Synthetic Microstructures in Biological Research

Synthetic Microstructures in Biological Research

Edited by

Joel M. Schnur
Center for Bio/Molecular Science and Engineering
Naval Research Laboratory
Washington, D.C.

and

Martin Peckerar
Nanoelectronics Processing Facility
Naval Research Laboratory
Washington, D.C.

Technical Editor
Helen M. Stratton

Springer Science+Business Media, LLC

Library of Congress Cataloging-in-Publication Data

Synthetic microstructures in biological research / edited by Joel M.
Schnur and Martin Peckerar ; technical editor, Helen M. Stratton.
p. cm.
"Proceedings of the Third International Conference on Synthetic
Microstructures in Biological Research, held September 9-12, 1991,
in Williamsburg, Virginia"--T.p. verso.
Includes bibliographical references and index.

1. Biosensors--Congresses. 2. Piezoelectric polymer biosensors-
-Congresses. 3. Biomaterials--Congresses. I. Schnur, Joel M.
II. Peckerar, Martin Charles, 1946- . III. Stratton, Helen M.
IV. International Conference on Synthetic Microstructures in
Biological Research (3rd : 1991 : Williamsburg, Va.)
[DNLM: 1. Biocompatible Materials--congresses. 2. Biomedical
Engineering--congresses. QT 34 S9957 1991]
R857.B54S975 1992
610'.28--dc20
DNLM/DLC
for Library of Congress 92-48767
CIP

Proceedings of the Third International Conference on Synthetic
Microstructures in Biological Research, held September 9–12, 1991,
in Williamsburg, Virginia

ISBN 978-1-4899-1632-7 ISBN 978-1-4899-1630-3 (eBook)
DOI 10.1007/978-1-4899-1630-3

INTRODUCTION

The Third International Conference on Synthetic Microstructures in Biological Research (SMIBR) was held in Williamsburg, Virginia, September 9-12, 1991. This book represents the compilation of many of the papers and posters presented at the meeting. Publications resulting from previous SMIBR meetings held at Airlie House, Airlie, Virginia (March 24-26, 1986 and March 20-23, 1988) can be obtained from the Institute of Electrical and Electronic Engineers, Inc. (IEEE) Service Center (908-562-5418).

The purpose of these conferences is to provide an interface between the engineering community and those at the cutting edge of biological and related material science. The overriding motive for assembling representatives from these diverse disciplines is clear. Engineers, even in 1986, could manipulate materials over dimensional scales on the order of those critical to biological systems. The devices that emerge from these manipulations could then be tailored to monitor system function more directly with significantly more localization than ever before. Thus, one important goal of the meetings is to provide the detailed specifications required for effective interfacing of devices with biological systems under investigation.

The basic theme of the conference has evolved over the years. In early discussions, it became quite clear that biologists and biochemists had as much to say as engineers about how materials could be effectively manipulated. The conference has become a forum for facilitating dialogue between these communities on the subject of material modification for biological applications. This dialogue consistently took place in the 1991 conference and is well represented in the text of this book.

In the past decade a number of advances in fabrication and characterization techniques have led to important improvements in the application of small structures to medical as well as other applications. Advances in microlithography, self-assembly, and advanced characterization techniques are leading the way for the fabrication of ever smaller devices. After we learn how to make and manipulate these "ultra small" structures, we must learn how to interface them with the outside world, e.g. biological systems, in order to evaluate their potential applications. All of this research and development will require a great deal of interdisciplinary interactions. The conference in Williamsburg was organized to provide a forum to encourage this type of interdisciplinary interaction.

Expert scientists from a number of diverse fields were invited to participate in the meeting. The scientific areas represented include medicine, biology, chemistry, physics, and engineering. The format of the meeting was designed to encourage communication between people with backgrounds as diverse as neurosurgery and quantum physics. A number of the more renowned speakers were asked to prepare posters rather than formal presentations in order to encourage lively interactions. A majority of the time at the meeting was spent in collegial discussions at the poster sessions rather than listening to a large number of formal presentations. Each speaker was encouraged to give a tutorial "Scientific American" level introduction and a significant amount of time was set aside for informal questions and discussion after each presentation.

The structure of the talks and the posters was developed along four general themes. The first theme was **Fabrication**. The premise was that one has to make something before one can study or use it. The session focused on some novel ways to make small structures which might have some useful properties. Recently, the process of self-assembly has been discussed.[1-3] The shape, size, and electronic properties of a molecule, coupled with the nature of its environment, can cause it to aggregate to form microstructures. Liposomes are examples of lipids self-assembling to form hollow spherical microstructures. By appropriately choosing the molecule and conditions, the nature of the structures formed can be studied and perhaps controlled. The session on Fabrication was devoted to this topic as well as more conventional microlithographic approaches to making microstructures.

As we make smaller and smaller objects, it becomes more difficult to see the structure that has been made! Fortunately, new techniques have been developed which now let us examine objects at even the sub-nanometer scale. The **Characterization** session examined the development of these techniques as well the novel use of more traditional ones for the study of sub-micron structures. The invention of scanning tunneling, atomic force, confocal, and near field microscopies have opened new vistas for the examination of microstructures. The potential, as well the limitations, of these techniques were closely examined during the conference. It is clear that no matter how powerful these new techniques may be, more traditional techniques such as optical and electron microscopy and vibrational spectroscopic techniques are still extremely important. In fact, it is only by collating all the information from a number of diverse techniques that one can begin to gain a meaningful understanding of these microstructures.

The process of characterizing synthetic microstructures should not be done in a vacuum. It is extremely important to consider the world in which these structures will be used. Thus, the session on **Biological Interfaces** should be seen to be closely coupled to the tools and methods of characterization. Critical issues facing the scientist in this area include compatibility of synthetic structures with biological ones, the ability of the synthetic structure to modify function in a rational and reproducible manner, as well as the question of modification of function because of the interface. The latter is of particular importance in applications involving micro probes in, for example, neuromedical applications. The talks and posters at the conference in this area covered a range of topics from the question of incorporating proteins in self-assembled films to the question of the interface between the ear and eye with the external world and the brain.

After the problems of biological interface have been addressed, **Applications** become feasible. The driving force for many at the meeting was in areas associated with brain function. A good deal of what was discussed was associated with the use of ultra small probes to report neurological function. Because of the synergy that developed between the neurologists, physicists, chemists, and engineers at the meeting, a number of new applications for small devices were actively considered. This book contains articles involving applications for biosensors, neural networks, and advanced neurosurgery.

There is a significant problem in organizing a conference and editing a book that covers such a diverse number of scientific disciplines. In order to maintain a consistent level of high scientific quality we have required that all papers published in this book be critically reviewed by suitable experts in the specific discipline. We consulted with a number of our colleagues and friends of these colleagues in hospitals, medical schools, universities, national laboratories, and industry in the formulation of the program and in the selection of the posters and oral presentations. We wish to thank all of those who helped us in this regard, especially those who attended the conference and made it such a memorable experience.

Special thanks and acknowledgments go to Mrs. Helen M. Stratton who has provided her special talents to the issues and hard work involved in organizing the conference and producing this book. The financial support of the National Science Foundation is gratefully acknowledged. Dr. Robert Welleck and Dr. Charles Hauer at the NSF provided important advice to us about the conference as well as advice (C.H.) about French automobiles. The Office of Naval Research provided financial support for the conference as well as significant advice. Dr. Steve Zornetzer provided important counsel with respect to the neurobiology topics covered at the meeting. The IEEE, Electron Devices Society has long been a supporter of this meeting. We thank that organization for its continued support and interest. Dr. Vickie Hallmark of the IBM Research Center wrote the preface to the Characterization section of this volume and her efforts are greatly appreciated.

August 1992 *Joel M. Schnur*

REFERENCES

1. J. Prost and F. Rondelez, Structures in colloidal physical chemistry, *Nature* 350:11-23 (1991).
2. A. H. Heuer et al., Innovative materials processing strategies: a biomimitic approach, *Science* 255:1098-1105 (1992).
3. G. Whitesides, et al., Molecular self-assembly and nanochemistry: a chemical strategy for the synthesis of nanostructures, *Science* 254:1300-1342 (1992).

CONTENTS

FABRICATION

CHARACTERIZATION

BIOLOGICAL INTERFACES

APPLICATIONS

FABRICATION

INTRODUCTION

In order to use microstructures (no matter what the application) one must first be able to make them. Microlithographic techniques have been the traditional way in which this has been done. The article by **Peckerar and Rebbert** reviews these classical techniques and shows examples of microelectronic processes used to fabricate bioprobes. Recently, the potential value of self-assembly has raised excitement. In this section there are two articles, **Hui et al. and Markowitz et al.**, that explore the way in which phospholipids can be modified to rationally control the nature of microstructures self-assembled from them. Hui describes fascinating structures formed in two dimensions, while Markowitz delves into the way the lipids can be induced to form sub micron size "soda straws" called tubules. These tubules have potential applications in areas such as controlled release, micro-separations, and high density microelectronics.

Controlled pores of very small sizes also offer potential advantages in separation science. **Krishnasastry et al.** describes ways in which the techniques of recombinant engineering might be used to form arrays of nanometer sized pores for this and several other potential applications. It is clear that the manipulation of proteins will be very important if we are going to be able take advantage of genetic engineering for nanofabrication applications. **Seeger et al.** elucidates techniques for attaching antibodies to piezoelectric substrates.

After one has the ability to fabricate an assortment of microstructures, one must still be able to assemble them into an integrated assembly that has function. **Ligler and Rabbany** describe ways biomolecules may be incorporated as recognition elements to construct "biosensors".

ADVANCED MATERIALS PROCESSES FOR BIO-PROBES

Martin Peckerar and Milton Rebbert

Nanoelectronics Processing Facility, Naval Research Laboratory
Washington, DC 20375

With the advent of the integrated circuit, our ability to form and to manipulate structures using a variety of materials has improved dramatically. We can currently fabricate objects with micron and submicron features out of silicon, polymers and metal. The geometries of these structures can be tailored to interface directly with living cells. In this paper, we review key processing techniques which may be applicable to biological research. We then present some examples of possible applications.

INTRODUCTION

The area of integrated circuit process technology is broad. It includes tools as diverse as particle accelerators, gas phase plasma reactors, chemical vapor deposition chambers, electron beam machines and complex optical systems. Many excellent overviews of these procedures have appeared in the literature.[1,2] Our interest here, though, is not in the fabrication of integrated circuits. But rather, we are concerned with the fabrication of microstructures capable of interfacing directly with cells and other elements which constitute the "fine-structure" of a biological system. In this paper we concentrate on those elements of process technology which are of interest in biological research.

In particular, this paper deals with the fabrication of probes of bio-system function. These probes may be "passive" in the sense that they record the state of electrical, mechanical or chemical activity of a system; or, they may be "active" in that they stimulate these types of activity. In all cases, probe dimension must be of a size suitable for interfacing with the system probed. Also, the probe must create minimum perturbation in the target system. "Per-turbation" implies insertion damage and/or long-term toxicity.

Our goal here is not to make the reader an expert in microstructure process technology. But rather, we are attempting to provide a general description of the key processes in current use. In addition, the "limits" of the technology are discussed. That is, we address such questions as: "How thin can I make a film of a given

Synthetic Microstructures in Biological Research, Edited by
J.M. Schnur and M. Peckerar, Plenum Press, New York, 1992

material? How fine can I pattern features in a given design? What dimensional tolerances can I expect?" Hopefully, this will allow the reader to better understand other publications in the field and will stimulate proposals of new applications of the technology.

The paper is organized in the following way. First, key areas of process technology are reviewed. The areas selected are those which, based on the author's experience, are useful in the fabrication of bio-probes. This usefulness must either have been demonstrated or it must (in the author's opinion) show promise of near-term demonstration. Detailed examples of processes currently in use and processes which may find use in biological research are presented.

PROCESS TECHNOLOGY FOR PROBING BIOLOGICAL SYSTEMS

The Basic Tool Set

In integrated circuit manufacture, materials are patterned into pre-determined shapes using a process similar to that of photo- graphic printing. In order to do this, the workpiece (the substrate on which the pattern is to be generated) must be turned into a photoplate. This is done by coating the work surface with a light-sensitive polymer called photoresist. A shadow image of the desired pattern is cast on the processing surface.[3] This "casting" may be accomplished either by holding a shadow mask in contact with the resist (contact printing) or at a small distance from it (proximity printing). In state-of-the-art printers, projection lenses are used to form images of the mask on the resist.[3]

Currently, optical systems are becoming available which can "print" features as small as 0.25 μm.[3] Ultra-high resolution (producing features smaller than 0.1 μm) is obtained by writing the pattern in photoresist with e-beams;[4] or, the pattern can be printed using short wavelength radiation, such as x-rays. The short wavelength avoids the diffraction limits on resolution.

There are two types of photoresist currently in use.[5] These are referred to as positive and negative systems. In the positive case, incident light renders the resist soluble in "developers," usually basic solutions such as tetramethylammonium hydroxide. In the negative case, light causes the resist to be less soluble in organic solvent developers, although newer negative systems also use basic aqueous developers. Both processes are illustrated in Figure 1.

The resist is applied to the substrate in the following way. The polymer is dissolved in its "spinning solvent," creating a viscous fluid. The fluid is "puddled" on the surface of the workpiece and the workpiece is spun about its central axis at speeds in the 2,000 - 7,000 revolutions per minute range. The centrifuge effect spins most of the resist off the surface. Surface tension and surface adhesion effects retain a thin film coating of resist. Resist thickness varies inversely with spin speed. Films thicknesses in the 0.5 - 2 μm range can be obtained using the spin speeds listed above. Thicker layers are obtained using more viscous materials. The remaining spinning solvent is driven off in a post-development bake. This creates a cured, solid film. More will be said on this process later. Other materials, such as polyimides, can be formed into films in this way. These thin polymer films are useful components in bio-probes.

Now let us consider the problems associated with image formation in photoresist. In the case of contact or proximity printing, resolution limit is set by diffraction. An expression for the minimum resolved feature size, r, is:

$$r = k\sqrt{s\lambda} \qquad (1)$$

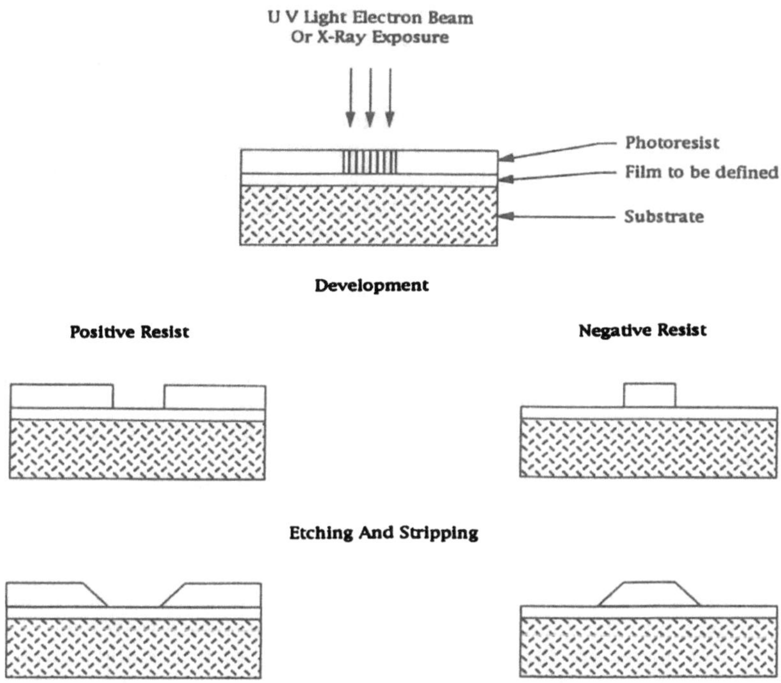

Figure 1. Positive and Negative photoresist processes.

where s is the separation of the mask from the wafer (usually taken as the resist thickness in contact printing) or, in proximity printing, this is the proximity offset plus the resist thickness. The λ term is the wavelength of exposing light (usually around 400nm); k is a process-related constant (ranging from 1 - 3). In a projection printer, resolution is set by the space-frequency filtration effects of the exit aperture of the lens system.[1] In projection printing, resolution is usually estimated by the following formula:

$$r = k_1 \lambda / NA \tag{2}$$

where k_1 is a process dependent parameter (ranging from 0.5 - 1) and NA is the numerical aperture of the optical system. In cutting edge lithographic systems, NA's of 0.4 - 0.5 are becoming available.

In the fabrication of most bio-probes, ultimate resolution is of little importance. Most feature sizes are well above a micron. Such dimensions do not strain existing lithographic capabilities. The main problem encountered in this field is printing on irregular surfaces. Think about the difficulty of projecting slide photographs on a wavy screen. The image goes in and out of focus. The maximum travel of the image plane over which acceptable resolution is obtained is known as depth of focus. Concepts developed above are useful in understanding this problem. In contact or in proximity effect printing, the "blur" can be estimated by assessing the degradation in resolution through equation 1. In projection printing, depth-of-focus, D, is given by the following relationship:

$$D = k_2 \lambda / (NA)^2 \tag{3}$$

where k_2 is a process related constant (usually about 1).

Depth-of-focus also impacts the aspect ratio of the resist exposure pattern. The image will "blur" as it traverses thick photoresist. In very thick resists, optical absorption may prevent full exposure. However, by using short wavelength, penetrating radiation, these limits can be overcome. Using sources emitting in the 10 Å wavelength range, we were able to obtain aspect ratios such as those shown in Figure 2.[6] Here we see tall cylinders (≥100 μm tall), whose diameter is ≤ 10 mm. Printing was done using a specialized shadow mask with a gold absorber pattern placed on a thin (≤2 μm thick) x-ray transparent silicon membrane. The exposure source was a synchrotron particle accelerator. Many of the processes used to make this mask are important in bio-probe fabrication and are discussed below. Admittedly, the synchrotron source is currently somewhat exotic. But they are available at universities (such as University of Wisconsin and Louisiana State University). Many industries (such as IBM) are currently buying such tools for circuit production.

An interesting paradox arises on comparison of equations 2 and 3. For high resolution, a large NA is required (equation 2); for a large D, small NA is required (equation 3). Thus demands for high resolution and large depth-of-focus are, in fact, contradictory.

Patterns in resist are important, but they are usually of little use in the finished device. The pattern must be transferred to underlying material through some etch process, and then the resist must be removed. Resists will withstand attack from most common acids and many bases. Consider, for a moment, the case of slots patterned into photoresist spun on glass. Dipping the glass/resist system into hydrofluoric acid will etch grooves into the glass where the glass is not protected by resist.

Most wet etches are "isotropic." That is they will etch laterally as well as in a direction perpendicular to the surface etched. As a rule-of-thumb the resist undercut is the same as the depth etched. To get around this problem, etching can be done using a gas-phase plasma.[7] In this process, gas is fed into an etching chamber and a glow discharge is struck using a radio-frequency source. The workpiece is surrounded by this plasma and an ionic space-charge develops over the work surface. Electric fields in the space charge accelerated reactive ions from the gas onto the surface where they chemically erode the unprotected regions. The electric field imparts a directionality to the etch, minimizing undercut. The etch rate is anisotropic - greatest in the direction perpendicular to the surface. For the purposes of most bio-probe fabrication projects, such etches are viewed as completely anisotropic.

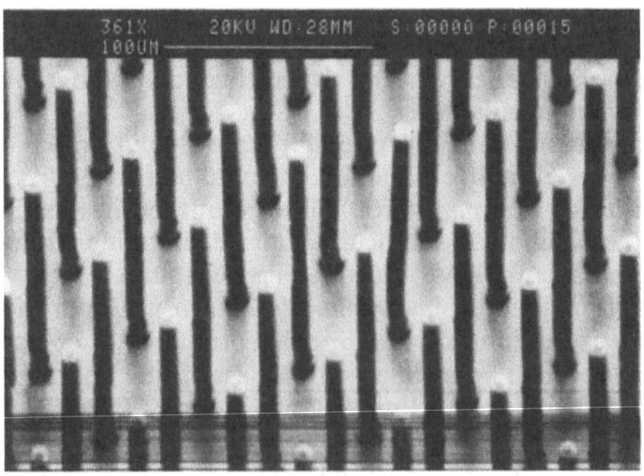

Figure 2. High aspect ratio resist structures, made using x-ray lithography.

Some wet etches can exhibit anisotropic properties.[1] Water-KOH mixtures as well as ethylene diamine pyrocatechol (EDP) will etch certain crystallographic planes of silicon faster than others. In particular, the (111)-face of silicon is almost unetchable to this solution, while other faces are. Consider the following process. Pattern circles of a suitable etch mask onto a silicon surface. Photoresist by itself is attacked by the KOH or EDP etchants. Perhaps we might use nitride or oxide films discussed below. If we start with a (100) face, the etch will undercut the etch mask and stop when the (111) surfaces are reached. This creates pyramids, as shown in Figure 3. If we start with a (110) cut, we will etch very deep slots. Some possible applications for this are discussed below.

Next, let us consider the types of films which can be placed on surfaces. These films can remain on the original surface or they may be "floated off" to create free standing membranes. Examples of such applications are shown in the next section of this paper.

Of course, photoresist is a surface film. It is usually installed on the surface by the spinning process described above. The spun thickness of resist, t, can be calculated from the following empirical formula:

$$t = \eta^{b}\omega^{-1/2}/t^{*} \qquad (4)$$

where η is the resist viscosity, b is a constant (usually about 2), ω is the spin speed, and t^{*} is the time-to-spin (the spin time needed to achieve final thickness, on the order of seconds). Thickness uniformity is better than 1% - on the order of hundreds of Ångstroms. Surface tension tends to pull the film surfaces taut, planarizing the structure.

Plastic films of polyimide, polystyrene and polyethylene can also be made in this way. These films are poorly adherent to many surface. Spinning polyimide on bare silicon, curing, and soaking in water will result separation of the film from the substrate.

Metal films are frequently used in the fabrication of microelectrode arrays. These films are usually deposited by vacuum evaporation techniques.[8] Electron beams impinge on a target of the metal to be deposited, evaporating a portion of this target.

Figure 3. Pyramids made using an anisotropic liquid etch.

The vapor streams out isotropically into the surrounding vacuum and deposits on surfaces within the vacuum chamber. The thickness of the deposited film is measured during the process using a quartz-crystal microbalance. This is simply a quartz oscillator whose surface is exposed to the depositing vapor. As mass builds on the exposed quartz surface, the quartz resonant frequency changes. The resonance is electrically measured and converted to deposited thickness. Film of gold and aluminum can be deposited as thin as 20 Å in this manner. Thickness control is on the order of 1%. Films thicker than 1 mm are difficult to do by evaporation. Stress mismatch (and thermally induced stress mismatch if the sample is heated during evaporation) will cause the films to flake away.

Thicker films can be generated by up-plating techniques.[9] Here, a thin (< 1000Å) plating base (usually of the material to be plated) is deposited on the surface. An electrode is attached to the surface and the sample is submerged in a bath containing ions of the plating metal. Applying electrical bias to the electrode will initiate the plating process. Films thicker than 25 μm can be deposited in this way. Free-standing metals foils can also be created by etching away the substrate after deposition, as shown below. Thick films of gold, nickel, platinum or chrome can be made with this process. Metals can be plated around resist features and resists can be used as plating guides. Many interesting and important microstructures can be made in this way, as shown below.

Many plating baths contain additives, or "brighteners" which affect the grain size of the metal plate or aid in achieving thickness uniformity. Some baths contain chemicals which will drive deposition reactions on the plating surface. Such "electroless" baths do not require external bias through electrodes.

Insulating films can be grown or deposited on substrates.[1] Silicon can be oxidized at temperatures in excess of 700°C. The oxidizing ambient can be either pure oxygen in a nitrogen carrier, in steam or in oxygen/nitrogen/HCl mixtures. The latter combination is used when surface cleanliness is required. The chlorine can remove harmful impurities from the surface during oxidation.

Oxidation of silicon "works" because the oxide film that forms matches the substrate in its thermal and mechanical properties. Many materials, such as most metals, do not exhibit this matching property. Blanket insulators must be deposited over these substrates. Usually, chemical vapor deposition (CVD) is used. For example, silicon dioxide can be deposited on a sample by placing the sample in a furnace and blowing a mixture of Silane gas (SiH_4) and oxygen over it. The surfaces in the furnace catalyze the following reaction:

$$SiH_4 + O_2 \rightarrow SiO_2 + 4\,H_2 \tag{5}$$

Such blanket insulators can be used to passivate individual electrode leads in microelectrode arrays, as discussed below.

Polycrystalline films of silicon can be chemically vapor deposited by surface-catalyzed thermal decomposition of silane. These films can be made conductive by doping with impurities such as phosphorous or boron. Doping is accomplished by exposing the films at high temperatures (>900°C) to phosphene or to diborane gas. The gas molecule breaks apart on the film surface and the doping impurity diffuses into the bulk of the film. The resulting film can be patterned and treated as a conductor. The polysilicon films can be heated to temperatures far higher than those that metals can be exposed to. Metals can rarely be exposed to temperatures higher that 500°C in completed devices. Interdiffusion and intermetallic formation generally destroys the devices.

It is frequently necessary to minimize the temperature in a given process to make it compatible with certain classes of material. The CVD process is usually accomplished at temperatures much lower than those of thermal oxidation. Most CVD processes are done in the 400 - 700°C temperature range. Reaction rates can be enhanced by breaking apart the molecules in a gas-phase plasma. Thus, plasma enhanced CVD (PECVD) can be done at lower temperatures still (perhaps as low as 250°C). Such low temperatures are compatible with some of the plastic films, such as polyimide, described above.

This concludes our presentation of processing fundamentals needed to understand basic concepts in bioprobe fabrication. We now proceed with examples of how these processes can be combined to form devices. In the next section, advanced processes groupings which may be useful in biological research are presented.

Integrated Process Flows

In order to fabricate a useful device, sequences of the processes outlined above must be employed. For example, as mentioned above, photoresist can be used as a plating guide for making high-aspect ratio structures. Figure 4 shows an example of

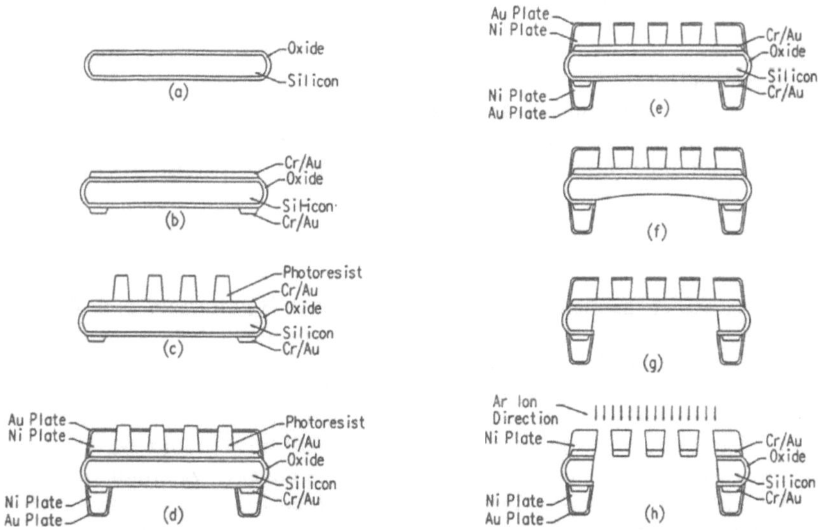

Figure 4. Illustration of the mesh fabrication steps. (a) 2A thermal oxide is grown on both sides of the wafer. (b) a chrome-gold film is deposited over the oxide, on both sides of the wafer. (c) A layer of photoresist deposited, exposed through a mask, and developed. (d) Nickel is up-plated using the resist template to form nearly perpendicular sidewalls. (e) The resist template is removed, leaving a nickel mesh on a gold film, all on a silicon wafer. (f) The back side of the silicon wafer is thinned from the center to the edge, putting a dimple in the backside of the wafer. (g) The silicon is etched away, leaving a full chrome-gold membrane and nickel mesh. (h) Ion milling is used to mill the gold membrane between the nickel wires, from the back side of the mesh, leaving a smooth, gold coating on the front side of the free-standing mesh.

how this process can be used to make a free-standing mesh (or screen) structure mounted on a silicon ring. Such meshes can be made with tailored aperture dimensions. Apertures as small as 1 micron are obtainable with current technology. The x-ray process described above[6] can be used to make free standing sub-micron meshes. The completed structure is useful in separation processes and in cell sorting.

Processes can be integrated to make movable structures which are tethered to the silicon substrate. This process flow (frequently called the surface relief process) is central to the fabrication of micromotors and actuators. These structures are made to move by application of external electrostatic bias.

The whole flow shown in Figure 5 involves two oxidation steps, two polysilicon depositions and three photolithography steps. First, CVD oxide is laid on the silicon surface. Polysilicon is deposited over the oxide. This layer of polysilicon is patterned into the moving structure. A second layer of oxide is placed over the patterned polysilicon and a hole is cut in the center of the moving structure through both layers of deposited oxide. A second layer of polysilicon is deposited and patterned to form the tethering pin for the movable structure. The whole wafer is soaked in hydrofluoric acid which dissolves the oxide and leaves the silicon and polysilicon untouched. This frees the first polysilicon layer for movement.

DETAILED EXAMPLES OF BIOPROBE FABRICATION

In this section, we present two examples of microstructure technology applied to neurophysiological research. Both examples are of multi-element electrode arrays. The first is a rigid implantible array which can be used for both recording and stimulation.[10] The second is an example of a flexible "blanket" array of electrodes. This blanket can be draped over exposed regions of cortex for measurements of evoked potential.

An Implantible Neural Probe

First, consider the rigid probe. The probe body is patterned as a "dagger" with recording sites on one side and contact pads to external circuitry on another (see Figure 6). The electrodes are patterned metal thin film wires on a thin silicon substrate. Electrode leads are insulated from the neuronal medium either by a blanket of CVD deposited SiO_2 or by polyimide. Contact holes are cut in the SiO_2 at either end of an individual electrode pattern. A small hole (about 1-2 mm across) is used as the recording site on the "blade" side of the dagger. A larger hole is cut on the "handle" side for contacts. The blade is embedded in the neuropile under study.

As this is an implantible array, the blade must be very thin (< 50 μm) to minimize damage to surrounding tissue. In the past[11] this was done as follows. A silicon wafer was oxidized and a hole in the shape of the dagger was lithographically defined using the processes described above. The wafer was than put in a furnace and diborane gas was bled into the furnace. The diborane would dissociate on the wafer surface and boron would diffuse into the silicon where the hole was cut. Electrodes, insulator blankets and contact windows were installed on the wafer surface and the surface was coated with an EDP-resistant film. The wafer was soaked in EDP and etched away. As discussed in reference 1, EDP will not etch the boron-doped regions. Thus, the shape of the dagger was cut out of the parent substrate. The EDP etch barrier was removed and appropriately bonded.

The process described below is similar. Here, the boron etch stop is not used to define the final thin dagger shape. But rather, the substrate wafer was pre-thinned

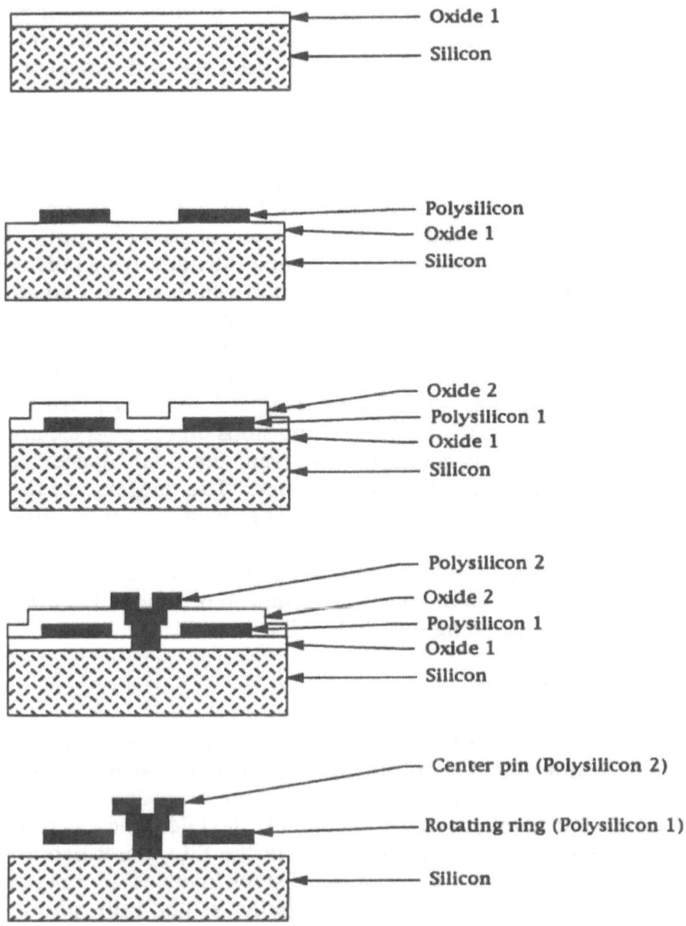

Figure 5. A process flow for fabricating a freely rotating section tethered to a silicon wafer.

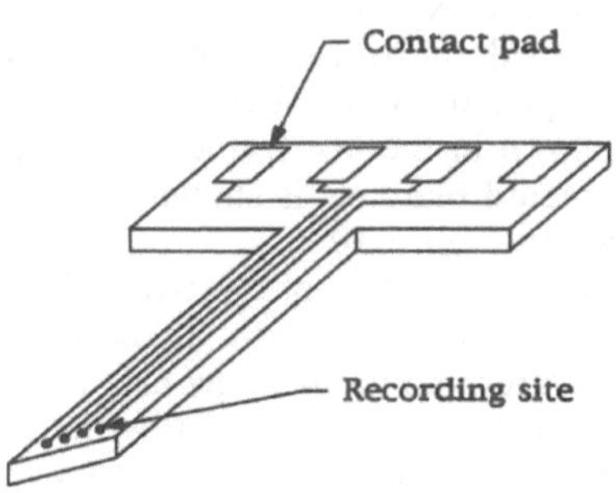

Figure 6. Silhouette of "Dagger" microelectrode array.

using a polishing process. Such thinned wafers are commercially available (through Virginia Semiconductors, Inc., Fredricksburg, VA). The thinned wafer is attached to another wafer to give it mechanical support during processing. In the end, the support wafer is removed. The daggers are mounted in a package and connected to external leads, thus completing the device.

Figure 7 discloses the process sequence used. The thin silicon wafer is attached to a thicker wafer which serves as a carrier. The "glue" used to attach the wafers was positive photoresist. Polyimide is spun on the surface of the workpiece creating an insulating barrier to the underlying silicon. A thin layer of chrome (<100Å) is evaporated on the surface, followed by a thicker layer of gold (0.4 μm). The chrome acts to promote good adhesion between the gold and the underlying polyimide surface. The metal films are patterned into electrode arrays which are subsequently covered by polyimide and CVD-deposited oxide. The oxide is patterned to act as a subsequent etch mask for the plasma etch process to follow. This patterning opens holes in the oxide which will be used to form the recording sites and contact pads. The oxide is also removed at the boundaries of the array "dagger." An SF_6 plasma etch is than employed. This etches through the polyimide and stops on the gold electrode surfaces in the active array. The etch is continued to reveal the final "dagger" shape of the electrode body in the field of the thin silicon wafer. The daggers are floated off the thick silicon substrate using an acetone soak. The daggers can be turned over for further thinning in SF_6plasma.

In any multi-layer structure, delamination is possible in use. This delamination is usually due to poor surface cleanliness or to improper polyimide cure. Soaking in saline solution is useful in weeding out early failures.

A Flexible Neural Blanket Array

The second example is also of a multi-unit electrode array. In this case, though, the substrate is a flexible blanket of polyimide which is draped across an exposed region of cortex (see Figure 8). Evoked potentials are recorded under conditions of cortical stimulation. The blanket array is completely flexible. The electrode insulating material is another layer of polyimide, rather than the SiO_2 described above.

With reference to Figure 8, the first set of steps involves spinning polyimide on an oxide-coated wafer. The blankets are individually defined using photolithography. Note that the photodefinition is done on partially cured polyimide. Fully cured polyimide will not dissolve in the positive resist developer. Effectively, the developer acts as a polyimide etch. The whole processing surface is metallized and patterned into electrodes. The structure is than coated again with polyimide and electrical contact windows and probe tips are exposed. The blanket is removed from the silicon substrate by etching away the oxide in hydrofluoric acid. The recording sites may be elevated to contact the neuropile directly. This is accomplished by up-plating the exposed recording sites, as shown in the last frame of the figure.

A picture of the completed blankets is shown in Figure 9. The contact pads and recording sites are shown in Figure 10.

This paper has focused on recording of electrical signals using microstructures which interface directly with the cellular medium. It should be pointed out that other sensors are possible. The literature contains many useful reviews describing the use of microstructure technology in the fabrication of flow meters, pressure sensors, Ph sensors, and temperature sensors. Field effect transistors can be configured to give information on ion concentrations in body fluids. The same basic processing tools are used to make all these devices. Reference 2 contains surveys of many of these sensor approaches.

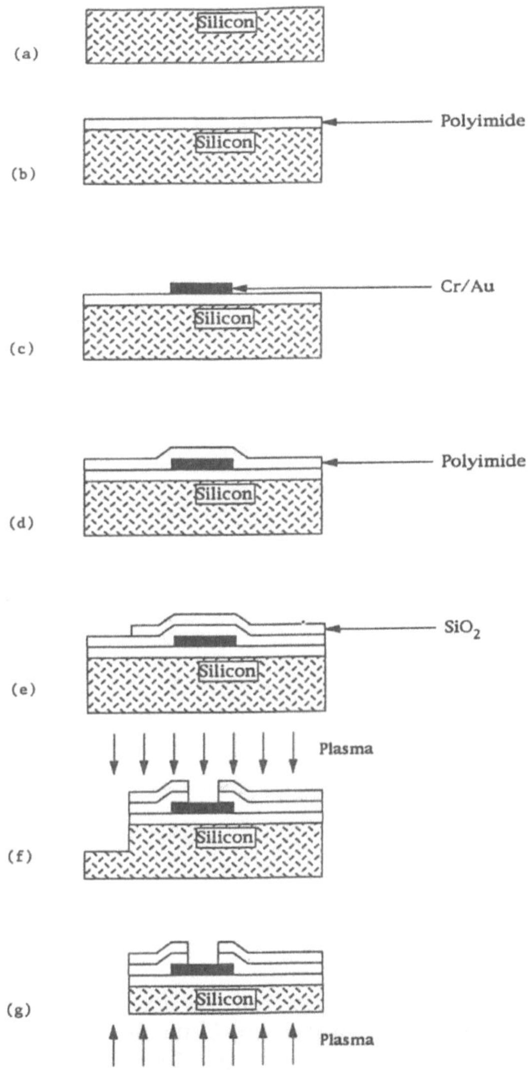

Figure 7. Illustration of the rigid silicon probe fabrication steps. (a) The starting material is a clean silicon wafer. (b) A layer of polyimide is applied to the wafer and is fully cured. (c) Metal for the electrodes is deposited and defined. (d) A top layer of polyimide is applied and cured. (e) A layer of SiO_2 is deposited and patterned on the polyimide. (f) The wafer is plasma etched on the front side to cut out the shape of the probe. (g) The wafer is plasma etched from the backside to separate the probe from the wafer.

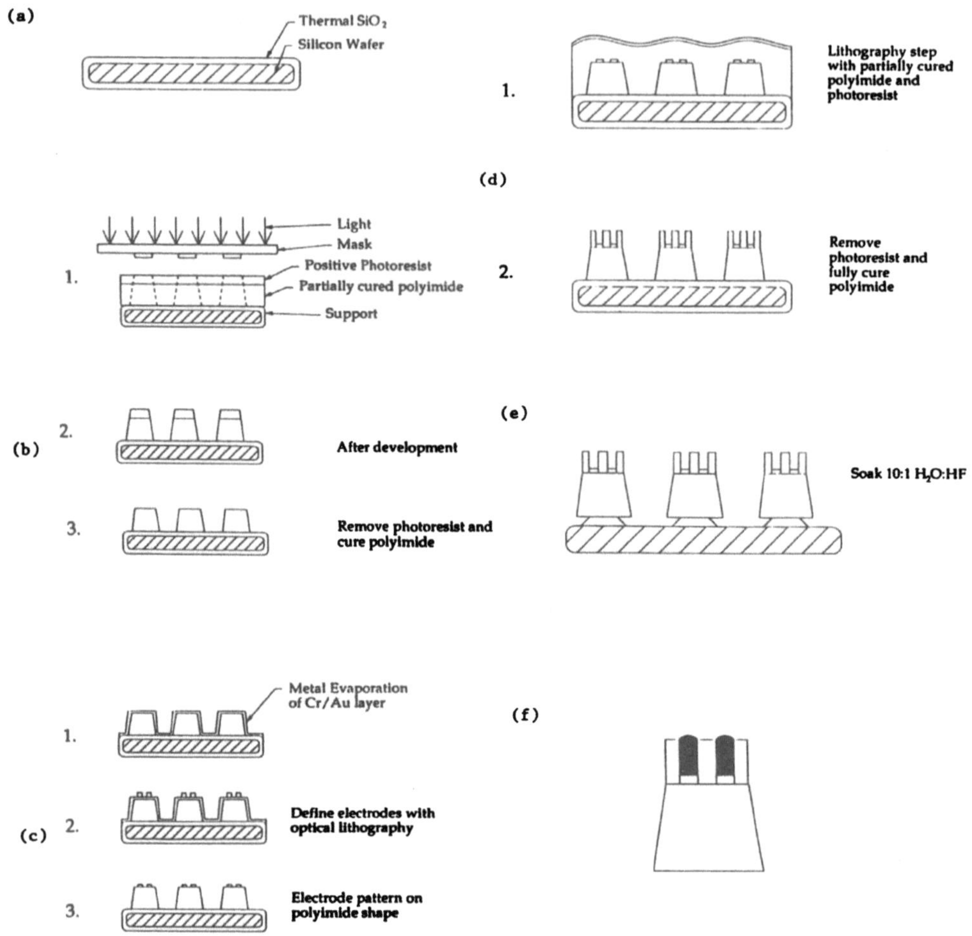

Figure 8. Illustration of the blanket probe fabrication steps. (a) A thermal oxide is grown on a silicon wafer. (b) Using the first mask level, the shape of the base polyimide is defined. (c) Metal for the electrodes is deposited and using the second mask level, the electrode pattern is defined. (d) Using the third mask level, the top polyimide is defined and contact windows to the electrodes are opened. (e) The probes are lifted from the support wafer by a soak in a dilute HF solution. (f) The electrodes contact pads are plated up through the contact windows to rise above the polyimide surface.

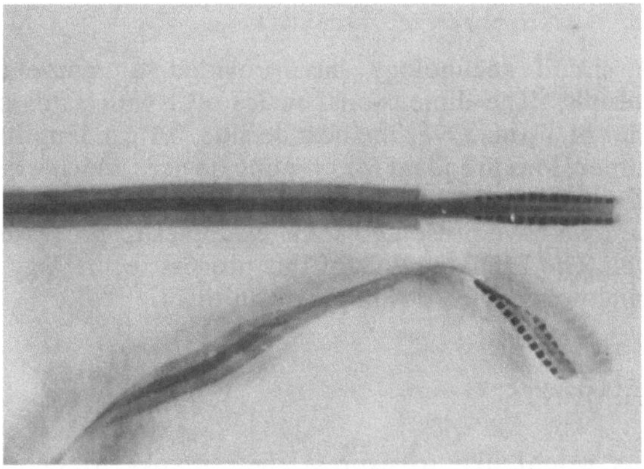

Figure 9. Completed blanket electrodes.

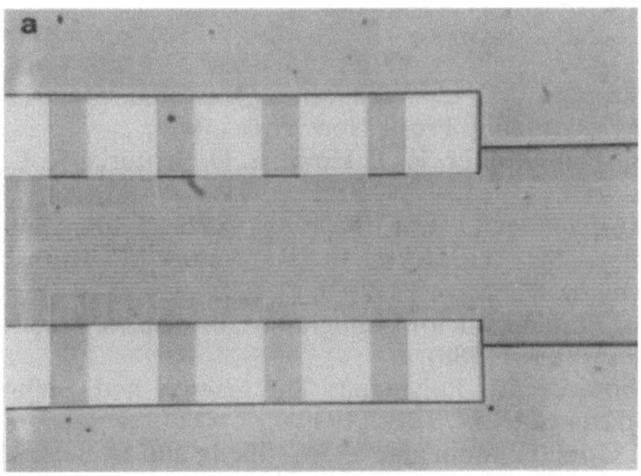

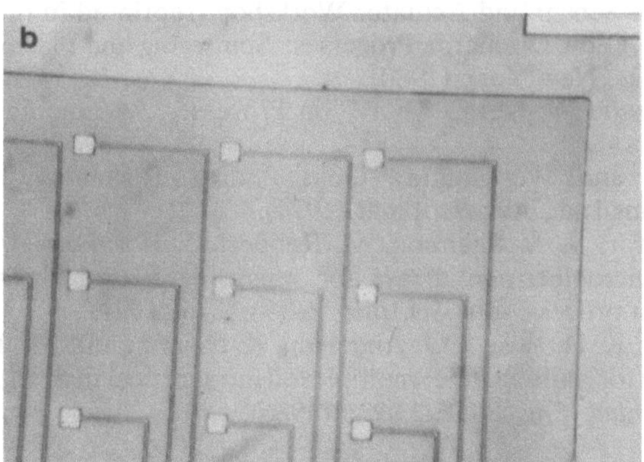

Figure 10. Contact pads (a) and recording sites (b) for blanket probes.

CONCLUSIONS

Integrated circuit technology has provided a marvelous capability for manipulating materials. The dimensional scales over which this manipulation takes place is on the order of 1 μm. Over the next decade, 0.1 μm length scales will become possible. These dimensions are ideal for creating devices which interface with cellular and sub-cellular components of living systems.

In this paper, we have presented some examples of processes already in use in biological research. The fundamentals of this process technology were presented in the hope of spurring new device development in this field.

ACKNOWLEDGEMENTS

The authors would like to express their thanks to Peter Isaacson and to Hope Ishii for helpful discussions relating to the work summarized in this manuscript. The Office of Naval Research and Technology are gratefully acknowledged for partial support of some of the research described in this paper.

REFERENCES

1. S. P. Murarka and M. C. Peckerar, "Electronic Materials: Science and Technology," Academic Press, New York (1989).
2. "Microsensors," R. S. Muller, R. T. Howe, S. D. Senturia, R. L. Smith, and R. M. White, eds., IEEE Press, New York (1991).
3a. Jain and Kanti, Excimer Laser Lithography, SPIE, Bellingham WA (1990).
3b. M. A. Hartney, M. W. Horn, R. R. Kunz, M. Rothschild, D. Shaver, *Microlithography World*, vol. 1(2):16 (1992).
4. W. M. Moreau, "Semiconductor Lithography: Principles, Practices and Materials," Plenum, New York (1988).
5. A. Reiser, "Photoreactive Polymers: The Science and Technology of Resists," Wiley-Interscience, New York (1989).
6. W. Ehrfeld, F. Gotz, D. Munchmeyer, W. Shelb, and D. Schmidt, "LIGA Process, Sensor Construction Techniques Via X-Ray Lithography," Record of the IEEE Solid-State Sensor and Actuator Workshop (reprinted in reference 2.) (1988).
7. B. Chapman, "Glow Discharge Processes: Sputtering and Plasma Etching," Wiley-Interscience, New York (1980).
8. J. L. Vossen and W. Kern, "Thin Film Processes," Academic Press, New York (1978).
9. F. H. Reid and W. Goldie, "Gold Plating Technology," Electrochemical Publications Ltd., Ayr, Scotland (1974).
10. M. C. Peckerar, S. A. Shamma, M. Rebbert, J. Kosakowski, and P. Isaacson, Passive microelectrode arrays for recording neural signals: a simplified fabrication process, *Rev. Sci. Instr.* 62(9):2276 (1991).
11. S. L. BeMent, K. D. Wise, D.J. Anderson, K. Najafi, and K. L. Drake, Solid state electrodes for multichannel multiplexed intracortical neuronal recording, *IEEE Trans. Biomed. Eng.* BME-33:230 (1986).

MICRODOMAINS IN POLYMERIZABLE DIACETYLENIC PHOSPHATIDYLCHOLINE MONOLAYERS

Sek Wen Hui[#a], Hao Yu[#], Zhenchun Xu[*], and Robert Bittman[*]

[#]Biophysics Department, Roswell Park Cancer Institute, Buffalo NY 14263; [*]Department of Chemistry and Biochemistry, Queens College of the City University of New York, Flushing NY 11367

INTRODUCTION

Monolayers of phospholipids may be used as substrates to stabilize membrane proteins on solid supports, in the fabrication of biosensors and other implantable devices. The stability of phospholipid monolayers can be significantly improved by using polymerizable lipids such as photo-polymerizable diacetylenic lipids. These lipids, once polymerized, form extremely stable structures which may be used as surface coating for biocompatible materials, and carrier vehicles for drugs, among other applications.[1-3] The function and reliability of the surface coating depend to a large extent on the microstructure of the monolayer used. Recent studies have shown that monolayers of lipids on the air-water interface are by no means uniform in structure.[4-7] Phase separated domains exist in most monolayers under certain surface pressure. The microscopic domains in monolayer become increasingly important as more microdevices are fabricated. One may even make use of the micro textures in monolayers to delineate spatially separated functional areas on a device surface. To this end, one must understand the physical chemistry of the formation of these microdomains to ascertain the uniformity and stability of the surface coating or deposit.

Although many structural studies have been made on common phospholipid monolayers,[4-7] as well as on the formation of tubules and bilayers of polymerized lipids,[8-11] very little is known about the microstructure of monolayers of polymerized lipids. Diacetylenic phospholipids and their analogs, including phosphoesters, alcohols, acids and amides containing diacetylenic hydrocarbon chains form monomolecular layers on the air-water interface.[12-13] The pressure-area curves of these analogs and some phospholipids (phosphatidylcholine, phosphatidylethanolamine, phosphatidic acid) containing hexacoxa-10,12-diynoic acids have been measured. It was reported that the diacetylenic phospholipids behave similarly to their saturated chain counterparts, but tend to have lower critical transition temperatures, and collapse at a larger area per molecule.[13] Recent advance of the use of fluorescence microscopy

[a]To whom correspondence should be addressed

Synthetic Microstructures in Biological Research, Edited by
J.M. Schnur and M. Peckerar, Plenum Press, New York, 1992

to study monolayers[4-7] enables us to make a thorough study of the microstructure of a polymerizable diacetylenic lipid on the air-water interface, prior to its deposition on a solid substrate. A good knowledge of the characteristics of this polymerizable lipid as a Langmuir monolayer will help us to optimize the solid surface coating technique.

We report here our measurement of the pressure-area relationship of a diacetylenic phosphatidylcholine, before and after the lipid is exposed to polymerizing UV light. We present our observation of reversible phase separated domains of unpolymerized lipid monolayers at relatively high surface pressures, and of irreversible domains formed by polymerized lipids. We have successfully transferred these monolayers to a solid film support, and obtained electron diffraction from selected domains.[14] Factors governing the size and shape of solid domains are compared with previous published theories.[15-17]

MATERIALS AND METHODS

1,2-bis-(octadeca-4,6-diynoyl)-phosphatidylcholine (PC-A) (Figure 1) was synthesized. The diacetylenic fatty acids were synthesized using butadiyne synthons as starting materials,[18] and then were coupled to glycerol-3-phosphocholine-cadmium-chloride complex in the presence of dicyclohexylcarbodiimide and 4-(dimethylamino) pyridine. The lipid was sealed in argon gas and stored at -70°C prior to use.

1-palmitoyl-2-oleoyl-phosphatidylcholine (POPC) and 1,2-dipalmitoyl-3-N-dimethyl-phosphatidylethanolamine [DP(Me)$_2$PE] were purchased from Avanti Polar Lipids (Birmingham, AL). The fluorescence probe phosphatidylethanolamine-rhodamine (PE-Rh) was purchased from Molecular Probes (Eugene, OR), and was mixed with the lipids at 1 mole percent. The probe is known to preferentially partition in the fluid domain of phospholipid monolayers, and has only a slight contribution to the pressure-area characteristics of the monolayer.[21]

All lipid samples were dissolved in chloroform at a concentration of 0.5 mg/ml prior to spreading on a Langmuir trough. A custom-built, four-compartment, environmentally controlled Langmuir trough was maintained within an enclosed chamber with thermostat wall, floor, and ceiling. The chamber was filled and continuously flushed with nitrogen gas. The size of the final trough where measurements were made was 15 X 15 cm. Surface pressure versus surface area per molecule of monomolecular films (π-A isotherms) were measured by a π-A recording system. The area was adjusted by a motor-driven Teflon bar moving along the surface of the trough. The movement of the bar is either manual or servo-controlled (in the isobaric mode) by the output of the electrobalance. The position of the barrier, which moved at a rate of 1-10 mm^2/s (0.18 - 1.8 Å^2/molecule/min), was converted to a voltage signal which was connected to the X-input of an X-Y recorder. Surface pressure measurements were made by a Wilhelmy plate. The change of surface tension was measured by a CAHN RG electric balance. The surface pressure (π) is defined as:

$$\pi = T_0 - T$$

where T_0 is the surface tension of triple distilled water, and T is the surface tension of the film covering the water substrate. The amplified signal was connected to the Y-input of a LINSEIS LY 1600 X-Y recorder which plotted the π-A isotherms.

Epifluorescence microscopy images were made by a fluorescent microscope AO-2071 with a 40X, (NA=0.85) objective lens. A Teflon retaining ring with a small opening was attached to the objective lens, and immersed through the air-water

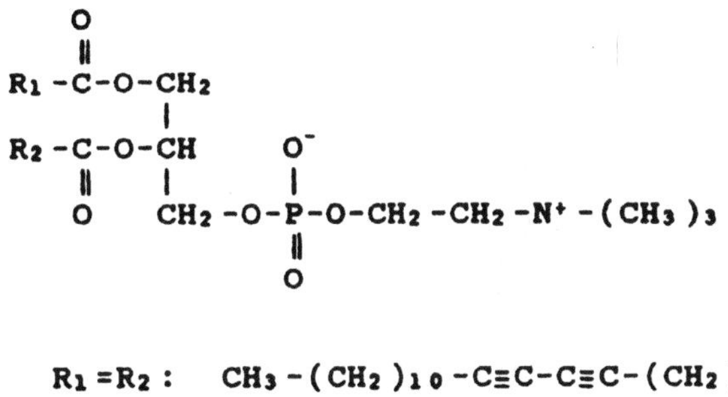

Figure 1. Chemical structures of PC-A.

interface. It was used to minimize monolayer drift. A Dage-MTI silicon intensified target (SIT) video camera MCP-SIT66 was attached to the microscope. A MAGNAVOX-80 monitor was used to observe the image, which was recorded by a Panasonic AG6050 VIDEO recorder. The processing of image from the video tape cassette was carried out in an IBM-PC-AT controlled Datacube Maxvision system.

Polymerization of the lipids in monolayers was by illuminating the monolayer with an 18.4 watt, 254 nm UV light (UVGL-58, UVP, San Gabriel, CA) for 15 min. The light source was about 1 cm from the monolayer. The temperature of the trough was maintained by circulating coolant during the UV exposure.

Phospholipid monolayers were deposited on Formvar covered marker-grids for electron microscopy. The Formvar-covered grids were coated with gold, and the coated side was pressed on a small piece of parafilm. A monolayer was deposited on the parafilm carrying Formvar-covered grids, by the transfer method described previously.[14] Successful transfer of a monolayer from the air-water interface to the Formvar film was verified by comparing the fluorescent domain patterns of the air-water interface and that of the deposit on a Formvar film. The domain patterns on selected locations on the marker grids were photographed using a fluorescence microscope (Olympus CH-5). The photomicrographs were used as guide maps for selected-area-electron-diffraction. The grids supporting the deposited monolayer were used in low dose electron diffraction experiments as previously described.[14] Briefly, a 5 μm condenser aperture was used to limit the already low beam current to below 50 $\mu A/cm^2$ at the specimen level to minimize radiation damage. Areas of the marker grid in the vicinity of the selected locations were moved to the beam, and diffraction conditions were set. The selected area was then quickly moved to the beam and the diffraction pattern was immediately recorded on Kodak DEF-5 diagnostic X-ray film.

RESULTS

1. Pressure-Area (π-A) Characteristics

The π-A isotherms of PC-A with 1 mole percent of PE-Rh are presented in Figure 2. The addition of 1 mole percent of the fluorescent dye has only a minor effect on the π-A characteristics. A transition point, equivalent to the "kink" of the onset of liquid expanded/liquid condensed (LE/LC) phase in most phospholipids, such as dipalmitoylphosphatidylcholine (DPPC) monolayers,[14] is observed in the isotherms at the region of 50-55 $Å^2$/molecule. However, this transition manifests as a "bump" or

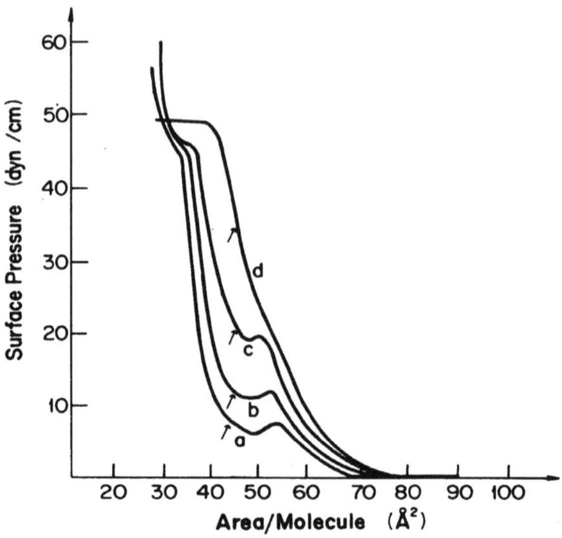

Figure 2. Pressure-area (π-A) isotherms at indicated temperatures on the first compression of PC-A at (a) 8.5°, (b) 10.5°, (c) 15.5° and (d) 21°C. The points marked by arrowheads indicate the onset of appearance of dark fluorescence domains.

overshoot rather than a "kink", especially when the monolayers are compressed slowly and for the first time. This transition point exists up to almost 21°C when it is no longer observable (Figure 2). The critical point is defined here as the transition point in the isotherm, above which the liquid expended phase and the liquid condensed phase are no longer distinguishable. This critical temperature is considerably lower than that of the corresponding saturated PC, namely distearylphosphatidylcholine (DSPC), in agreement with previous findings.[13] Unlike DPPC and other saturated PC's, PC-A has no well-defined point on the π-A curve that marks the transition from lipid condensed phase to solid condensed (SC) phase. The relatively incompressible solid phase area increases with temperature, from 35 Å^2 to 40 Å^2 as the temperature increases from 8.5°C to 21°C (Figure 2).

It is interesting to note that the overshoot "bump" is observed only in monolayers compressed for the first time. The expansion curve and the subsequent compression curves show no bumps but a kink and a relatively flat plateau, similar to the DPPC curves.[14] Upon subsequent recompressions, the curves shift slightly to the left, indicating some loss of lipids from the monolayer at each cycle.

UV induced polymerization at the high pressure range results in rendering the monolayer extremely inelastic. An abrupt decline of pressure was observed upon a slight increase of area (Figure 3). UV exposure has no effect on the π-A curves if applied at low pressure ranges where no domains are observed.

To examine the effect of possible "impurities seeding" as a cause for the disappearance of the "bump" during repeated or rapid compressions, 5 mole% of 1-palmitoyl-2-oleoyl-phosphatidylcholine (POPC) was mixed with PC-A. The resulting π-A curve shifts to a larger area per molecule at high pressure, indicating a looser packing with "impurity" molecules (Figure 3). The bump at the phase transition is much less prominent than that for pure PC-A.

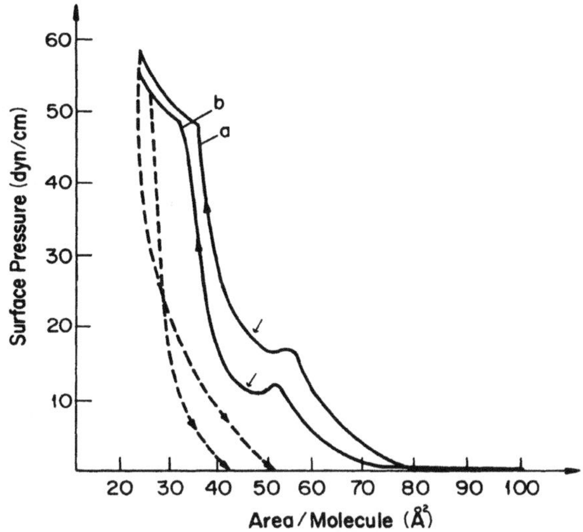

Figure 3. The initial compression (———) and decompression (after UV exposure ------) isotherms of (a) PC-A + 5 mole % of POPC at 11.5°C and (b) PC-A at 10.5°C.

2. Fluorescence Microscopic Imaging of Monolayer Domains

At near zero pressure and prior to compression, the fluorescence microscopic view of PC-A monolayers containing 1% PE-Rh is homogeneous. Upon continuous but slow compression, dark domains begin to appear as the monolayer area approaches about 45 Å^2/molecule. The domain distribution in the monolayer is not uniform; domains cluster into strips. Dark domains first appear as irregular dots. As the area is reduced, the domains show curved, feather-like shapes (Figure 4a). The thick ends of the feathers join together, and the thin ends invariably curve counterclockwise, resulting in a "phoenix tail" cluster.

As the monolayer is compressed, the phoenix clusters are pushed together but the thickness of each feather remain constant (Figure 4b). During the expansion cycle, the feather-shape domains disperse and separate (Figure 4c), and finally merge with the uniform brightness background. The feather-like domains are longer at lower temperatures (Figure 5a). Some feather-shape domains show branching at the thin end. If the monolayer pressure is reduced to near zero and the monolayer then recompressed, the feather-shape domains become shorter and more numerous, and the phoenix clusters are much smaller (Figure 4d).

When monolayers of PC-A are exposed to UV while compressed to less than 40 Å^2/molecule (about 50 dynes/cm), polymerization occurs as indicated by the inelastic behavior of the monolayer (Figure 3). Upon decompression, phoenix clusters remain and become thickened (Figure 5b). Further decompression leads to the breaking-up of feathers into fragments, but each fragment retains the elongated or spindle shape. In some areas, these domain fragments coalesce into irregular shape patches. Further expanding the area results in the transformation of single feathers and remaining phoenix clusters into large irregular shape patches. Recompressing these samples again results in the reformation of domains, similar to those shown in Figure 5b.

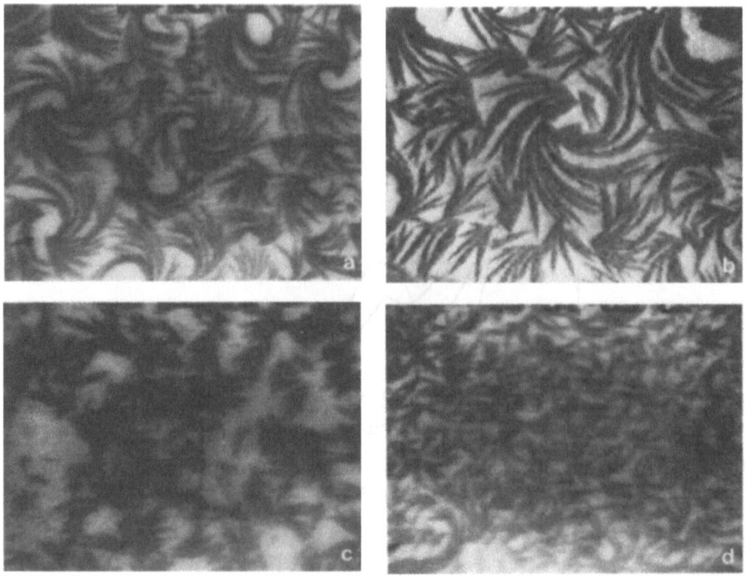

Figure 4. Solid domains in a PC-A (+1% PE-Rh) monolayer at 10.5°C, as revealed from a higher fluorescent background of PE-Rh labelled fluid lipids. The surface pressures are: a) 12 and b) 30 dynes/cm in first time compression, then reduced to c) 4 dynes/cm and recompressed to d) 40 dynes/cm.

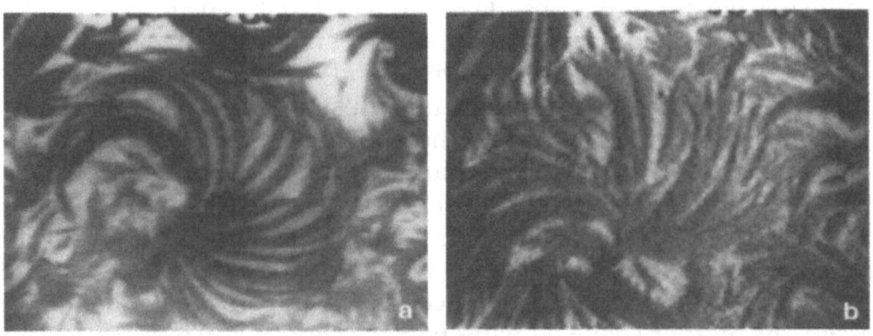

Figure 5. Solid domains in PC-A (+1% PE-Rh) monolayer at 8.5°C, as revealed from a higher fluorescent background of PE-Rh labelled fluid lipids. The surface pressures are: a) 13 dynes/cm in first time compression. The monolayer is then exposed to UV and decompressed to b) 10 dynes/cm.

3. *Electron Diffraction of Monolayer Domains*

Monolayers transferred to Formvar films show similar fluorescent patterns as those seen from monolayers at the air-water interface. The transfer is carried out under conditions for Figure 5a, either before or after exposure to UV. An example of fluorescence and bright field images of a selected area is shown in Figure 6a,b. Electron diffraction from dark domains or from bright interdomain space, as identified from marker-grid positions in fluorescence micrographs, gives only a diffuse ring (Figure 6c) typically of a liquid-crystalline bilayer.[14] Expanding the selected diffraction area to include many dark domains and interdomain space results in similar diffraction patterns. On the contrary, electron diffraction derived from dark and bright areas of a DP$(Me)_2$PE monolayer transferred to a Formvar film give distinct patterns (Figures 7 a,c). For patterns derived from dark domains, a hexagon diffraction pattern with a (100) spacing of 4.2Å is observed, as expected from a gel phase bilayer.[14]

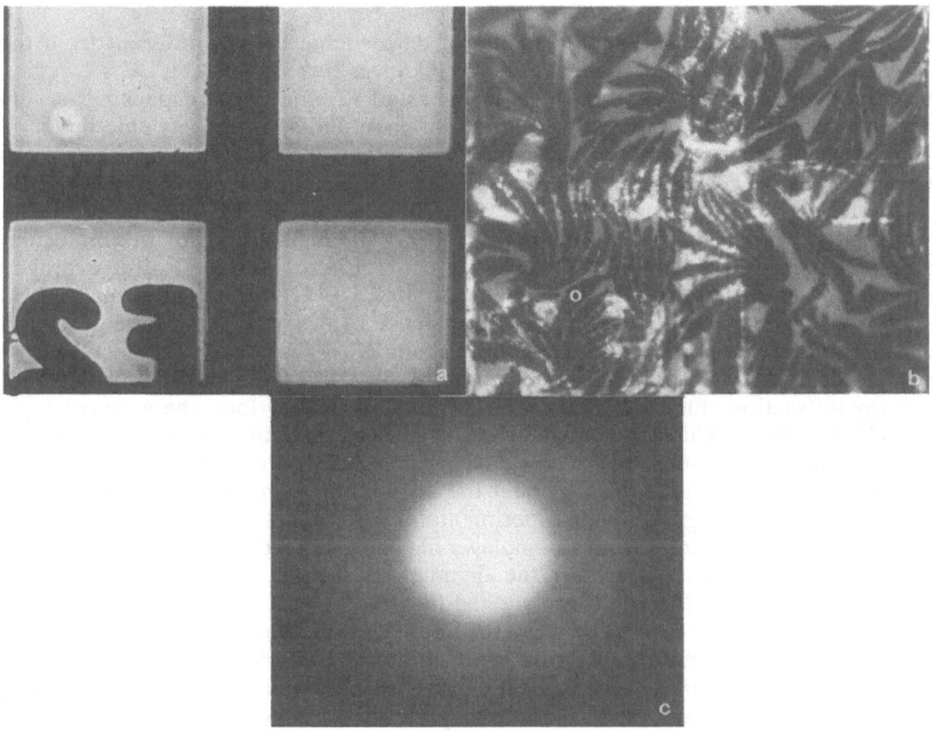

Figure 6. (a,b) A pair of bright field and fluorescence micrographs of a selected area on a marker grid which supports a monolayer of PC-A (+1% PE-Rh) on a Formvar film. (c) Electron diffraction from the circled area in (b). Diffraction patterns derived from bright inter-domain area are similar to (c).

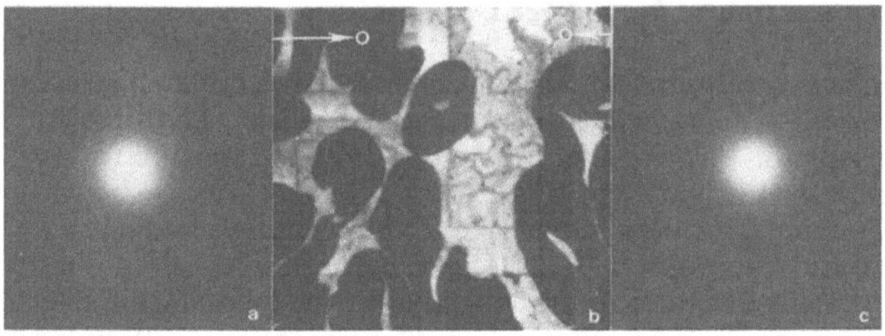

Figure 7. (a,c) Electron diffraction from circled areas of a monolayer of $DP(Me)_2PE$ transferred to a Formvar film. The selected areas on a marker grid are indicated on the fluorescence micrograph (b).

DISCUSSION

The monolayer π-A characteristics of diacetylenic PC are different from the well-studied saturated PC's, in that (1) the well known kink and plateau LE/LC region is replaced by a "bump" and a steep incline, and (2) the S-phase packing is very sensitive to temperature. Previous monolayer π-A studies of diacetylenic lipids[12,13] included some measurement of hexacosa-10,12-diynoyl-PC, but these characteristics were not noted. The difference is perhaps due to different diacetylenic PC's used.

The critical temperature is lower for the diacetylenic PC (~21°C for PC-A versus 41°C for DPPC and even higher in DSPC), indicating the intermolecular cohesive energy is less than its saturated counterpart. However, the averaged molecular area of solid phase diacetylenic lipids (35-45 $Å^2$) is smaller than DPPC (45-50 $Å^2$), given that the temperature dependence of the former is wide. This is in contrast to a previous report using a longer chain PC.[13] The smaller molecular area may be accommodated only if the packing of the subcell is different. This is depicted by x-ray diffraction studies of liposomes and tubules of very long chain diacetylenic lipids.[8,9] For PC-A which has only 18 carbons in its acyl chains, and has diacetylenic bonds at the 2,4-carbons, the molecular packing seems to be dominated by the alignment of the diacetylenic bonds near the head group. The remainder of the acyl chains are not well aligned at our experimental temperatures to give a gel-phase-like electron diffraction pattern from a monolayer deposit, even if the deposited monolayer contains dark domains. Apparently, the alignment of the diacetylenic moiety alone in the dark domains is sufficient to exclude the fluorescence probe from these domains.

The marked differences between the initial and subsequent compression curves and their domain geometries indicate a possibility of lack of nucleation in the first compression. The "bump" in the first time compression in the π-A isotherm is similar to a supercooling phenomenon, where extra pressure is needed to overcome the lack of nuclei for crystal formation. The nucleation in this case involves the alignment of diacetylenic bond planes of neighboring fatty acids. Similar interpretations are given for "bumps" in π-A curves.[19,20] This hypothesis is supported by the fact that the first time "crystallized" domains are larger, and several of these feather-shaped "crystals" are initiated from a single center, forming a phoenix cluster. The thin, elongated domains indicate that well oriented dipoles exist in these domains against the line tension which tends to favor round domains.[16,17] The fact that the domains remain thin and become even thinner at high surface pressure is in accordance with the theoretical prediction[16,17] that the higher the dipole density, the stronger is the mutual repulsion which keeps domains narrow against line tension. Lower tempera-

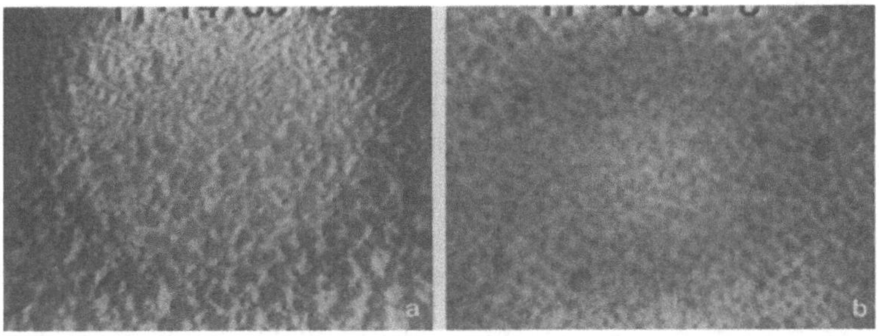

Figure 8. Solid domains in a monolayer of PC-A + 5 mole% of POPC at 11.5°C, as revealed from a higher fluorescent background of PE-Rh labelled fluid lipids. The surface pressure is (a) 48.5 dynes/cm in first time compression. The monolayer is then exposed to UV and decompressed to 0 dyne/cm (b).

tures favor the stabilization of molecular and dipole alignment, therefore longer and thinner domains are found (Figures 4,5). The strong dipole alignment we found in PC-A is likely to be a result of the alignment of the diacetylene bonds at the same positions along the chains.

After the first compression, apparently some molecules become interlocked and provide nucleation for subsequent domain formation on later compressions. Therefore the second and later compression isotherms show no "supercooling" bumps, and the domains are more numerous and smaller. The nucleation hypothesis agrees with the fact that faster compression rate abolishes the bump, since the rate of compression may not concur with the rate of nucleation. This hypothesis is also endorsed by the result of doping the PC-A monolayer with POPC, which acts as an impurity, and provides seeds for crystal growth. The seeding provided by POPC is not very effective, since most of the impurity molecules, including the fluorescence label, prefer fluid lipids. Even so, the doped PC-A has a less prominent bump in the first compression isotherm (Figure 3), and the domains are much smaller (Figure 8), either before or after exposure to UV. A mixture of both elongated and more round shape domains indicate that there is some degree of phase separation. The PC-A we used is apparently of high chemical and optical purity.

The polymerization of PC-A was not completed in our experiment, as indicated by the reformation of domains upon decompression of the monolayer. At reduced pressures when the molecular dipole density was lowered, unpolymerized molecules rearranged to form irregular-shaped domains which screened the polymerized domains. Recompressing the monolayer re-introduced the feather-like domains in a manner similar to unpolymerized lipids. However, for partially polymerized monolayers, the loss of elasticity and the persistence of solid domains showed that a considerable portion of monolayer lipid was polymerized.

In conclusion, even at high surface pressure, polymerized lipid monolayers are not as homogeneous as one expects. The heterogeneous domain structure should be taken into consideration when Langmuir-Blodgett films are made from these lipids for application purposes.

ACKNOWLEDGEMENT

This work is supported by a grant GM-28120 (to S.W.H.) and HL-16660 (to R.B.) from the National Institutes of Health.

REFERENCES

1. B. P. Gaber, J. M. Schnur, and D. Chapman, eds., "Biotechnological Applications of Lipid Microstructures," Plenum Press, New York (1988).
2. S. L. Regan, Polymerized phosphatidylcholine vesicles as drug carriers. *Ann. N.Y. Acad. Sci.* 446:296-307 (1985).
3. D. S. Johnston, L. R. McLean, M. A. Whittam, A. D. Clark, and D. Chapman, Spectra and physical properties of liposomes and monolayers of polymerizable phospholipids containing diacetylene groups in one or both acyl chains. *Biochem.* 22:3194-3202 (1983).
4. H. M. McConnell, Structures and transitions in lipid monolayers at the air-water interface. *Annu. Rev. Phys. Chem.* 42:171-195 (1991).
5. H. Möhwald, H., Phospholipid and phospholipid-protein monolayers at the air/water interface. *Annu. Rev. Phys. Chem.* 41:441-76 (1990).
6. R. M. Weis, R. M., Fluorescence microscopy of phospholipid monolayer phase transitions. *Chemistry and Physics of Lipids* 57:227-239 (1991).
7. R. M. Weis, R. M. and H. M. McConnell, Cholesterol stabilizes the crystal-liquid interface in phospholipid monolayers. *J. Phys. Chem.* 89:4453-4459 (1985).
8. P. Yager, P. E. Schoen, C. Davies, R. Price, and A. Singh, Structure of lipid tubules formed from a polymerizable lecithin. *Bioph. J.* 48:899-906 (1985).
9. M. J. Caffrey, J. Hogan, and A. S. Rudolph, Diacetylenic lipid microstructures: structural characterization by x-ray diffraction and comparison with the saturated phosphatidylcholine analogue. *Biochem.* 30:2134-2146 (1991).
10. A. L. Plant, D.M. Benson, and G. L. Trusty, Probing the structure of diacetylenic phospholipid tubules with fluorescent lipophiles. *Biophys. J.* 57:925-933 (1990).
11. R. Treanor and M. D. Pace, Microstructure, order and fluidity of 1,2-bis(tricosa-10,12-diynoyl)-*sn*-glycero-3-phosphocholine($DC_{8,9}PC$), a polymerizable lipid, by ESR and NMR. *Biochimica et Biophysica Acta.* 1046:1-11 (1990).
12. H. H. Hub, B. Hupfer, H. Koch, and H. Ringsdorf, Polymerization of lipid and lysolipid like diacetylenes in monolayers and liposomes, *J. Macromol. Sci.-Chem.* A15:701-715 (1981).
13. B. Hupfer, B. and H. Ringsdorf, Spreading and polymerization behavior of diacetylenic phospholipids at the gas-water interface, *Chemistry and Physics of Lipids* 33:263-282 (1983).
14. S. W. Hui, S. W. M. Cowden, D. Papahadjopoulos, and D. F. Parsons, Electron diffraction study of hydrated phospholipid single bilayers: effects of temperature hydration and surface pressure of the "precursor" monolayer. *Biochimica Biophysica Acta.* 382:265-275 (1975).
15. A. Fischer, M. Lösche, H. Möhwald, and E. Sackmann, On the nature of the lipid monolayer phase transition *J. Physique Lett.* 45:L-785-L-791 (1984).
16. D. J. Keller, H. M. McConnell, and V. T. Moy, Theory of superstructures in lipid monolayer phase transitions. *J. Phys. Chem.* 90:2311-2315 (1986).
17. H. M. McConnell and V. T. Moy, Shapes of finite two-dimensional lipid domains. *J. Phys. Chem.* 92:4520-4525 (1988).
18. Z. Xu, Z., H.-S. Byun, and R. Bittman, Synthesis of photopolymerizable long-chain conjugated diacetylenic acids and alcohols from butadiyne synthons. *J. Org. Chem.* 56:7193-7198 (1991).
19. T. Thuren, T., J. A. Virtanen, P. Vainio, and P. K. J. Kinnunen, Hydrolysis of 1-triacontanoyl-2-(pyren-1-yl)hexano-*sn*-glycero-3-phosphocholine by human pancreatic phospholipase A_2. *Chemistry and Physics of Lipids* 33:283-292 (1983).

20. H. Matuo, D. K. Rice, D. M. Balthasar, and D. A. Cadenhead, Mixed monolayer studies of 12-hydroxyoctadecanoic acid and its esters, *Chem. Phys. Lipids* 30:367-380 (1982).
21. H. Yu and S. W. Hui, Microdomain structures of methylation effects of phosphatidylethanolamine monolayers, *Chem. Phys. Lipids* (in press).

THE EFFECT OF MODIFICATION OF THE POLAR HEADGROUPS OF ACIDIC DIACETYLENIC PHOSPHOLIPIDS ON THEIR BILAYER MORPHOLOGY

Michael A. Markowitz, Joel M. Schnur and Alok Singh

Center for Biomolecular Science and Engineering, Naval Research Laboratory, Washington, DC 20375

ABSTRACT

The polar headgroup of acidic diacetylenic phospholipids has been modified synthetically and by systematic changes in their aqueous environment. The effect of these modifications on bilayer morphology have been investigated. For this study, the acidic diacetylenic phospholipids 1,2-Bis(10,12-tricosadiynoyl)-sn-glycero-3-phosphohydroxyethanol ($DC_{8,9}PEtOH$), -phosphohydroxypropanol ($DC_{8,9}PPrOH$), and -phosphohydroxybutanol ($DC_{8,9}PBuOH$) were prepared by means of a phospholipase D catalyzed transphosphatidylation. The diameter of tubule-like cylinders which were formed in aqueous media was influenced by headgroup size, pH, and the counteranion present in the aqueous medium. The results are discussed with regard to the manner in which the modifications affect the ability of the metal ion to bind to the headgroup and affect headgroup mobility and the consequent affect on acyl chain packing.

INTRODUCTION

In recent years, a great deal of research has been directed towards characterizing and understanding the mechanism of tubule formation from diacetylenic phosphocholines. Other investigations have examined the utility of tubules in technological applications.[1-26] Many of the potential technological applications which take advantage of the controlled release properties of tubules or their aspect ratio would be enhanced by the ability to rationally control tubule diameters over a wide diameter range. Our earlier studies have focussed on the role of the diacetylenic acyl chains in tubule formation. The results of these studies have demonstrated that the remarkably rigid diacetylenic acyl chains are required for tubule formation.[27,28] Various studies have shown that the length of the acyl chains and the placement of the diacetylene group within the chains do not have a major effect on tubule properties, i.e., tubules of similar diameter are still formed. Given these results, we decided to study how the headgroup modification would affect tubule formation and morphology.

Synthetic Microstructures in Biological Research, Edited by J.M. Schnur and M. Peckerar, Plenum Press, New York, 1992

We have recently reported that the formation characteristics of tubules from an acidic diacetylenic phospholipid, 1,2-bis(10,12-tricosadiynoyl)-sn-glycero-3-phosphohydroxyethanol ($DC_{8,9}PEtOH$), can be influenced by a metal counterion.[29,30] The range of tubule diameters extended down to 0.04 μm, a size not previously observed for those formed from diacetylenic phosphocholines. No dependence of tubule diameter or morphology on countercation size was observed. Studies of rod-like microstructure and tubule formation from diacetylenic chain surfactants with glutamate, oligopeptide, and cerebroside headgroups have been reported.[31-36] In particular, tubules with diameters of 0.03 μm have been formed from the oligopeptide headgroup-based surfactants. These results stimulated us to proceed further by systematically investigating the role of the headgroup geometry, pH, and counteranion on the bilayer morphology of the microstructures formed from this class of phospholipid. For this study, we prepared the following compounds by exchanging the choline functionality of ($DC_{8,9}PC$) with the appropriate glycol by means of a phospholipase D transphosphatidylation[37] (Figure 1).

The effect of surface charge and molecular geometry on the self-assembly behavior of surfactants has been thoroughly investigated.[38,39] Altering the headgroup structure of phosphatidyl cholines and phosphatidyl ethanolamines or the bulk pH and salt content of the aqueous medium has been shown to affect lipid bilayer surface charge and interfacial hydration.[40] These in turn profoundly affect the functional properties of the membrane. NMR studies have demonstrated that the binding of metal ions to the negatively charged phosphates of acidic phospholipids affects polar headgroup mobility and therefore, molecular motion and acyl chain packing.[41-44] In addition, the role of pH and Ca^{2+} on the phase behavior of phosphatidyl glycerol and in the formation of cochleate cylinders by phosphatidyl serine has been well characterized.[45,46]

By working with acidic diacetylenic phospholipids **1-3**, **n** could be incrementally increased. This would progressively alter headgroup size and, consequently, acyl chain packing and these changes should affect lipid hydration. Likewise, systematically altering the pH and counterions of the metal ion will also have an affect on the headgroup to headgroup interactions and the acyl chain interactions. As a result of altering headgroup size and environment, changes in bilayer morphology could occur. These changes were observed by electron microscopy.

MATERIALS AND METHODS

Materials

$CHCl_3$, CH_3OH, ether, and acetone were obtained from Burdick and Jackson (Muskegon, MI). The $CHCl_3$ and CH_3OH were used as is and the ether and acetone were stored over $CaCl_2$. $CaCl_2$, NH_4OH, glycerol, and HCl were obtained from Fisher (Columbia, MD). NaOAc, Tris, ethylene glycol, 1,3-propanediol, 1,4-butanediol, $CDCl_3$ and $CuCl_2$ were obtained from Aldrich Chemical Co (Milwaukee, WI). EDTA and Dragendorf's reagent were obtained from Sigma Chemical Co (St. Louis, MO). $DC_{8,9}PC$ was synthesized in our laboratory following the published procedures.[47-50] Water was three times distilled from glass using a Corning MP-12A apparatus. IR spectra were obtained using a Perkin-Elmer 1800 FT-IR. NMR spectra were obtained in $CDCl_3$ using a Varian EM 390 nuclear magnetic resonance spectrophotometer. FAB mass spectra were obtained using a Finnigan triple quadrupole mass spectrometer.

$CH_2OC(O)\text{-}(CH_2)_8\text{-}C{\equiv}C\text{-}C{\equiv}C\text{-}(CH_2)_9\text{-}CH_3$
|
$CHOC(O)\text{-}(CH_2)_8\text{-}C{\equiv}C\text{-}C{\equiv}C\text{-}(CH_2)_9\text{-}CH_3$
|
$CH_2OP(O_3)\text{-}(CH_2)_n\text{-}OH$

Figure 1. Acidic diacetylenic phospholipids.
1; n = 2 **2**; n = 3 **3**; n = 4

Synthesis of Phospholipids (1-3)

The lipids were prepared using an established phospholipase D catalyzed transphosphatidylation procedure.[37] The general procedure for the lipid preparation was as follows: 450 mg (0.44 mmols) of $DC_{8,9}PC$ was dissolved in 48 mL of ether with gentle warming. A 100 fold excess of the appropriate diol was added. Then, 24 mL of a 1/1 (v/v) mixture of the enzyme extract and 0.2 M aq. NaOAc-HOAc buffer (40 mM $CaCl_2$, pH 5.6) was added and the reaction mixture was stirred vigorously at 40°c for 10 hrs. During this time, the reaction mixture was protected from light. The progress of the reaction was followed by thin layer chromatography using two solvent systems (A 65/26/4 $CHCl_3/CH_3OH/H_2O$ (v/v/v), B 60/30/5 $CHCl_3/CH_3OH/NH_4OH$ (v/v/v)). The developed plates were analyzed using Dragendorf's reagent to monitor the disappearance of the PC and the appearance of the product lipid was monitored with phosphomolybdate reagent and iodine vapor. The ether was evaporated under reduced pressure and 25 mL of a saturated aq. EDTA solution was added to the remaining aqueous phase. The lipid was extracted successively with a 2/1 $CHCl_3/CH_3OH$ (v/v) mixture and $CHCl_3$. The organic fractions were combined and the solvent was evaporated under reduced pressure. The residue was redissolved in a minimum amount of $CHCl_3$ and the pure lipid was precipitated at 0°C with dry acetone. To ensure the absence of any ion the lipids were dissolved in $CHCl_3$ and treated with an ion exchange resin (Biorad AG 50W-X8). FAB mass spectral analysis established that the treated compounds were ion free. It should be noted that other methods, acidification and/or treatment with EDTA, were not sufficient to completely remove the calcium ion. The lipids were characterized by IR and NMR spectra and purity was routinely checked by TLC. The characterization data for each lipid are as follows: **1**, NMR ($CDCl_3$) 0.9 (distorted **t**, 6H, CH_3), 1.25 (**s**, 56H, $-CH_2-$), 2.2 (**m**, 12H, $-CH_2$-COO,$-CH_2$-C-), 3.8-4.3 (**m**, 8H, $-OCH_2-$), and 5.2 (broad **m**,1H, -O-CH-) ppm. Negative ion mass spectrum (using triethanolamine as matrix) revealed a mass peak (m/e) at 871.3 $(M\text{-}H)^-$; **2**, NMR ($CDCl_3$) a spectrum similar to that of **1** was obtained except that chemical shift at 1.25 was broad at the base due to the presence of methylene proton from propane diol fragment. Mass spectrum (m/e) 885.4 $(M\text{-}H)^-$; **3**, NMR contained additional chemical shift peaked at 1.65 ppm and the number of protons calculated under 1.65 and 1.25 ppm were 60. Mass spectra (m/e) 899.1 $(M\text{-}H)^-$.

Buffer Preparation

The pH was maintained using the following aqueous buffers: pH 2.0, 0.2 M KCl-HCl; pH 5.6, 0.2 M NaOAc-HOAc; pH 7.5 and 9.0, 0.2 M Tris-HCl. All buffers contained 1 mM $CuCl_2$ unless otherwise stated.

Calorimetry

Calorimetric studies were performed with a Perkin-Elmer DSC-7 differential scanning calorimeter. Samples were prepared by weighing a known amount of the lipid into the DSC pan and then injecting the appropriate amount of buffer. The lipid/metal ion ratio in each experiment was 2.7:1. The DSC pan was immediately sealed, kept at 80°C for 3 hrs, cooled to room temperature and transferred to the calorimeter. Samples were scanned at a rate of 2°/min until reproducible scans were produced. The melting transition (T_m) and cooling transition (T_c) values were determined from the peak temperature of the transitions. Enthalpies were determined from the area under the peak as determined by the calorimeter. Using this procedure, a T_m of 41.7°C and an enthalpy of 7.8 kcal/mole were obtained for DPPC. These values are in good agreement with those obtained previously. The phase transitions of each lipid under the different experimental conditions used in this study are given in Table 1.

Microstructure Preparation

The lipid microstructures were prepared using slow cooling. A $CHCl_3$ solution of the lipid was evaporated to dryness under a stream of argon, then placed in vacuo for 60 min. Buffer was added to the lipid. Generally, the concentration of the lipid was 2.1 mg/mL except for the samples used in freeze fracture experiments. No concentration effects were observed for the samples used in freeze fracture. The lipid was sonicated (Branson Sonifier 250, sonicating cup) above the phase transition temperature of each lipid until the solution became translucent. The temperature was maintained using a MGW Lauda RC6 circulating water bath. The sonicated solution was cooled to room temperature and then kept at 4°C for at least 3 hrs. Then, the lipid samples were heated at 80°C for 45 minutes. The temperature was slowly cooled to a few degrees below the exothermic transition temperature of the lipid at < 1°C/min. The final 10° of the cooling cycle proceeded at a rate of 1°C/5 min. The solution was maintained at this temperature for 14 hrs before electron microscopic analysis. In order to determine the extent to which the lipids were hydrolyzed as a result of this procedure, the lipid dispersion was lyophilized and redissolved in 1/1 $CHCl_3/CH_3OH$ (v/v). An aliquot of this mixture analyzed by thin layer chromatography (Silica Gel 60, 250 μm layer, Whatman, Hillsboro, OR). The plate was developed in (Solvent A). The results indicated that no more than 5% of the lipid was hydrolyzed except at pH 2.0 where significant hydrolysis occurred.

Microscopy

Electron microscopy was performed using a Zeiss EM-10. For transmission electron microscopy (TEM), samples were pipetted onto carbon-coated copper grids, air dried, and observed directly. All samples were unstained. For freeze-fracture, the cryoprotectant glycerol (10 wt.%) was added to samples immediately before freezing. Samples were then transferred to Balzers copper specimen plates, equilibrated at room temperature, and quickly frozen by plunging into melting nitrogen. They were then transferred to a Balzers 360 freeze-fracture device, fractured, and then replicated with a Pt-C film at -100°C and 10^{-6} Torr. The replicas were floated off onto triply distilled water then picked up on coated grids and examined.

Table 1

Melting and Crystallization Temperatures of Acidic Diacetylenic Phospholipids[1]

Compound	Salt	pH	T_m[2] °C	T_c[2] °C
1	$CuCl_2$	2.0	41.71	33.01
1	$CuCl_2$	5.6	51.12, 61.55	35.74
1	$CuCl_2$	7.5	55.65	45.75
1	$CuCl_2$	9.0	45.46, 49.91	[3]
1	CuF_2	5.6	49.86	36.20
1	$CuClO_4$	5.6	51.89, 62.19	40.02
1	$CuNO_3$	5.6	51.39, 61.07	39.22
1	$CuSO_4$	5.6	51.47, 61.86	39.41
2	$CuCl_2$	2.0	36.48, 53.47	27.06, 43.14
2	$CuCl_2$	5.6	47.95	38.96
2	$CuCl_2$	7.5	43.97	25.50
2	$CuCl_2$	9.0	41.23, 44.92	22.10
3	$CuCl_2$	2.0	34.90, 52.28	25.10, 41.91
3	$CuCl_2$	5.6	51.01	35.43
3	$CuCl_2$	7.5	45.47	28.25
3	$CuCl_2$	9.0	41.53, 45.66	[3]

1 Ionic strength of the buffer = 0.18

2 T_m and T_c refer to the melting and crystallization temperatures, respectively

3 No transition was observed

RESULTS

Microscopy

The electron micrographs of the microstructures formed by compounds **1-3** in 0.2 M aq. NaOAc - HOAc (pH 5.6) in the presence of $CuCl_2$ (1 mM) display the changes in morphology which occur as **n** is increased from 2-4 (Figure 2). In particular, the affect on morphology was revealed in differences in tubule diameter and features observed. Tubule diameters were measured from the center of the cylindrical structure. Unless otherwise noted, average diameters and standard deviations were based on 100 tubules. In the absence of Cu^{2+}, compound **2** did not form any cylindrical microstructures while compounds **1** and **3** did. Only a very small number of microstructures were observed for **1** (< 100) with diameters ranging from 1.00 μm - 1.20 μm while a much higher number of microstructures were observed for **3** with an average diameter of 0.88 μm ± 0.12 μm. In the presence of Cu^{2+}, **1** (Figure 2a) formed a bimodal population of small diameter tubule-like cylinders (average diameter of 0.12 μm ± 0.03 μm) which had a ropy appearance and larger diameter tubules (average diameter of 0.93 μm ± 0.15 μm). The addition of a single methylene to the headgroup alkyl chain resulted in the formation of a bimodal population of small and large tubules with average diameters of 0.30 μm ± 0.06 μm and 0.87 μm ± 0.18 μm, respectively (Figure 2b, the larger diameter tubules are not shown). These tubules resembled those formed from diacetylenic phosphocholines in appearance. A higher proportion of large diameter tubules were observed. When **n** was increased to 4, only a single population of large cylindrical microstructures which had a tapered appearance was observed. The average diameter of these cylinders was 0.80 μm ± 0.17 μm (Figure 2c). Altering of the pH of the dispersion medium produced significant changes in the diameter and observed yield of lipid cylinders. As the headgroup phosphate becomes increasingly ionized, the effective headgroup size increases. Based on the pK_a of simple alcohols and phosphatidyl glycerol, the hydroxyl group was expected to remain protonated. However, at the highest pH of this experiment, pH 9.0, partial deprotonation of this hydroxyl cannot be ruled out. At pH 2.0, none of the compounds dispersed but rather, aggregated into large precipitates. Thin layer chromatography analysis indicated that this was due to hydrolysis and/or polymerization. At pH 7.5, the compounds did not form any cylindrical microstructures in the absence of Cu^{2+}. In the presence of Cu^{2+}, homogeneous populations of tubule-like cylinders were observed from **1** (average diameter = 0.15 μm ± 0.08 μm, < 100 cylinders counted) and from **2** (average diameter = 0.38 μm ± 0.14 μm, < 100 tubules counted) while no cylindrical microstructures were observed for **3**. At pH 9.0, no microstructures were formed in the presence of Cu^{2+}. Thin layer chromatographic analysis of the lipids (Solvent A) indicated that only a small percentage (< 5%) of each of the lipids was hydrolyzed during microstructure formation at pH 5.6 - 9.0.

The replicas of freeze fractured samples of **1** (pH 5.6, 1 mM $CuCl_2$, ionic strength 0.18); **2** (1 mM $CuCl_2$, pH 7.5, ionic strength 0.18); and **3** (1 mM $CuCl_2$, pH 5.6, ionic strength 0.18) were examined by transmission electron microscopy (Figures 2d-f). The tubules formed from all three lipids consisted of a thin wall of lipid of similar thickness around an aqueous core.

The effect of counteranion on tubule formation from **1** was studied by substituting Cl^-, with various anions (0.2 M aq. NaOAc-HOAc, 1 mM Cu^{2+}, pH 5.6). In the presence of ClO_4^-, a bimodal population of tubule-like cylinders, similar in average diameters to that seen in the presence of Cl^-, was observed. In the presence of F^- a single population of tubules with average diameter of 0.78 μm ± 0.13 μm was

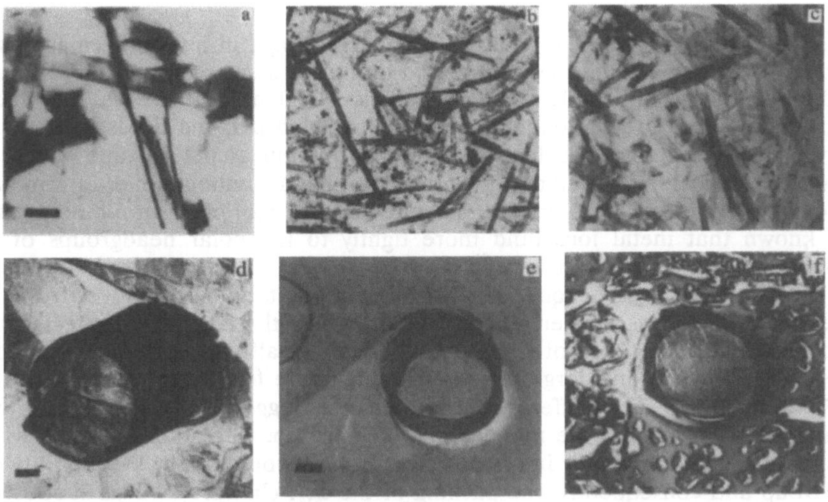

Figure 2. Transmission electron micrographs of microstructures formed from:
a) **1** (1 mM $CuCl_2$, pH 5.6, 0.18 IS)
b) **2** (1 mM $CuCl_2$, pH 5.6, 0.18 IS)
c) **3** (1 mM $CuCl_2$, pH 5.6, 0.18 IS)
d) freeze fracture of **1** (1 mM $CuCl_2$, pH 5.6, 0.18 IS)
e) freeze fracture of **2** (1 mM $CuCl_2$, pH 5.6, 0.18 IS)
f) freeze fracture of **3** (1 mM $CuCl_2$, pH 5.6, 0.18 IS)
For a, bar represents 1.0 μm.; for b and c, bar represents 2.0 μm.; for d and e, bar represents 0.1 μm.; for f, bar represents 0.2 μm.

formed. Tubules formed in the presence of SO_4^{2-} and NO_3^- from **1** had average diameters of 0.77 μm + 0.11 μm and 0.96 μm + 0.12 μm, respectively. Transmission electron microscopic analysis of a replica of a freeze fractured sample of tubules formed from **1** in the presence of 1 mM CuF_2 (pH 5.6, ionic strength 0.18) revealed that the thickness of the tubule walls was similar to that seen in Figures 2d-f.

DISCUSSION

The transmission electron microscopy (TEM) measurements provide clear evidence of a correlation between the size of the headgroup of the diacetylenic phospholipids and the resulting microstructure morphology. The diameters observed by TEM are external diameters. However, the freeze fracture results indicate that tubule wall thickness was not significantly affected by changes in headgroup size or environment. The ropy appearance of the cylinders formed from **1** may arise from their formation by the twisting of the larger diameter tubules or by some other mechanism which does not involve the larger diameter tubules. The precise mechanism cannot be determined based on the available experimental data. The effect of increasing the number of methylenes in the headgroup was demonstrated by the corresponding changes in tubule diameter. At pH 5.6 and pH 7.5, the average diameter of the smaller diameter cylinders observed for **2** was greater than that for **1** and no small diameter tubules were observed with **3**. Variation of the hydrogen association of these lipids produced changes in observed tubule yield consistent with a change in the way the headgroups are able to engage in intermolecular hydrogen bonding. At low pH, the fully protonated lipids are uncharged and therefore, their amphiphilicity is significantly decreased. This is reflected in the way in which the lipids agglomerate in the aqueous medium. At high pH, the lipids will be almost entirely deprotonated. The lack of tubule formation at pH 9.0 and the reduced amount of cylinders observed at pH 7.5 suggests that a combination of intermolecular hydrogen bonding and Cu^{2+}-lipid binding are necessary for efficient tubule formation. It is known that metal ions bind more tightly to the polar headgroups of acidic phospholipids as the pH increases.[51]

The effect of the aqueous environment on tubule formation from acidic diacetylenic lipids was further demonstrated by how the counterion affected tubule diameter. In the presence of Cl^- and ClO_4^-, both small and large diameter tubules were observed while only large diameter tubules were formed in the presence of F^-, NO_3^-, and SO_4^{2-}. The differences in diameter suggest that the nature of the counteranion may affect the manner in which the countercation interacts with the phosphate at the lipid-water interface. This in turn would affect the mobility of the headgroup and consequently, the packing of the acyl chains.[41-44] This supposition is supported by results which demonstrate that at $Ca^{2+} > 10^{-3}$ M, the ratio of metal ion to phosphatidyl serine approaches one and a lipid-Ca^{2+}-X complex is formed where X is the counteranion.[52-55] However, the actual ratio of metal ion - lipid was not determined in this study. The results of this study were not strictly influenced by counterion size which increases in the following manner: $F^- < Cl^- < NO_3^- < SO_4^{2-} < ClO_4^-$. Since SO_4^{2-}, ClO_4^-, and NO_3^- have different configurations, it is possible that the symmetry of each counterion plays a role in this process.[56] In addition, other factors such as the manner in which the anions influence the neighboring water molecules may also be a factor.

CONCLUSION

It is now clear that modification of the headgroup, either by synthesis or by changes in solvent condition, can profoundly affect the diameter of this class of cylindrical microstructures. The set of observations reported here suggest that the headgroup plays an important role in determining the magnitude of the curl or wrapping of the bilayer sheets which are thought to be important in the formation of the tubules. It is possible that a fulcrum effect exists in that headgroup modification, either synthetically or by a perturbation of the aqueous environment, affects the spacing between headgroups in the bilayer and thus, the tendency for curling.

At this time, we do not understand the mechanism of tubule formation from acidic diacetylenic phospholipids. One of the more interesting questions involves the coexistence of both small and large diameter tubules under certain conditions. A possible explanation is that the small diameter tubules are formed from a further twisting of the large diameter tubules. Another possible explanation is that the microstructure formation is more sensitive to temperature during formation than are the phosphocholine analogs. Perhaps the type of thermal treatment has a profound effect on tubule diameter.

Further studies of the effect of headgroup shape and environment on the bilayer morphology of microstructures formed from these lipids are in progress. A systematic altering of the ionic strength of the dispersion medium should give some insight into how the interaction between the metal ion and the headgroup phosphate affects tubule morphology. The effect of the various headgroup perturbations on the packing of the acyl chains at the air-water interface will be examined by monolayer studies. X-ray and electron beam studies are planned in order to directly observe the headgroup packing in tubules as a function of changes in size and environment. NMR spectroscopy will also be used to investigate how changes in the conformation of the lipid headgroups affects tubule morphology.

ACKNOWLEDGEMENTS

We wish to thank Dr. Bruce Gaber, Dr. Alan Rudolph, Dr. R. Shashidhar, and Dr. B. Ratna for helpful discussions, Dr. Jim Callaghan for mass spectral analysis, Mr. Ronald Price for the electron microscopy, and Dr. Li-I Tsao for preparing the phospholipase D extract. This research was conducted through the Naval Research Laboratory (NRL) program on the Molecular Design of Microstructures which was funded by the Office of Naval Research. Dr. Michael Markowitz is a recipient of an Office of Naval Technology postdoctoral fellowship.

REFERENCES

1. D. S. Johnston, S. Sanghera, M. Pons, and D. Chapman, Phospholipid polymers; synthesis and spectral characteristics, *Biochim. Biophys. Acta* 602:57 (1980).
2. H. Hub, B. Hupfer, H. Koch, and H. Ringsdorf, Polymerizable phospholipid analogues. New stable biomembrane and cell models, *Angew. Chem. Int. Ed. Engl.* 19:938 (1980).
3. D. F. O'Brien, R. T. Klingbiel, and T. H. Whitesides, The polymerization of lipid diacetylenes in bimolecular-layer membranes, *J. Polym. Sci., Polym. Lett. Ed.* 19:95-101 (1981).
4. A. Singh, R. R. Price, P. E. Schoen, P. Yager, and J. M. Schnur, Tubule formation by heterobifunctional polymerizable lipids: synthesis and characterization, *Polymer Preprints* 27:393 (1986).
5. B. Hupfer, H. Ringsdorf, and H. Schupp, Liposome from polymerizable phospholipids, *Chem. Phys. Lipids* 33:355 (1983).
6. P. Yager, P. E. Schoen, Formation of tubules by a polymerizable surfactant, *Mol. Cryst. Liq. Cryst.* 106:371 (1984).
7. J. H. Georger, A. Singh, R. R. Price, J. M. Schnur, P. Yager, and P. E. Schoen, Helical and tubular microstructures formed by polymerizable phosphatidylcholines, *J. Amer. Chem. Soc.* 109:6169 (1987).
8. P. Yager, P. Schoen, C. A. Davies, R. Price, and A. Singh, Structure of lipid tubules formed from a polymerizable lecithin, *Biophys. J.* 48:899 (1985).
9. A. L. Plant, D. M. Benson, and G. L. Trusty, Probing the structure of diacetylenic phospholipid tubules with fluorescent lipophiles, *Biophys. J.* 57:925 (1990).
10. M. F. Sonnenschein and R. G. Weiss, Depth profiles for permeation of water and oxygen in vesicles and tubule phases by fluorescence quenching studies, *Photochem. Photobiol.* 51:539 (1990).
11. A. Singh, T. G. Burke, J. M. Calvert, J. H. Georger, B. Herendeen, R. R. Price, P. E. Schoen, and P. Yager, Lateral phase separation based on chirality in a polymerizable lipid and its influence on formation of tubular microstructures, *Chem. Phys. Lipids* 47:135 (1988).
12. A. S. Rudolph, J. M. Calvert, M. E. Ayers, and J. M. Schnur, Water-free self-assembly of phospholipid tubules, *J. Amer. Chem. Soc.* 111:8516 (1989).

13. T. G. Burke, A. S. Rudolph, R. R. Price, J. P. Sheridan, A. W. Dalziel, A. Singh, and P. E. Schoen, Differential scanning calorimetric study of the thermotropic phase behavior of a polymerizable, tubule-forming lipid, *Chem. Phys. Lipids* 48:215 (1988).
14. A. S. Rudolph, P. E. Schoen, M. Nagumo, F. Behroozi, T. G. Burke, M. E. Ayers, A. Singh, and R. Treanor, Spectroscopic studies of tubule-forming polymerizable lecithins, *SPIE* 1057:57 (1989).
15. P. E. Schoen and P. Yager, Spectroscopic studies of polymerized surfactants: 1,2-bis(10, 12-tricosadiynoyl)-sn-glycero-3-phosphocholine, *J. Polym. Sci: Polym. Phys. ed.* 23:2203 (1985).
16. A. S. Rudolph, B. P. Singh, A. Singh, and T. G. Burke, Phase characteristics of positional isomers of 1,2 bis heptacosadiynoyl-sn-glycero-3-phosphocholines, *Biochim. Biophys. Acta* 943:454 (1988).
17. W. Helfrich, Helical bilayer structures due to spontaneous torsion of the edges, *J. Chem. Phys.* 85:1085 (1986).
18. P. G. de Gennes, Surface and interphase physics. Electrostatic buckling of chiral lipid bilayers, *C.R. Seances Acad. Sci.* 304:259 (1987).
19. P. Yager, R. R. Price, J. M. Schnur, P. E. Schoen, A. Singh, and D. Rhodes, The mechanism of formation of lipid tubules from liposomes, *Chem. Phys. Lipids* 46:171 (1988).
20. P. Schoen, P. Yager, J. P. Sheridan, R. Price, J. M. Schnur, A. Singh, D. G. Rhodes, and S. L. Blechner, Order in diacetylenic microstructures, *Mol. Cryst. Liq. Cryst.* 153:357 (1987).
21. P. Yager, P. E. Schoen, J. H. Georger, R. R. Price, and A. Singh, Two mechanisms for forming novel tubular microstructures from polymerizable lipids, *Biophys. J.* 49:320a (1986).
22. D. G. Rhodes, S. L. Blechner, P. Yager, and P. E. Schoen, Structure of polymerizable lipid bilayers.I. 1,2-Bis (10,12-tricosadiynoyl)-sn-glycero-3-phosphocholine, a tubule-forming phosphatidylcholine, *Chem. Phys. Lipids* 49:39 (1988).
23. A. S. Rudolph, J. M. Calvert, P. E. Schoen, and J. M. Schnur, Technological development of lipid based tubule microstructures, in: "Biotechnological Applications of Lipid Microstructures," B. P. Gaber, J. M. Schnur, and D. Chapman, eds., Plenum Press, New York (1988).
24. H. Ringsdorf, B. Schlarb, and J. Venzmer, Molecular architecture and function of polymeric oriented systems: models for the study of organization, surface recognition and dynamics of biomembranes, *Angew. Chem. Int. Ed. Engl.* 27:113 (1988).
25. J. M. Schnur, R. Price, P. Schoen, P. Yager, J. Calvert, and A. Singh, Lipid based tubule microstructures, *Thin Solid Films* 152:181 (1987).
26. T. G. Burke, A. Singh, and P. Yager, Entrapment of 6-carboxyfluorescein within cylindrical phospholipid microstructures, *Ann. N. Y. Acad. Sci.* 507:330 (1987).
27. J. P. Sheridan, Conformational order in lipid tubules formed from a diacetylenic lecithin: a raman spectroscopic study, *Memorandum Report 5975, Naval Research Laboratory* (1988).
28. A. Singh and M. A. Markowitz, Phospholipids and self-assembled microstructures: accomplishments, challenges and future prospects, in: "Membrane Structure and Functions-The State of the Art," B. P. Gaber K. R. K. Easwaran, eds., Adenine Press, New York, pp. 37-51 (1992).
29. A. Singh and S. Marchywka, Synthesis and characterization of headgroup modified 1,3 diacetylenic phospholipids, *Polym. Mat. Sci. Eng.* 61:675 (1989).

30. M. Markowitz and A. Singh, Self-assembling properties of 1,2-Diacyl-sn-glycero-3-phosphohydroxyethanol: a headgroup modified diacetylenic phospholipid, *Langmuir* 7:16 (1991).
31. K. Yamada, H. Ihara, T. Ide, T. Fukumoto, and C. Hirayama, Formation of helical superstructure from single-walled bilayers by amphiphiles with oligo-L-gutamic acid headgroup, *Chem. Lett* 1713 (1984).
32. W. Curatolo, Thermal behavior of fractionated and unfractionated bovine brain cerebrosides, *Biochemistry* 21:1761 (1982).
33. W. Curatolo, The effects of ethylene glycol and dimethyl sulfoxide on cerebroside metastability, *Biochim. Biophys. Acta* 817:134 (1985).
34. W. Curatolo and F. B. Jungalwala, Phase behavior of galactocerebrosides from bovine brain, *Biochemistry* 24:6608 (1985).
35. W. Curatolo and L. J. Neuringer, The effects of cerebrosides on model membrane shape, *J. Biol. Chem* 261:17177 (1986).
36. P. Yager, J. Chappell, and D. D. Archibald, When lipid bilayers won't form liposomes: tubules, helices, and cochleate cylinders, in: "Membrane Structure and Functions-The State of the Art", B. P. Gaber and K. R. K. Easwaran, eds., Adenine Press, New York, pp. 1-19 (1992).
37. A. Singh, M. Markowitz, and Li-I Tsao, Phosphatidylhydroxyalkanols as versatile intermediates in the synthesis of headgroup modified daicetylenic phospholipids, *Syn. Comm.* 22:000 (1992).
38. J. N. Israelachvili, "Intermolecular and Surface Forces: With Applications to Colloidal and Biological Systems," Academic Press, New York (1985).
39. H. Hauser and M. C. Phillips, Interactions of the polar groups of phospholipid bilayer membranes, *Progress in Surface and Membrane Science* 13:297 (1979).
40. G. Cevc, How membrane chain melting properties are regulated by the polar surface of the lipid bilayer, *Biochemistry* 26:6305 (1987).
41. H. Hauser and M. C. Phillips, Structures of aqueous dispersions of phosphatidyl serine, *J. Biol. Chem.* 248:8585 (1973).
42. G. H. Schnepel, D. Hegner, and U. Schummer, Influence of calcium on the molecular mobility of fatty acid spin labels in phosphatidyl serine and phosphatidyl inositol structures, *Biochim. Biophys. Acta* 367:67 (1974).
43. I. Haller and M. J. Frieser, Structural changes in bilayer membranes by multivalent ions, *Biochim. Biophys. Acta* 455:739 (1976).
44. H. Hauser, E. G. Finer, and A. Darke, Crystalline anhydrous calcium-phosphatidyl serine bilayers, *Biochem. Biophys. Res. Commun.* 76:267 (1977).
45. P. W. M. Van Dijck, B. De Kruijff, A. J. Verkleij, L. L. M. Van Deneen, and J. De Gier, Comparative studies of the effects of pH and Ca^{2+} on bilayers of various negatively charged phospholipids and their mixtures with phosphatidylcholine, *Biochim. Biophys. Acta* 512:84 (1978).
46. D. Papahadjopoulos, W. J. Vail, K. Jacobson, and G. Poste, Cochleate lipid cylinders: formation by fusion of unilamellar lipid vesicles, *Biochim. Biophys. Acta* 394:483 (1975).
47. D. Chapman, "Biocompatible Surfaces," U.S. Patent # 4348,329, Sept. 7 (1982).
48. J. Leaver, A. Alonso, A. A. Durrani, and D. Chapman, The physical properties and photopolymerization of diacetylene-containing phospholipid liposomes, *Biochim. Biophys. Acta* 732:210 (1983).
49. A. Singh, An Efficient Synthesis of Phosphatidylcholines, *J. Lipid Res.* 31:1522 (1990).

50. C. M. Gupta, R. Radhakrishan, and H. G. Khorana, Glycerophospholipid synthesis: improved general method and new analogs containing photoactivable groups, *Proc. Natl. Acad. Sci., USA* 74:4315 (1977).
51. P. J. Quinn and R. M. C. Dawson, pH Dependence of calcium adsorption onto anionic phospholipid monolayers, *Chem. Phys. Lipids* 8:1 (1972).
52. H. Hauser, A. Darke, and M. C. Phillips, Ion binding to phospholipids. interaction of calcium with phosphatidyl serine, *Eur. J. Biochem.* 62:335 (1976).
53. H. Hauser, D. Chapman, and R. M. C. Dawson, Physical studies of phospholipids. XI. calcium ion binding to monolayers of phosphatidyl serine and phosphatidyl inositol, *Biochim. Biophys. Acta* 183:320 (1969).
54. E. Rojas and J. M. Tobias, Membrane model: association of inorganic cations with phospholipid monolayers, *Biochim. Biophys. Acta* 94:394 (1965).
55. T. Seimiya and S. Ohki, Ionic structure of phospholipid membranes, and binding of calcium ions, *Biochim. Biophys. Acta* 298:546 (1973).
56. F. A. Cotton and G. Wilkinson, The electronic structures of transition metal complexes, in: "Advanced Inorganic Chemistry," Interscience Publishers, New York, pp. 620-644 (1972).

GENETICALLY ENGINEERED PROTEIN PORES AS COMPONENTS OF SYNTHETIC MICROSTRUCTURES

Musti Krishnasastry[#], Barbara Walker[#], Lynda Zorn[+], John Kasianowicz[§,a], and Hagan Bayley[#,+,a]

[#]Worcester Foundation for Experimental Biology, 222 Maple Avenue, Shrewsbury, MA 01545, [+]Graduate School of Biomedical Sciences, University of Massachusetts Medical Center, 55 Lake Avenue North, Worcester, MA 01655, [§]NIDDK-LBM, National Institutes of Health, Bethesda, MD 20892

INTRODUCTION

Proteinaceous, nanometer-scale pores might be used as components of new materials.[1] For example, such pores could be incorporated into thin films to confer novel permeability properties upon them. These films in turn would have several technological applications, which are discussed below, including acting as molecular filters in sensors or as components of microelectronic devices. The development of genetically engineered pores for this purpose is timely as it can be based on three recent advances in molecular biology: the use of recombinant DNA technology to alter the structures of polypeptides at will;[2,3] a rapidly increasing body of knowledge about the mechanisms of channel and pore proteins;[4-6] and progress on the formation and structural analysis of 2D protein lattices.[7,8] With this in mind, we have chosen to explore the properties of the α-hemolysin (αHL) from *Staphylococcus aureus*.[9] This surprisingly hydrophilic 293 amino acid polypeptide[10] is secreted by *Staphylococcus* as a monomer and assembles into a hexameric pore in target membranes such as those of rabbit red blood cells. Single channel recordings indicate that the pore is 1.1-1.2 nm in internal diameter.[11] The monomer and the pore have similar secondary structures predominantly comprising ß-sheet.[12,13]

There are several advantages of working with αHL that are not available with many other channel or pore proteins. The staphylococcal protein (s-αHL) is readily purified in large amounts as is recombinant protein from *Escherichia coli* (r-αHL, see below). The pore formed by αHL is robust; for example, it survives treatment with the detergent sodium dodecyl sulfate under conditions which are generally denaturing. The polypeptide chain is relatively short and therefore highly amenable to mutagenesis. The pore is a homooligomeric hexamer of defined tertiary structure,

[a] To whom correspondence should be addressed

Synthetic Microstructures in Biological Research, Edited by
J.M. Schnur and M. Peckerar, Plenum Press, New York, 1992

while other channel or pore proteins may have several types of subunits (e.g. the acetylcholine receptor[14]) or a variable number of subunits (e.g. the pore formed by the C9 subunit of complement[15]). The hexamer self-assembles in solution when treated with an inexpensive detergent, sodium deoxycholate.[16] αHL oligomers have been induced to form sheets[17,18] that might be the basis for the production of extended monolayers.

In this paper, we review our work on the expression of αHL in *E. coli*, which potentially allows the production of gram amounts of r-αHL, including mutated forms. We also outline recent data on the properties of various mutants, with which we are attempting to define the functional domains of αHL, exemplified by deletions that affect pore assembly. Finally, we sketch our future goals which include the production of pores with different conductances, selectivities and gating properties and the manufacture of monolayers containing these pores. Detailed descriptions of this work will appear elsewhere.

AMPLIFICATION OF THE α-HEMOLYSIN GENE AND EXPRESSION IN E. COLI

The αHL gene had previously been isolated and sequenced by Gray and Kehoe.[10] This permitted the design of polymerase chain reaction (PCR) oligonucleotide primers with which the protein coding sequence has been directly amplified from *S. aureus* genomic DNA (Figure 1). As we wished to express the mature protein, the primers were chosen to exclude the 5' region of the gene that encodes the hydrophobic leader sequence of the preprotein. The leader sequence is presumably required for secretion from *S. aureus*, but might not have been correctly processed in *E. coli*.

Several schemes for obtaining r-αHL are outlined in Figure 1. For expression in *E. coli*, the gene was inserted behind the T7 promoter in the expression plasmid pT7f1A.[19,20] This allows expression of αHL at high levels in *E. coli* strains harboring the T7 polymerase gene (Figure 2A). Utilizing a promoter encoded by the 5' PCR primer,[21] ^{35}S-labeled α-hemolysin of high specific radioactivity has also been obtained using an *E. coli* transcription-translation (IVTT) system (Figures 1 & 2B).

RECOMBINANT α-HEMOLYSIN IS INDISTINGUISHABLE FROM THE WILD-TYPE PROTEIN

Recombinant αHL has been purified to homogeneity from *E. coli* extracts (data not shown) and is closely similar to the hemolysin isolated from *S. aureus* as judged by several criteria. 1) Protein chemistry: the recombinant and staphylococcal polypeptides have indistinguishable electrophoretic mobilities, peptide maps, amino acid analyses and N- & C-terminal sequences. 2) Binding to red blood cells: in a kinetic assay the two forms compete similarly with [^{125}I]s-αHL for binding sites on rabbit red blood cells. Both the recombinant and staphylococcal polypeptides bind less avidly to human cells. 3) Hemolysis: the two forms lyse rabbit red cells with similar potencies in a quantitative assay. Both are far less effective against human cells. 4) Hexamer formation: both forms assemble into hexamers on rabbit red cells or when treated with deoxycholate in solution. 5) Pore formation: both forms make pores of well-defined conductance in planar lipid bilayers and the characteristics of the pores formed by r-αHL are indistinguishable from the characteristics of s-αHL pores, including: the pH-dependent rate of pore formation, the pH-dependence of conductance, rectification characteristics, the weak anion selectivity, and the pH- & voltage-dependence of inactivation.

These results are illustrated with two examples particularly relevant to the use of αHL in biomaterials. Hexamer formation, which is induced when the monomeric recombinant protein is treated in solution with the detergent sodium deoxycholate (Figure 2C), could be readily performed on a multigram scale. These hexamers could then be induced to form sheets or otherwise introduced into materials (see Goals and Applications). The properties of pores formed by r-αHL in a planar lipid bilayer are illustrated in Figure 3. Discrete conducting states of ~180 pS (V = 10 mV, 0.1 M KCl, 5 mM citric acid, pH 4.5) are apparent. These represent current flow through pores ~1 nm in diameter.[11]

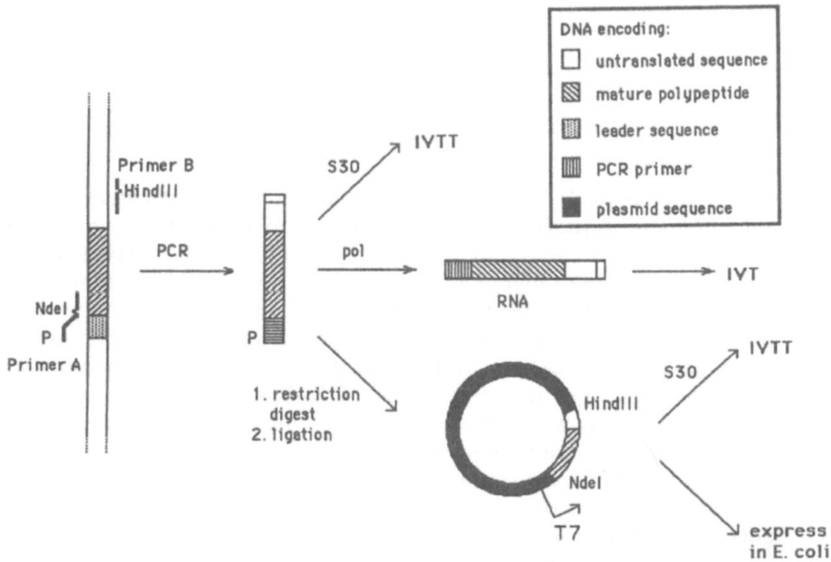

Figure 1. Methods for heterologous expression of αHL. The coding region of the αHL gene is amplified from *S. aureus* genomic DNA using the polymerase chain reaction (PCR). Primer A contains a phage RNA polymerase promoter (P: T7, T3 or SP6) and mismatches generating an NdeI restriction site at the initiator ATG. Primer B contains mismatches generating a new HindIII site 3' of the natural stop codon. The PCR product is used in three ways. First, αHL can be produced directly by coupled in vitro transcription/ translation (IVTT) using an *E. coli* S30 extract. Although hemolytic concentrations of αHL are not readily achieved, this procedure is useful for obtaining radiolabeled αHL (Figure 2B). Second, RNA can be synthesized from the PCR product using the phage RNA polymerase (pol) that recognizes the built-in promoter (P). The RNA is translated in vitro (IVT) using a reticulocyte or wheat-germ system. Third, the PCR product is trimmed with the appropriate 5' and 3' restriction enzymes (NdeI and HindIII) and inserted into the plasmid pT7f1A for expression in *E. coli* (see the text and Figure 2A). In addition, IVTT was generally more efficient when supercoiled plasmid DNA was used as a template rather than than the linear PCR product (Figure 4B).

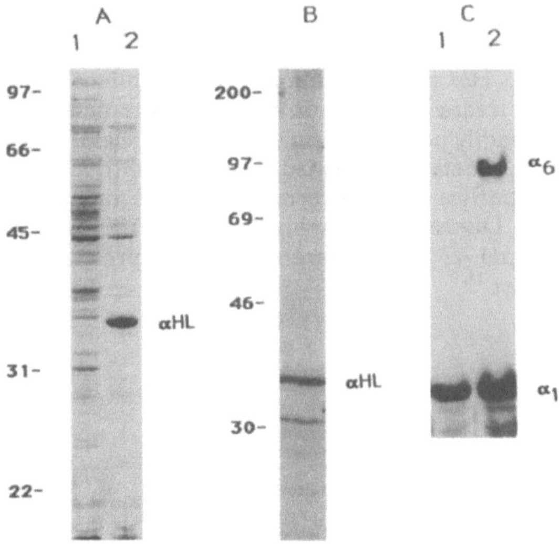

Figure 2. Expression of αHL in *E. coli* cells and extracts, and conversion of the recombinant polypeptide to hexamer with deoxycholate. A: Expression in *E. coli*. The PCR product was transferred to the expression vector pT7f1A as shown in Figure 1. The resulting plasmid was used to transform JM109(DE3), which bears a chromosomal copy of the phage T7 RNA polymerase gene. After growth under conditions optimal for expression, proteins in cell pellets were subjected to sodium dodecyl sulfate-polyacrylamide gel electrophoresis (SDS-PAGE). A Coomassie Blue stained gel is shown. Lane 1: lysate of untransformed *E. coli*; lane 2: lysate of *E. coli* containing the expression plasmid pT7f1A-αHL. The recombinant αHL was subsequently purified to homogeneity (see the text). B: Direct synthesis of radiolabeled αHL by IVTT. A PCR amplified αHL gene containing a T7 promoter (Figure 1) was transcribed and translated in vitro in the presence of [^{35}S]Met. The S30 extract was supplemented with T7 RNA polymerase. The IVTT products were subjected to SDS-PAGE. An autoradiogram of the gel is shown. C: Deoxycholate induces the formation of hexamers from monomeric r-αHL. Recombinant αHL was treated with 6.33 mM deoxycholate at room temperature for 1 h.[16] The products were subjected to SDS-PAGE, omitting the customary heat denaturation of the electrophoresis samples. A Coomassie Blue stained gel is shown. Lane 1: Untreated r-αHL; lane 2: r-αHL treated with deoxycholate.

MAPPING FUNCTIONAL DOMAINS OF α-HEMOLYSIN BY DELETION MUTAGENESIS

By analogy with other channel or pore proteins, the properties of αHL may map to different segments of the polypeptide chain. These properties, which must be at least partly interdependent, have been enumerated in the previous section and include the ability to bind to certain eukaryotic cells, to form hexamers both on cell membranes and in the presence of detergent, to lyse cells, and to form pores in lipid

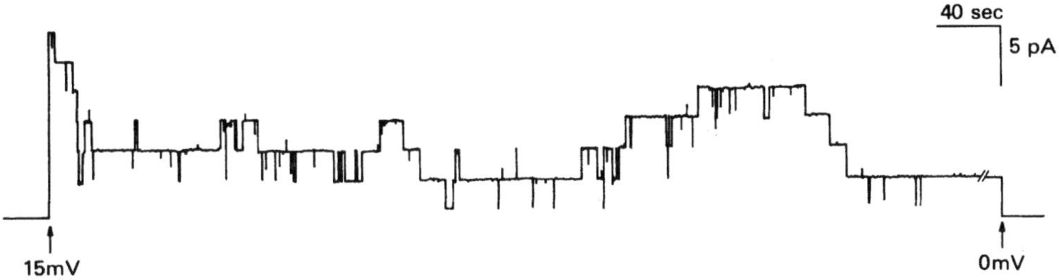

Figure 3. The current flowing through pores formed by r-αHL in a planar lipid bilayer of diphytanoyl phosphatidylcholine. The major conductance transitions are rapid and nearly equal with an average conductance of $<\Delta G> = -180$ pS. The membrane, which contained six r-αHL channels, was bathed by symmetric solutions of 0.1 M KCl, 5 mM citric acid at pH 4.5. The channels formed spontaneously after adding r-αHL (1 μL, 25 μg/mL) to one side of the bilayer. The potential applied across the membrane is denoted at the arrows. The time between transitions is a function of applied potential (data not shown) and varies over a wide range.

bilayers. Polypeptides of the colicin E1 class, for example, contain separate domains for binding to a receptor on gram negative bacteria, for passage through the outer bacterial membrane and for pore formation in the inner membrane.[6,22] Eukaryotic potassium channels apparently contain separate domains for controlling voltage-activation, for forming the channel itself and for mediating inactivation.[4]

To map the functional domains of αHL, deletion mutants have been generated using PCR (Figure 4A). Promoter-containing PCR products have been used directly for IVTT (cf. Figure 2B) allowing preliminary measurements of the properties of mutants without subcloning of the amplified DNA into expression vectors (data not shown). Similar approaches have been developed for examining malaria antigens[23] and *E. coli* transcription factors.[24] The PCR primers were also designed with appropriate restriction sites so that the mutant coding sequences could be introduced into the pT7f1A expression vector described above. One of our findings is illustrated: thirty seven amino acids at the C-terminus of the polypeptide are required for hexamer formation on red cell membranes (Figure 4B). In addition, the polypeptide missing these residues has no hemolytic activity in an assay that would detect one one-hundredth of the activity of the unaltered polypeptide. Further preliminary studies have revealed that the deletion of only three residues at the C-terminus prevents both hexamer formation and red cell lysis. It would be surprising if such a small alteration upset the entire folded structure of the polypeptide.[25] Preliminary experiments indicate that a deletion of 39 amino acids at the N-terminus of the polypeptide has the same effect. These results suggest that virtually the entire sequence of αHL is required for assembly of the pore. By contrast, only the C-terminal one-third[6] or perhaps less[26] of the pore-forming colicins is required.

FURTHER MUTAGENESIS OF α-HEMOLYSIN AND PORE ASSEMBLY

The findings described above suggest that point mutagenesis may ultimately be a more profitable way to determine the relationships between structure and function in αHL. For example, in the case of pore assembly, we already know that subtle

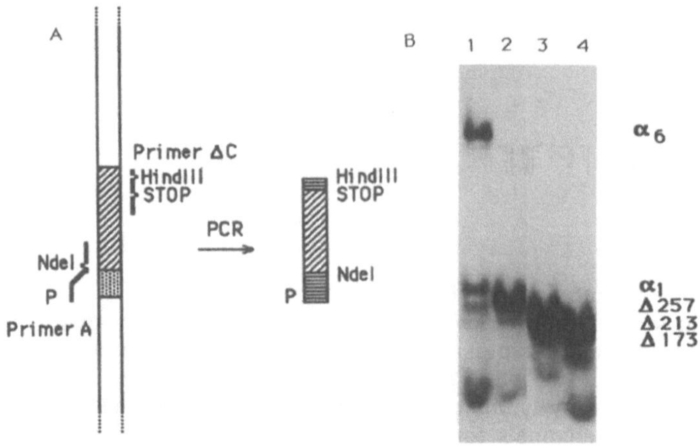

Figure 4. Deletion mutants of αHL. A: Scheme for forming a C-terminal deletion mutant using PCR. Key: see Figure 1. The method is the same in principle for an N-terminal deletion. Nested series of N- and C-terminal deletions have been generated. B: C-terminal deletions preventing the formation of hexamers in rabbit red cell membranes. ^{35}S-labeled r-αHL subunits prepared by IVTT using plasmid DNA as template were allowed to bind to rabbit red cells at 0°C. Then, a lytic concentration of s-αHL was added at 37°C. The cell membranes were recovered by centrifugation and the proteins in them subjected to SDS-PAGE, without heat denaturation of the samples. An autoradiogram of the gel is shown. Lane 1: full-length r-αHL is incorporated into hexamers; lane 2: the deletion mutant ΔC257, which lacks 37 amino acids at the C-terminus remains monomeric; lane 3: the deletion mutant ΔC213, which lacks 81 amino acids at the C-terminus; lane 4: the deletion mutant ΔC173, which lacks 121 amino acids at the C-terminus. The lanes containing the mutant polypeptides were intentionally overexposed.

changes at the N- and C-termini are tolerated as two points mutants, Ser 3 → Cys and Thr 292 → Cys, and the corresponding double mutant all form hexamers on rabbit red cells and have hemolytic activity. Wild-type αHL contains no cysteine residues and these "cysteine mutants" have been made so that αHL chemically modified at selected sites (see below) can be prepared for various biophysical studies.

Educated guesswork based on protein chemistry will be used to choose additional regions of the polypeptide as primary targets in point mutagenesis experiments designed to elucidate the assembly process. For example, we earlier identified a glycine-rich hinge or loop near the midpoint of the polypeptide chain that is sensitive to mild proteolysis in the monomer but occluded in the hexamer. This finding and biophysical data led us to propose a working model for assembly of the pore (Figure 5A).[1,12] Thelestam and colleagues[9,27] have suggested that the glycine-rich loop, which is indeed one of the least charged regions of the generally hydrophilic polypeptide, takes part in the initial contact with the lipid bilayer. Based on the finding that αHL cleaved with trypsin in the hinge region forms a hexamer but not a functional pore,[28] they have also proposed that at least two forms of the hexamer can exist in association with the cell membrane.[27] In Figure 5B, these ideas are combined with our earlier proposal. Of course, both of these models must be oversimplifications and the existence of further intermediates in assembly have been proposed by many investigators.[11,12,18,29] Because these speculations about αHL assembly revolve around the glycine-rich hinge, we are currently generating numerous mutants in this region by segment-directed saturation mutagenesis.[2,30,31]

GOALS AND APPLICATIONS

Goals

A short term goal is to use mutagenesis to produce αHL polypeptides that yield pores with altered properties including: conductance, ion selectivity and gating. Preliminary work on the assembly of the pore (see above) suggests that it may be impossible to track down the residues that project into the pore by deletion analysis because all deletion mutants tested so far cannot assemble. It will therefore be necessary to build a collection of point mutants (or perhaps small internal deletion or insertion mutants) to find these vital amino acids. In doing this it is important not to disrupt the tertiary structure of the protein or its segments.[25] One approach has been to make substitutions throughout a polypeptide with the amino acid alanine, which is thought to be the residue least generally disruptive of secondary structure and hence protein folding.[32] As a collection of mutants is built, the locations of residues that determine pore properties will be mapped allowing us to make simple models for the pore structure and then plan further mutations. It will also be possible to screen mutants, for example, by assaying the ability of bacterial colonics to lyse red cells in blood agar plates or by electrophysiological analysis of pores formed in bilayers from crude r-αHL in cell extracts or in IVTT reaction mixes. Thus, an iterative approach (mutations→ functional changes→ model→ mutations &c.) utilizing a combination of guesswork and logic will yield interesting new pore molecules.

Ultimately, the information gathered by this approach will allow the rational design of pores with defined properties. To alter pore conductance two basic strategies will be explored. The first will be to expand or reduce the size of the pore by substituting the amino acids whose side-chains project into it. As the pore is a hexamer, a small change in the volume of an amino acid side-chain might produce an appreciable change in conductance. A second and more speculative idea is to alter the number of subunits in the αHL oligomer by changing amino acids located at the

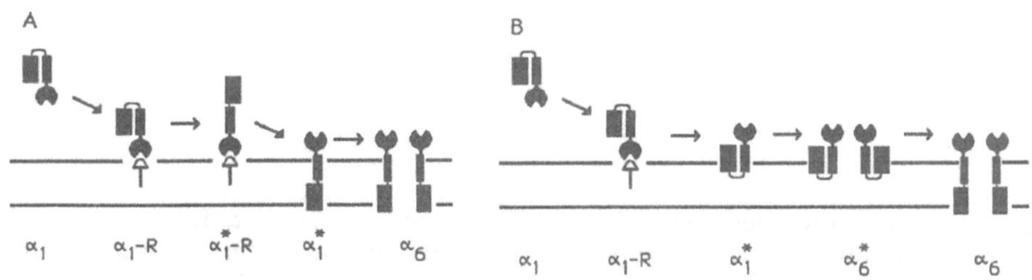

Figure 5. Models for the assembly of αHL. A: Model based on the insertion of monomeric αHL polypeptides into the lipid bilayer followed by oligomerization. B: Model based on the formation of hexameric αHL at the membrane surface followed by insertion into the lipid bilayer. α_1: monomer; α_6: hexamer; R: receptor; *: unstable intermediate. For details see the text.

interfaces between the subunits. The selectivity of the pore, for anions versus cations for example, might also be changed by making substitutions in the residues that line the pore.[5,33,34] The most difficult task will be to introduce useful gating characteristics. One speculative idea is to place a photoisomerizable group, such as azobisbenzene,[35] into the pore or at its mouth (Figure 6). This group would be attached at a unique site defined by a cysteine substitution. A related idea has been attempted recently with the channel-forming antibiotic gramicidin.[36] Another possibility will be to use new knowledge about voltage[4] and mechanically-gated channels[37] to build gating structures into the polypeptide sequence.

Applications

Perhaps the most compelling proposed application of proteinaceous, nanometer-scale pores is in monolayer (or perhaps multilayer) films bonded to support matrices for use in membrane separation systems, sensor technology and electronic devices. The need for defect-free ultrafiltration devices with uniform pores remains unfulfilled.[38,39] In sensor technology a coating with a semipermeable ultrafiltration membrane might be used to select species of predetermined size and charge for access to an electrode surface. Pores gated by voltage, light, or pressure might be incorporated into microelectronic devices (e.g. Figure 6). A calculation based on the outer diameter of the αHL hexamer[40] suggests that 1 g of αHL would yield ~300 m^2 of a cubic monolayer lattice (Figure 6). Lattices of αHL oligomers have been known for some time,[17] but the surprising recent studies of Olofsson and colleagues[18] suggest that, in one case at least, the constituents are tetramers lying on the surface of a lipid bilayer. Therefore it may be necessary to make films using preformed hexamers (made for example by the deoxycholate method) and Langmuir-Blodgett techniques. By using mutant polypeptides with aptly introduced residues such as

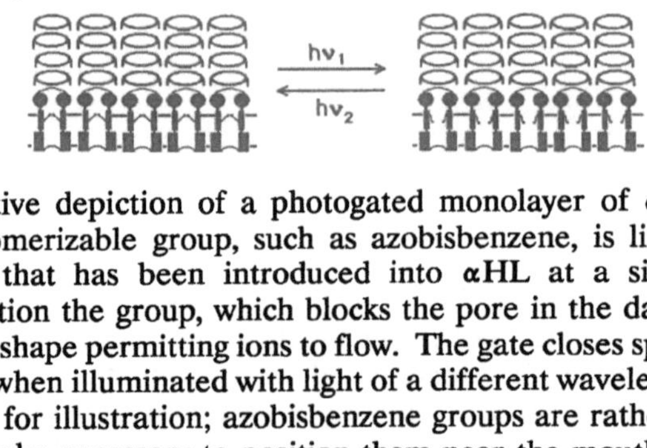

Figure 6. Speculative depiction of a photogated monolayer of αHL pores. A: A photoisomerizable group, such as azobisbenzene, is linked to a cysteine residue that has been introduced into αHL at a single site. B: On illumination the group, which blocks the pore in the dark, isomerizes and changes shape permitting ions to flow. The gate closes spontaneously in the dark or when illuminated with light of a different wavelength. The diagram is solely for illustration; azobisbenzene groups are rather bulky and it will probably be necessary to position them near the mouth of the pore or to devise ways of introducing one group per hexamer.

cysteine it should be possible to strengthen the films by chemical crosslinking. αHL hexamers are also readily incorporated into lipid bilayers[16] (H. Bayley, unpublished work). Again, such bilayers might be strengthened, in this case by polymerization of their lipid components.[41] For certain applications, such as facilitated or active transport systems, αHL molecules could be incorporated into bilayers with one or more additional components.

In addition, regular arrays of αHL might be used as templates for forming metallic nanostructures.[42] It is possible that the repeat distance or metal deposition characteristics could be manipulated by altering the constituent polypeptides. αHL lattices might also be used as substrates to orient other proteins, which would be attached by "single" cysteine residues introduced into αHL by mutagenesis. The pore cavity itself might be used to orient smaller molecules. As it will eventually be possible to alter the side-chains that project into the cavity in a predictable fashion, it is also conceivable that proteinaceous pores in solution or in lattices could, like antibodies,[43] be endowed with catalytic properties.

ACKNOWLEDGEMENTS

We are indebted to our colleagues at the Worcester Foundation and the National Institutes of Health for their help and guidance, especially Rekha Panchal for help with DNA sequencing and Stephen Cheley for advice on protein expression. This study was supported by grants from the DOE (to H.B.) and the ONR (to V.A. Parsegian), and by a postdoctoral fellowship (to J.J.K.) from the NIH (IRTA).

REFERENCES

1. H. Bayley, Monolayers from genetically engineered pores, *MRS Symp.Proc.* 218:69 (1991).
2. J. D. Hermes, S. C. Blacklow, and J. R. Knowles, Searching sequence space by definably random mutagenesis: improving the catalytic potency of an enzyme, *Proc. Natl. Acad. Sci. USA* 87:696 (1990).
3. W. Bloch, A biochemical perspective of the polymerase chain reaction, *Biochemistry* 30:2735 (1991).
4. C. Miller, 1990: annus mirabilis of potassium channels, *Science* 252:1092 (1991).
5. E. Blachly-Dyson, S. Peng, M. Columbini, and M. Forte, Selectivity changes in site-directed mutants of the VDAC ion channel: structural implications, *Science* 247:1233 (1990).
6. M. W. Parker, A. D. Tucker, D. Tsernoglou, and F. Pattus, Insights into membrane insertion based on studies of colicins, *Trends Biochem. Sci.* 15:126 (1990).
7. R. Henderson, J. M. Baldwin, T. A. Ceska, F. Zemlin, E. Beckman, and K. H. Downing, Model for the structure of bacteriorhodopsin based on high-resolution electron cryomicroscopy, *J. Mol. Biol.* 213:899 (1990).
8. S. A. Darst, A. M. Edwards, E. W. Kubalek, and R. D. Kornberg, Three-dimensional structure of yeast RNA polymerase II at 16Å resolution, *Cell* 66:121 (1991).
9. M. Thelestam and L. Blomqvist, Staphylococcal alpha toxin- recent advances, *Toxicon* 26:51 (1988).
10. G. S. Gray and M. Kehoe, Primary sequence of the α-toxin gene from *Staphylococcus aureus* Wood 46, *Infect. Immun.* 46:615 (1984).

11. G. Menestrina, Ionic channels formed by *Staphylococcus aureus* alpha-toxin: voltage-dependent inhibition by divalent and trivalent cations, *J. Membrane Biol.* 90:177 (1986).
12. N. Tobkes, B. A. Wallace, and H. Bayley, Secondary structure and assembly mechanism of an oligomeric channel protein, *Biochemistry* 24:1915 (1985).
13. H. Ikigai and T. Nakae, Conformational alteration in alpha-toxin from *Staphylococcus aureus* concomitant with the transformation of the water-soluble monomer to the membrane oligomer, *Biochem. Biophys. Res. Commun.* 175:181 (1985).
14. J.-L. Galzi, F. Revah, A. Bessis, and J.-P. Changeux, Functional architecture of the nicotinic acetylcholine receptor: From electric organ to brain, *Annu. Rev. Pharmacol. Toxicol.* 31:37 (1991).
15. H. Sauer, L. Pratsch, J. Tschopp, S. Bhakdi, and R. Peters, Functional size of complement and perforin pores compared by confocal laser scanning microscopy and fluorescence microphotolysis, *Biochim. Biophys. Acta Bio-Membr.* 1063:137 (1991).
16. S. Bhakdi, R. Füssle, and J. Tranum-Jensen, Staphylococcal α-toxin: oligomerization of hydrophilic monomers to form amphiphilic hexamers induced through contact with deoxycholate micelles, *Proc. Natl. Acad. Sci. USA* 78:5475 (1981).
17. J. H. Freer, J. P. Arbuthnott, and A. W. Bernheimer, Interaction of staphylococcal α-toxin with artificial and natural membranes, *J. Bacteriol.* 95:1153 (1968).
18. A. Olofsson, U. Kavéus, I. Hacksell, M. Thelestam, and H. Hebert, Crystalline layers and three-dimensional structure of *Staphylococcus aureus* α-toxin, *J. Mol. Biol.* 214:299 (1990).
19. J. Kuret, K. E. Johnson, C. Nicolette, and M. J. Zoller, Mutagenesis of the regulatory subunit of yeast cAMP-dependent protein kinase: isolation of site-directed mutants with altered binding affinity for catalytic subunits, *J. Biol. Chem.* 263:9149 (1988).
20. P. J. Bergold, S. A. Beushausen, T. C. Sacktor, S. Cheley, H. Bayley, and J. H. Schwartz, Identification of a regulatory subunit of the cAMP-dependent protein kinase downregulated in Aplysia sensory neurons during long-term sensitization, *Neuron* 8:387 (1992).
21. G. Sarkar and S. S. Sommer, Access to a messenger RNA sequence or its protein product is not limited by tissue or species specificity, *Science* 244:331 (1989).
22. H. Benedetti, M. Frenette, D. Baty, M. Knibiehler, F. Pattus, and C. Lazdunski, Individual domains of colicins confer specificity in colicin uptake, in pore-properties and in immunity requirement, *J. Mol. Biol.* 217:429 (1991).
23. K. C. Kain, P. A. Orlandi, and D. E. Lanar, Universal promoter for gene expression without cloning: expression PCR, *BioTechniques* 10:366 (1991).
24. S. A. Lesley, M. A. D. Brow, and R. R. Burgess, Use of *in vitro* protein synthesis from polymerase chain reaction-generated templates to study interaction of *Escherichia coli* transcription factors with core RNA polymerase and for epitope mapping of monoclonal antibodies, *J. Biol. Chem.* 266:2632 (1991).
25. P. Schimmel, Hazards and their exploitation in the applications of molecular biology to structure-function relationships, *Biochemistry* 29:9496 (1990).
26. Q. R. Liu, V. Crozel, F. Levinthal, S. Slatin, A. Finkelstein, and C. Levinthal, A very short peptide makes a voltage-dependent ion channel: the critical length of the channel domain of colicin E1, *Proteins* 1:218 (1986).

27. M. Thelestam, A. Olofsson, L. Blomqvist, and H. Hebert, Oligomerisation of cell-bound staphylococcal α-toxin in relation to membrane permeabilisation, *Biochim. Biophys. Acta Bio-Membr.* 1062:245 (1991).
28. L. Blomqvist and M. Thelestam, Oligomerization of ^{3}H-labelled staphylococcal alpha-toxin and fragments on adrenocortical Y1 tumour cells, *Microbial Pathogen.* 4:223 (1988).
29. S. Harshman, P. Boquet, E. Duflot, J. E. Alouf, C. Montecucco, and E. Papini, Staphylococcal α-toxin: a study of membrane penetration and pore formation, *J. Biol. Chem.* 264:14978 (1989).
30. A. Peterson and B. Seed, Monoclonal antibody and ligand binding sites of the T cell erythrocyte receptor (CD2), *Nature* 330:842 (1987).
31. A. R. Oliphant and K. Struhl, An efficient method for generating proteins with altered enzymatic properties: application to ß-lactamase, *Proc. Natl. Acad. Sci. USA* 86:9094 (1989).
32. B. C. Cunningham, D. J. Henner, and J. Wells, Engineering human prolactin to bind to the human growth hormone receptor, *Science* 247:1461 (1990).
33. A. J. Yool and T. L. Schwarz, Alteration of ionic selectivity of a K channel by mutation of the H5 region, *Nature* 349:700 (1991).
34. M. P. Anderson, et al. and M. P. Welsh, Demonstration that CFTR is a chloride channel by alteration of its anion selectivity, *Science* 253:202 (1991).
35. H. Bayley, F. Gasparro, and R. Edelson, Photoactivatable drugs, *Trends Pharm. Sci.* 8:138 (1987).
36. C. J. Stankovic, S. H. Heinemann, and S. L. Schreiber, Photo-modulated ion channels based on covalently linked gramicidins, *Biochim. Biophys. Acta Bio-Membr.* 1061:163 (1991).
37. F. Sachs, Mechanical transduction by membrane ion channels: A mini review, *Mol. Cell. Biochem.* 104:57 (1991).
38. R. W. Baker, E. L. Cussler, W. Eykamp, W. J. Koros, R. L. Riley, and H. Strathman. "Membrane Separation Systems: A Research & Development Needs Assessment," U.S. Department of Energy, Springfield, VA (1990).
39. J. Beard and A. Coghlan, Filters provide traps for catching viruses, *New Sci.* 131(1778):20 (1991).
40. A. Olofsson, U. Kavéus, M. Thelestam, and H. Hebert, The projection structure of α-toxin from Staphylococcus aureus in human platelet membranes as analyzed by electron microscopy and image processing, *J. Ultrastruct. Mol. Struct. Res.* 100:194 (1988).
41. S. L. Regen, Polymerized phosphatidylcholine vesicles as drug carriers, *Ann. N.Y. Acad. Sci.* 446:296 (1985).
42. K. Douglas, N. A. Clark, and K. J. Rothschild, Nanometer molecular lithography, *Appl. Phys. Lett.* 48:676 (1986).
43. R. A. Lerner, S. J. Benkovic, and P. G. Schultz, At the crossroads of chemistry and immunology: catalytic antibodies, *Science* 252:659 (1991).

PREPARATION AND CHARACTERIZATION OF ANTIBODY FILMS ON LITHIUM NIOBATE SURFACES

S. Seeger, K. Bierbaum, R. Dahint, C. L. Feng*, M. Mantar, and M. Grunze

Institut für Angewandte Physikalische Chemie, Universität Heidelberg, Im Neuenheimer Feld 253, 6900 Heidelberg, Germany
*permanent address: Faculty of Chemistry, University of Liaoning, Huanghe Lu 850, 116022 Dalian, China

ABSTRACT

Lithium niobate is a piezoelectric substrate of interest for applications as an electro-acoustic immunosensor. The activation of lithium niobate substrates with antibody films is carried out by using the standard immobilization technique, i.e. silanization with 3-aminopropyltriethoxysilane prior to linkage of antibody molecules via a glutaraldehyde crosslinker. The cleaning of the substrate prior to silanization was achieved by microwave and ozone/uv cleaning. X-ray photoelectron spectra show a significant reduction of impurities compared to cleaning with organic solvents only. Silanization was performed in toluene solution, toluene vapour, and by chemical vapour deposition. The differences between the various silanization processes for antigen binding examined by enzyme linked immunosorbent assay are small. Using protein G as an immobilized binding molecule for antibodies the binding capacity was enhanced by 40%.

INTRODUCTION

The immobilization of biological active molecules on surfaces has become a field of great interest in several analytical processes, for example affinity chromatography,[1,2] immobilization of enzymes,[3] and solid phase analytical techniques.[4] In biochemical and immunological analysis, e.g. for applications in biosensors, a very high reproducibility of film characteristics such as homogeneity, thickness, long term stability, and activity is required. Immobilization of biomolecules has been achieved on several substrates.[5] However, covalent linking to lithium niobate surfaces has not been studied in detail to our knowledge.

Acoustic waves in piezoelectric materials can be used for sensing ions in solutions with high sensitivity.[6,7] $LiNbO_3$ was also applied to detect an antibody-antigen reaction[8] and DNA hybridization[9] in the liquid phase. Stimulated by these results, we investigated several techniques to covalently bond antibodies to a $LiNbO_3$ crystal surface.

Synthetic Microstructures in Biological Research, Edited by
J.M. Schnur and M. Peckerar, Plenum Press, New York, 1992

A standard technique for covalent immobilization of proteins, e.g. antibodies, is the silanization of the substrate after cleaning prior to the reaction with a crosslinker and the protein. After activation of the $LiNbO_3$ surface with NaOH, 3-aminopropyltriethoxysilane reacts with the hydroxygroups of the crystal surface. In a third step a crosslinker is linked to the aminogroups of the aminosilane. The remaining electrophilic group of the crosslinker molecule reacts subsequently with antibody molecules forming a covalent bond to free nucleophilic groups of the aminoacids (Figure 1).

The treatment with reactive silanes has been performed on several substrates, for example silicon.[10] Jönsson et al.[11] prepared silane films on silica for subsequent antibody immobilization. Deposition of silanes on surfaces was usually done in the liquid phase. To avoid accompanying difficulties such as inhomogeneity of the films and codeposition of polymeric silane globules, the silanization has also been performed in the gas phase. Haller[10] demonstrated the formation of a monolayer coverage on silicon and gallium arsenide crystals after refluxing the substrates in a vapour of 3-aminopropyltriethoxysilane in toluene and xylene. As shown by Mittal and O'Kane,[12] the chemical vapour deposition of several silane derivates on metal substrates results in film thicknesses from 0.6 to 11.5 nm.

The quality of the immobilized antibody film depends on the purity of the surface. Impurities like small particles and hydrocarbons inhibit covalent linking between the surface and the silane. Usually, surfaces are cleaned before silanization with alcohol like propanol. To improve the purity of the surface oxygen radicals produced by ozone photolysis and microwave discharge were used for the cleaning process. The surfaces were analysed before and after silanization by x-ray photoelectron spectroscopy (XPS) and atomic force microscopy. The present studies describe the application and a comparison of the different deposition techniques to silanization of lithium niobate surfaces.

MATERIALS AND METHODS

1. Cleaning of the $LiNbO_3$ Surface

Cleaning of the crystal surface was performed by four different procedures: In *Cleaning procedure I*, the substrate was washed in 2-propanol and subsequently air-dried 3 times. In the second cleaning procedure, the substrate was first washed in 2-propanol as described in I. Subsequently the crystals were treated in a ultrasonic bath with chloroform at 65°C for 30 minutes. This was followed by cleaning of the surface in a mixture of hydrogenperoxide (stock solution 30%), concentrated ammonia (stock solution 25%) and bidistilled water (1:1:5). After washing with bidistilled water the substrates were dried at 105°C for 20 minutes.

During cleaning process III and V, oxygen radicals were produced either by uv-cleavage of ozone or by microwave discharges in an oxygen flux, respectively. In the third cleaning procedure the substrates were treated after performing procedure I with an oxygen/ozone mixture (ozone concentration 0.1%) for 6.5 hours during UV-irradiation of the surface in the experimental set-up described in Figure 2. The microwave cleaning (process IV) was performed as follows: At first the surface is cleaned with 2-propanol as described for procedure I. After channelling the substrate in the reaction chamber of the experimental set-up oxygen radicals were produced by microwave discharge. The surface was treated with the radicals for 2.5 hours.

The cleaning of surfaces with uv-ozone and the microwave cleaning was performed in a quartz cell. In order to perform gas phase silanization without exposing

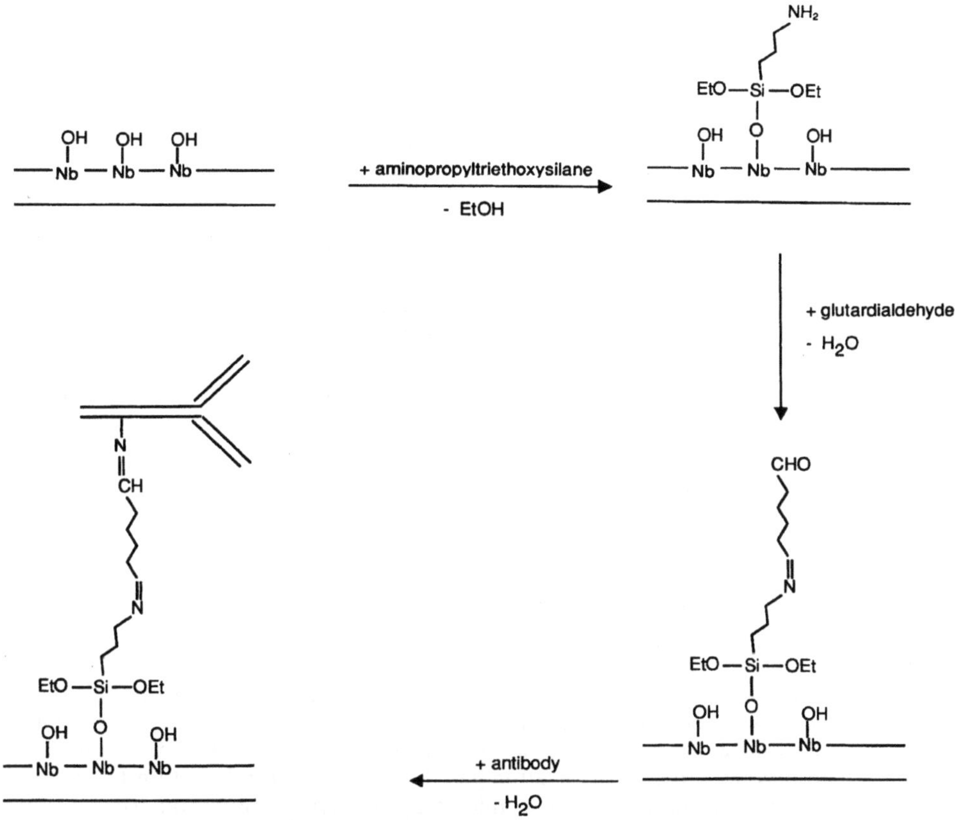

Figure 1. Covalent immobilization of antibodies. After activation of the $LiNbO_3$ surface aminogroups were introduced. Subsequently, a crosslinking agent like glutaraldehyde offers an aldehyde function for linkage of antibody molecules.

the surface to the ambient after cleaning, a small vacuum system was assembled using standard KF parts. The quartz cell consists of a cylindrical tube containing a sample holder and was connected to an ozone generator and an additional quartz tube for the microwave discharge. A mercury lamp was fixed outside the quartz tube for the production of oxygen radicals from ozone.

The KF system for the uv-ozone and microwave cleaning and silanization of the lithium niobate surfaces is shown in Figure 2. The sample holder was placed in a KF-cross supplied with a quartz window necessary for uv-irradiation. A quartz tube connects to the KF-cross with the oxygen reservoir, the ozone generator and the carrier gas reservoir (carrier gas: argon). The quartz tube is surrounded by a resonator coupled with a microwave generator. The fourth port of the KF-cross is coupled to the aminosilane reservoir.

The microwave generator was a Microtron 200 generator, 60 W, 2.456 Hz. Ozone was produced by an Argentox generator operating at 1 l/min oxygen flow. At 155 V it produced 1000 ppm ozone. As a light source a mercury lamp working at wavelength greater than 275 nm was used.

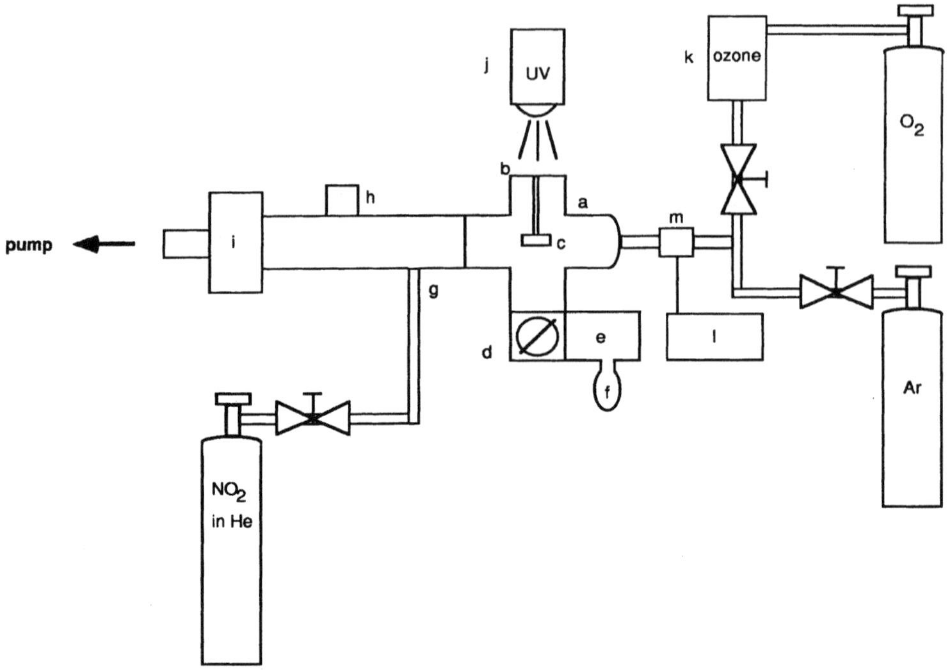

Figure 2. The cleaning and silanization set-up is used for uv/ozone cleaning, microwave cleaning, and silanization of substrates. **a** KF-cross; **b** quartz window; **c** sample holder; **d** viton-sealed valve; **e** KF-Tee; **f** aminosilane reservoir; **g** input for titration gas; **h** TC gauge; **i** adsorption trap; **j** uv-lamp; **k** ozone generator; **l** microwave generator; **m** resonator

2. Silanization

Linking of the 3-aminotriethoxypropylsilane to the surface, cleaned by procedure I - IV was performed by three different methods: In the *liquid phase silanization process* the surface was activated prior to the silanization by creating hydroxylgroups in 5n NaOH at room temperature for 15 minutes. Subsequently the crystal was washed three times for 5 min with bidistilled water. The completely dried substrates were dipped into a solution of 2% 3-aminopropyltriethoxy-silane in ethanol. After 10 minutes the crystals were removed and rinsed carefully with ethanol. The last step involved curing at 105°C for 1 hour.[13]

The set-up used for vapour phase silanization consists of a modified 100 ml flask, connected for a vacuum pump and a reflux cooler heated in an oil bath. A small porous container keeps the substrate above the solution. The flask contains 14 ml of a 5% 3-aminopropyltriethoxysilane solution in toluene under a nitrogen atmosphere. The solution is refluxed at 140°C for 13 hours. After removing, the substrate was rinsed with toluene.[10]

The chemical vapour deposition of 3-aminotriethoxypropylsilane was performed in the uv-ozone and microwave cleaning apparatus at room temperature (Figure 2). The pressure during silanization was held constant at $2 \cdot 10^{-3}$ bar, the silane reservoir contained 4 ml 3-aminopropyltriethoxysilane. Deposition was varied from 15 minutes to 41 hours.

3. Antibody Immobilization

The silanized substrates were treated with a 2.5% solution of glutaraldehyde in PBS for 30 minutes at room temperature, followed by washing three times with PBS/Tween-20. Subsequently the crystals were treated with a solution of 10 μg antibody (mouse-anti-horseradish peroxidase (POD)-antibody or goat-anti-rabbit-antibody) in 1 ml PBS for 1 hour at room temperature and dipped in a solution of 50μl monoethanolamine in PBS for 15 minutes at room temperature. The last step consisted of washing the samples carefully for three times in PBS/Tween-20.

4. Enzyme-Linked Immunosorbent Assay (ELISA)

To quantify the amount of horseradish peroxidase linked to the crystal surface via antigen-antibody reaction, the prepared substrates were put into a cuvette with 2.5 ml substrate buffer solution. The reaction was started by adding 1 μl 30% H_2O_2 solution while vigorously stirring with a small teflon rod. For at least 5 minutes every 30 seconds the absorption at 405 nm was measured.[14]

The calibration was performed by measuring the absorption of solutions of 1.0, 0.75, 0.5, 0.1 and 0.0 ng horseradish peroxidase in 2.5 ml substrate solution starting the reaction with 1 μl 30% H_2O_2 solution. As buffers we used a *substrate buffer* (3.25 mM (50 mg)) sodiumperborate, 39.8 mM (835.9 mg) citric acidmonohydrate, and 60 mM (1.068 g) di-sodiumhydrogenphosphate-dihydrate dissolved in bidistilled water (total volume: 100 ml). As a *substrate solution* 95.3 mg 2.2'-Azino-di[3-ethylbenzthiazoline-sulfonate(6)] (ABTS) (Sigma, Deisenhofen, FRG) were dissolved in 100 ml of the substrate buffer.

Absorption spectroscopy measurements were performed at 405 nm with a Lambda 7 spectrometer (Perkin-Elmer). X-ray photoelectron spectra were recorded with a Max 200 system (Leybold, FRG). The α-step measurements were carried out at Max-Planck Institut für Polymerforschung, Mainz, FRG. The atomic force images were taken in air with a Park Instruments AFM in a standard configuration.

RESULTS

Cleaning and Activation

The homogeneity of the silane films is essential to produce homogeneous antibody films. This requires that the substrate surface prior to silanization is free from impurities. Small particles and hydrocarbon contaminations can be washed off by organic solvents. The effect of several solvents was examined by XPS and it was found, that 2-propanol was most effective. Table 1 shows relative photoemission intensities obtained from XP-spectra of $LiNbO_3$ substrate after cleaning with 2-propanol. Compared to the uncleaned surface a significant removal of carboneous impurities is observed, although a small C1s signal remains. The data show a distinct reduction of the carbon 1s peak after ozone/uv exposure for 6.5 hours following initial cleaning with 2-propanol compared to the only 2-propanol-cleaned surface. The small remaining carbon signal is probably due to air borne contaminants adsorbing during transportation of the substrate from the reactor to the XPS spectrometer.

Using the quartz cell cleaning and activation with oxygen radicals $O(^3P)$ generated by a microwave discharge result in a similar surface quality as the ozone cleaning process. Both gas phase cleaning processes show the successful removal of carbon from the surface. The relative peak intensities in Table 1 reflect the effectiveness of microwave and ozone/uv generated oxygen radicals.

With the KF cell the relative intensity of the C1s peak was slightly higher after microwave cleaning compared to the same procedure in the quartz cell. A fluorine signal appears in the XPS data indicating a contamination of the surface by hydrocarboneous material originating from the teflon seals. After the KF cell was modified to avoid exposure of the teflon seals to the microwave plasma, a very efficient cleaning was found in the XPS data.

Silanization of the Lithium Niobate Surfaces

Incomplete silanization is expected to reduce the coverage of antibodies on the surface and hence the biological activity of the film. Also the nature of binding may play an important role for the application of these films, e.g. physical adsorption rather than covalent bonding may reduce the stability of the film. It is also known that alkoxysilanes show a tendency to polycondensation which results in forming polymeric globules on the surface. The homogeneity of the film, however, is required for reproducible biological activity and for a defined interaction of the acoustic plate modes in the $LiNbO_3$ substrate with the film.

The comparison of the experimental C:Si-ratios from XPS data (Table 2) with the expected stochiometries (see Table 3) indicates the extent of polycondensation in the silane film. By the formation of covalent bonds between the silane molecules ethoxygroups are released and the percentage of carbon decreases. The N(1s) intensities in the XP spectra of the aminosilane films can not be used to infer on the degree of condensation, because it was found that the N(1s) intensities decrease with X-ray exposure as will be described elsewhere.

Silanization for 18 hours of the microwave cleaned surface results in a C(1s):Si(2p) ratio consistent with physisorbed silane. Longer silanization times (approximately 40 hours) of the microwave cleaned surface seems to result in partial

Table 1

Relative intensities of C1s signals after several cleaning processes: a) cleaning with 2-propanol; b) cleaning with uv/ozone in a quartz cell; c) cleaning with uv/ozone in a KF cell; d), e), f) cleaning with microwave technique in a quartz cell and a KF cell, respectively.

cleaning process	rel. C(1s) intensities [%]
a) 2-propanol	16,8
b) uv/ozone (quartz cell)	9,6
c) uv/ozone (KF cell)	9,1
d) microwave (quartz cell)	11,4
e) microwave (KF cell)	14,3
f) microwave (modified KF cell)	5,5

Table 2

Experimental relative peak intensities at difference cleaning procedures. In rows 2 and 3, two different samples were analyzed.

cleaning process	silanization period [h]	peak intensities [%]	
microwave	18	C(1s) 92,5 Si(2p) 7,5	
microwave	41	C(1s) 80,8 Si(2p) 19,2	C(1s) 87,0 Si(2p) 13,0
ozone/uv	40	C(1s) 88,9 Si(2p) 11,1	C(1s) 89,9 Si(2p) 10,1

Table 3

Expected relative peak intensities at different substitution of ethoxygroups.

kind of bonding	peak intensities [%]
physical adsorbed	C(1s) 90 Si(2p) 10
substitution of one ethoxy group	C(1s) 87,5 Si(2p) 12,5
substitution of two ethoxy groups	C(1s) 83,3 Si(2p) 16,7
substitution of three ethoxy groups	C(1s) 75,0 Si(2p) 25,0
average	C(1s) 84,3 Si(2p) 15,7

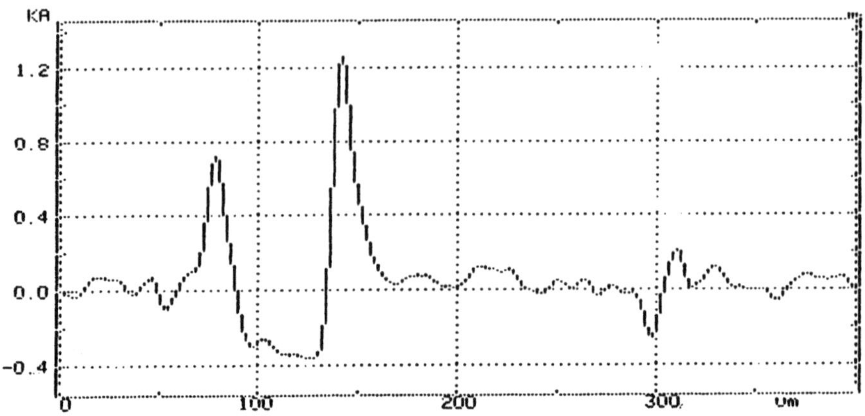

Figure 3. α-step measurement of a liquid phase prepared antibody film (see text).

conclusion, where the deviation between the two experiments does not allow to specify the degree of crosslinking. The same silanization time for the uv/ozone cleaned sample again gives higher C(1s):Si(2p) ratios suggesting a lower degree of substitution of the ethoxy groups.

Preliminary atomic-force microscopic images (not shown here) of a silanized microwave cleaned surface and an ozone/uv cleaned surface reveal polymeric globules on the surface. Whereas the globules on the microwave cleaned surface are small (maximum height of 10 nm), the size of the globules on the ozone/uv cleaned surface is more than twice as large. The average film thickness determined by the attenuation of the Nb3d signal in the XPS experiments was 0.5-0.7 nm and 8.4-11.2 nm for the microwave cleaned and uv/ozone cleaned surfaces, respectively. Angle resolved XP measurements further show, that the films on the microwave treated substrates are not homogeneous, whereas a substantial better homogeneity is found on the uv-ozone cleaned substrates. A comparison and discussion between the XPS derived film thickness and the AFM data will be given elsewhere.

ANTIBODY FILMS

After silanization the surface is activated by glutardialdehyde. The addition of antibodies results in covalent bonds of the remaining aldehyde groups to amino groups of the antibody molecules (see Figure 1). Film thickness, specific antigen binding capacity, unspecific adsorption, and long term stability were examined.

The film thickness of liquid phase silanized and antibody covered films were examined by α-stepping experiments. Figure 3 shows the profile of the antibody film. The thickness of the film consisting of the silane film crosslinker and the antibody molecules is approximately 38 nm. With respect to the size of an antibody molecule of approximately 10 nm only a few monolayers (1 - 4) cover the surface. However, these measurements give no indication about the orientation of the antibody molecules linked to the surface and the homogeneity of the film.

The binding capacity of immobilized antibodies to the antigens was determined by an immunological method, ELISA. The principle is shown in Figure 4. The immobilized antibody binds an enzyme-labeled antigen or the enzyme itself. Here we used anti-peroxidase antibodies for immobilization. The enzyme "horseradish-peroxidase" (POD) is used as antigen, which catalyzes the reaction of ABTS (2.2'-Azino-di[3-ethylbenzthiazol-inesulfonate(6)] with H2O2 to a green coloured dye, whose concentration can be determined by absorption 1 spectroscopy. Kinetic studies using the Michaelis-Menten-kinetics of enzyme catalysed reactions yield the amount of trapped POD molecules, and hence the number of active antigen-binding sites.

Cleaned and silanized lithium niobate substrates were treated with 2.5% glutaraldehyde in phosphate-buffered saline (PBS) for 30 minutes. After washing with PBS the substrates were incubated with an antibody solution. After this step the ELISA-absorption measurements with POD (antigen) and ABTS as a substrate were carried out. Due to the different sizes of the $LiNbO_3$ crystals the kinetic data had to be normalized to the substrate surface area. The antibody activity per area was calculated from the kinetic data of the POD catalysed dye production (Figure 5). The absolute number of binding sites is obtained from calibration curves. The mass of immobilized POD per mm^2 lithium niobate surface is shown in Table 4. It can be clearly seen, that the activation of lithium niobate with 5N NaOH before silanization in the liquid phase improves the immobilization result by about a factor of 6. Using cleaning procedure II only a weak increase in antibody immobilization can be observed.

As well the vapour phase silanization as the cleaning of the substrate before silanization with ozone/uv obviously has no measurable advantages about the liquid phase silanization.

The theoretically maximum binding capacity can be calculated as follows: One antibody molecule of the immunoglobuline-G-type covers a surface of approximately 30 nm^2. Sincc two binding sites are provided by one antibody molecule, the required surface area for one binding site is 15 nm^2. If all antibody molecules are ideally attached at the Fc region on the crystal surface, that means all binding sites are free for antigen binding, $6.6 \cdot 10^{11}$ antigen molecules can be linked to 1 mm^2 of the surface. Because of the molecular weight of the antigen HRP (40000 D), the binding capacity of a regular monomolecular anti-HRP-antibody film is approximately 1.67 ng/mm^2. The average binding site capacity of the produced antibody films is about 0.1 ng/mm^2. This means that under the conditions applied here only 6.0% of the theoretically maximum coverage can be achieved.

Using a purified goat-anti-rabbit-antibody the specific activity increases about ten-fold to an average binding of 1.1 ng rabbit-anti-horse-peroxidase-conjugate per mm^2 (Table 5). The theoretical maximum binding capacity calculated in the same was described above for horseradish peroxidase, is approximately 8.3 ng/mm^2. This indicates, that 13.3% of the area are covered with the antigen. Note the high unspecific binding observed for the rabbit-anti-horse-peroxidase-conjugate.

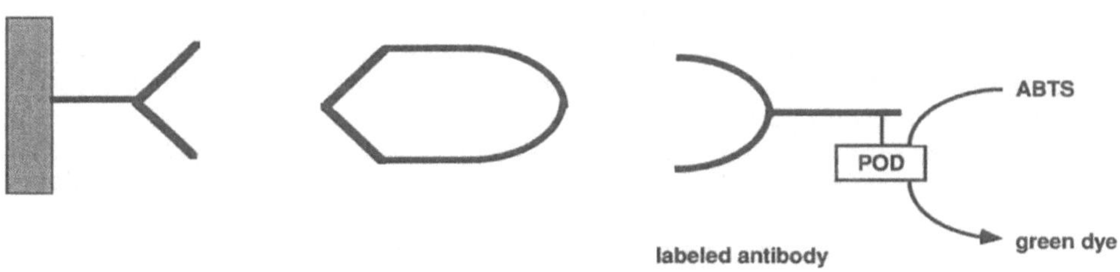

Figure 4. Principle of enzyme-linked immunoassay.

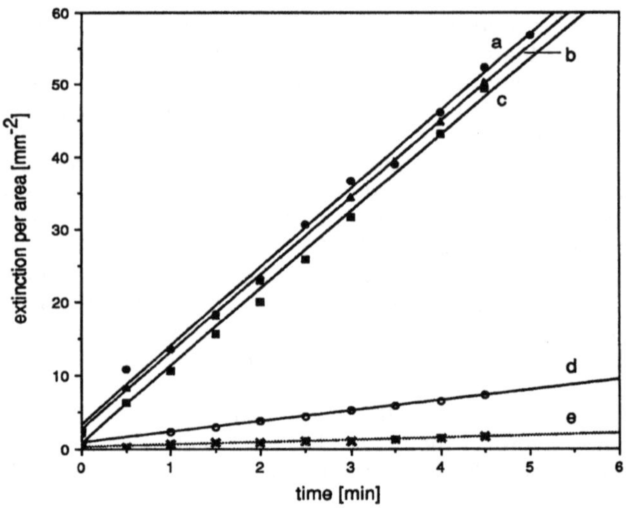

Figure 5. Absorption kinetics of the enzyme catalysed dye formation normalized to crystal size. a) propanol-2 cleaned, NaOH treatment, liquid silanization; b) microwave cleaned, gas phase silanization; c) ozone/uv cleaned, gas phase silanization; d) propanol-2 cleaned, gas phase silanization; e) propanol-2 cleaned, liquid phase silanization, control experiment.

Protein G, a 35000 Dalton protein is known to bind the Fc fragment of the antibody, i.e. the antigen binding sites are free for antigen binding.[15-17] It has been demonstrated,[18] that immobilization of protein G and subsequent specific binding of antibodies at the Fc part results in a higher antigen binding capacity per area than direct linkage of the antibody to the surface. Using this technique, the binding capacity was increased to 38% compared to antibody linkage without protein G (Table 5).

Unspecific Adsorption

The unspecific binding of proteins at surfaces is a general problem in immunoassay technique. Unspecific adsorption usually is one of the reasons for a high noise level and hence a reduced sensitivity of the immunotest. The unspecific interaction of proteins with the sensor surface depends strongly on the surface material and the microstructure of the film.

Several substrate surfaces were incubated with a solution of the antigen horseradish peroxidase. Using completely untreated lithium niobate surfaces, the unspecific adsorption was found to be higher than on a surface with an immobilized rabbit-anti-goat antibody not recognizing horseradish peroxidase. This implies, that the antibody film itself operates as an blocking agent.

In all test series the control experiments representing the unspecific adsorption show a surface coverage of HRP of $0.07 \cdot 10^{-2}$ to $1.22 \cdot 10^{-2}$ ng/mm^2 lithium niobate. The specific/unspecific binding ratio in our experiments therefore was larger than 10. The high unspecific binding using the anti-rabbit/rabbit system possibly is due to an enhanced cross-reactivity.

Table 4

Specific activity of immobilized mouse-anti-horseradish peroxidase-antibodies of the immunoglobuline-G-type achieved from the slope of the lines as shown in Figure 5. * = control experiments.

	cleaning process	silanization process	specific activity [10^{-2}ng/mm]
I	propanol-2 no NaOH treatment	liquid phase	1,67
I	propanol-2 NaOH treatment	liquid phase	10,76
I	propanol-2 NaOH treatment	liquid phase	9,62
I	propanol-2 NaOH treatment	liquid phase	0,33*
I	propanol-2	gas phase	1,47
I	propanol-2	gas phase	0,67
II	H_2O_2/NH_3	liquid phase	8,24
II	H_2O_2/NH_3	liquid phase	10,64
III	ozone/uv	gas phase	10,54
III	ozone/uv	gas phase	8,06
III	ozone/uv	gas phase	1,22*
IV	microwave	gas phase	10,58
IV	microwave	gas phase	1,30*

Stability of Antibody Films

Application of immunosensors in routine diagnostics and analysis requires a stability of antibody films of at least several weeks. This is one of the major problems of immunosensor application. We examined the binding capacity of the prepared films after 22 hours incubation in PBS/Tween-20 buffer (pH 7.4) and in the ELISA substrate buffer (pH 4.4). The data show a dramatic decrease in activity for both buffer solutions. The buffer with the lower Ph value results in a decrease of approximately 80% of the initial value in a time period of 22 hours.

DISCUSSION

The cleaning of lithium niobate surfaces with conventional techniques (i.e. using only solvents) is not sufficient for a reproducible film deposition on the surface.

Table 5

Specific activity of immobilized goat-anti-rabbit antibodies of the immunoglobuline-G-type. As antigen a rabbit-anti-horse-POD-conjugate was used. * = control experiment.

	cleaning process	silanization process	antibody immobilization	specific activity [10^{-2}ng/mm]
II	H_2O_2/NH_3	liquid phase	directly	109,6
II	H_2O_2/NH_3	liquid phase	directly	101,1
II	H_2O_2/NH_3	liquid phase	directly	100,9
II	H_2O_2/NH_3	liquid phase	directly	30,0 *
II	H_2O_2/NH_3	liquid phase	via protein G	151,0
II	H_2O_2/NH_3	liquid phase	via protein G	134,1

A substantial decrease in carboneous surface contamination can be achieved using microwave produced oxygen radicals or oxygen radicals produced by UV-cleavage of ozone.

For a successful silane film deposition from the liquid phase, activation of the $LiNbO_3$ surface with 5N NaOH is a necessary step. The ELISA data in Table 4 show the very different binding capacity of NaOH activated and non activated substrates. With the microwave and uv/ozone cleaning technique, a NaOH activation was not required for silanization. The oxygen radicals obviously not only clean the surface, they apparently also activate the $LiNbO_3$ for covalent silane adsorption. The non-NaOH activated gas phase cleaned surfaces show a comparable binding capacity to NaOH activated covered $LiNbO_3$ substrates (see Table 4).

It was shown by Jönsson et al.[19] that gas phase deposition of aminosilanes on silica results in thin and stable films, whereas after liquid phase silanization formation of globules due to the well known self condensation of alkoxysilanes were observed. In our experiments, however, silanization of uv-ozone and microwave cleaned $LiNbO_3$ surfaces with the CVD technique show polycondensated globules in the preliminary AFM data, whereas the averaged film thickness obtained from XPS data indicates for the microwave cleaned surface a monomolecular film. Angular resolved XP-data show, that the silane film on the microwave cleaned surface is inhomogeneous, i. e. part of the surface is not silanized. Hence, the average film thickness of approximately one monolayer refers to an inhomogeneous film structure. A comparison of the experimental C:Si ratios with expected data show incomplete condensation. Furthermore, a different condensation state can be seen using uv/ozone or microwave cleaning and activation, respectively. Using microwave cleaning the average condensation reaction appears higher than uv/ozone treatment. The reason for this is presently not known.

Fluorine signals in the XP-spectra of the microwave cleaned substrate indicated deposition of teflon material on the substrates. This disadvantage does not appear using uv/ozone. Assuming a correlation between the teflon impurities at the surface and the smaller silane film thickness compared to uv/ozone cleaned surfaces the set-up was changed in a way that teflon could not be adsorbed on the substrate. This was confirmed by XP-data. Subsequent vapor phase silanization, however, produced the same silane film thickness as on the teflon contaminated surfaces. Hence, the film thickness does not depend on the teflon impurities.

The differences between the gas phase cleaning procedures are not evident in the ELISA measurements on the binding capacities of silanized and subsequently antibody-linked surfaces. The antigen binding capacity of all surfaces studied is nearly the same, irrespective of the silane film quality. The average binding capacity of 0.1 ng horseradish peroxidase/mm^2 indicates a maximum coverage of 6-8% of the theoretical value. Using a purified antibody the binding capacity increases to about 13% (Table 5).

An increase in binding capacity of the film was found, when the antibody was linked via protein G to the surface. The binding capacity increases by a factor of 1.4 compared for using conventional techniques. This observation is in agreement with the results of Hilpert et al.[18]

The silanization results in an unstable film consisting of self-condensated globules, which rapidly hydrolyses in an aqueous solution. Hence, together with hydrolysed silane molecules antibody molecules were washed off and the binding capacity of the film decreases. In addition, binding between the surface and the antibody occurs randomly, i.e. the antigen binding site may be involved in surface linkage in a way that binding of antigen molecules is no more possible. Using protein G, the surface coverage of the antibodies is adjusted to the Fc fragments of the antibodies and hence a loss of antigen binding capacity is reduced.

The purity of the antibody is also important for antigen binding capacity. In the experiments described here, we used an anti-horseradish peroxidase antibody of ascites fluid without further purification. Preliminary immobilization experiments with highly purified antibodies resulted in a significantly higher binding capacity.

Immobilization of antibodies on $LiNbO_3$ substrates is hampered by similar difficulties as observed on other substrates. The instability of the films is one of the main problems to solve before a reproducible and routine application in an electro-acoustic immunosensor is possible.

ACKNOWLEDGMENT

The authors would like to thank S. Schwiegk, Max-Planck-Institut für Polymerforschung, Mainz, FRG for performing the α-step measurements; W.N. Unertl, University of Maine, USA, for taking the atomic force microscopic images; the Max-Buchner-Stiftung and the Verein zur Materialforschung for financial support; F. Josse, University of Milwaukee, USA, for helpful discussions and acoustic sensor design.

REFERENCES

1. P. Cuatrecasas, M. Wilchek, and C. B. Antinsen, Selective enzyme purification by affinity chromatography, *Proc. Nat. Acad. Sci. USA* 61:636 (1968).
2. P. Cuatrecasas, Protein purification by affinity chromatography, *J. Biol. Chem.* 245:3059 (1970).
3. K. Mosbach, Immobilized enzymes, *Trends Biochem. Sci.* 5:1 (1980).
4. J. L. Guesdon and S. Avrameas, Solid phase enzyme immunoassays, *Appl. Biochem Bioeng.* 3:207 (1981).

5. W. H. Scouten, A survey of enzyme coupling techniques, *Meth. Enzymol.* 135:30 (1987).
6. F. Josse, Z. A. Shana, D. T. Haworth, S.. Liew, and M. Grunze, On the use of ZX-$LiNbO_3$ acoustic plate mode devices detectors for dilute electrolytes, *Sens. and Act.*, to appear (1992).
7. R. Dahint, M. Grunze, F. Josse, and J. C. Andle, Probing of strong and weak electrolytes with acoustic wave fields, *Sens. and Act.*, submitted (1992).
8. J. Andle, J. Vetelino, R. Lec, and D. McAllister, An acoustic plate mode immunosensor, *Proc. IEEE Ultras. Symp.* pg 579 (1989).
9. J. Andle, J. Vetelino, M. Lade, and D. McAllister, Detection of nucleic acid hybridization with an acoustic plate mode sensor, to be published.
10. I. Haller, Covalently attached organic monolayers on semiconductor surfaces, *J. Am. Chem. Soc.* 100:8050 (1978).
11. U. Jönsson, M. Malqvist, and I. Rönnberg, Immobilization of immunoglobulines on silica surfaces, 227:363 (1985).
12. K. L. Mittal and D. F. O'Kane, Vapor deposited silanes and other coupling agents, *J. Adhes.* 8:93 (1976).
13. H. H. Weethall and M. Lynn, "Immobilized Enzymes, Antibodies, Cells, Peptides: Preparation and Characterization," M. Dekkar, Inc., New York (1975).
14. J. H. Peters, M. Schulze, and M. Grol, "Monoklonale Antikörper," Springer Verlag, Heidelberg, New York (1990).
15. L. Björck and G. Kronvall, Purification and some properties of streptococcal protein G. a novel IgG binding reagent, *J. Immunol.* 133:969 (1984).
16. B. Åkerström, T. Brodin, K. Reis, and L. Björck, Protein G: a powerful tool for binding and detection of monoclonal and polyclonal antibodies, *J. Immunol.* 135:2589 (1985).
17. B. Åkerström and L. Björck, A physicochemical study of protein G., a molecule with unique immunoglobulin G-binding properties, *J. Biol. Chem.* 261:10240 (1986).
18. R. Hilpert, F. Binder, J. Ritter, S. Drost, C. Klößinger, S. Koch, E. Müller, I. Ruge, and M. H. Zenk, Biosensoren zur Gewässerüberwachung, 9. Dechema Jahrestagung der Biotechnologen, Berlin 1991, p. 257
19. U. Jönsson, G. Olofsson, M. Malmqvist, and I. Rönnberg, Chemical vapour deposition of silanes, *Thin Solid Films* 124:117 (1985).

BIOLOGICAL MICROSTRUCTURES IN BIOSENSORS

Frances S. Ligler[1] and Sina Y. Rabbany[2]

[1]The Center for Bio/Molecular Science and Engineering, Naval Research Laboratory, Washington, DC 20375; [2]Bioengineering Program, Department of Engineering, Hofstra University, Hempstead, NY 11550

INTRODUCTION

To construct ultrasensitive detection systems, biomolecules are being incorporated as recognition elements into devices called biosensors. The choice of the biomolecule, its method of incorporation into the system, and the mechanism of signal transduction determine the specificity, speed, and the sensitivity of the biosensor. However, the molecular nature of the microstructure at the interface and the events during analyte recognition and signal generation are poorly understood. The interaction of the microstructure at the surface of the device with a complex environment yields a signal. The biochemical and physical interactions of these biological microstructures at the solvent interface are of critical importance to the behavior of the biosensor, but are yet to be fully defined. Figure 1 outlines some of the issues relevant to understanding the behavior of molecular microstructures at the solid-liquid interface. This chapter provides a conceptual framework for understanding the molecular complexity of the microstructures which are responsible for biosensor function. Alteration of antibody kinetics caused by change from an equilibrium to a nonequilibrium system is used to exemplify how variation in a single physical parameter can drastically alter the behavior of the biomolecule.

BACKGROUND

The incorporation of biomolecules into optoelectronic devices is becoming an accepted methodology for the fabrication of ultrasensitive detection systems. In the development of these biosensors, careful attention is given to the choice of molecules, the mechanism of signal transduction, and the device. Table I illustrates how various biological recognition elements are immobilized on different devices.

Biosensors are constructed by immobilizing biomolecules to provide a molecular-level recognition for the compound being detected. Biomolecules frequently used include antibodies, enzymes, chelators, DNA probes, and ionophores. Antibodies

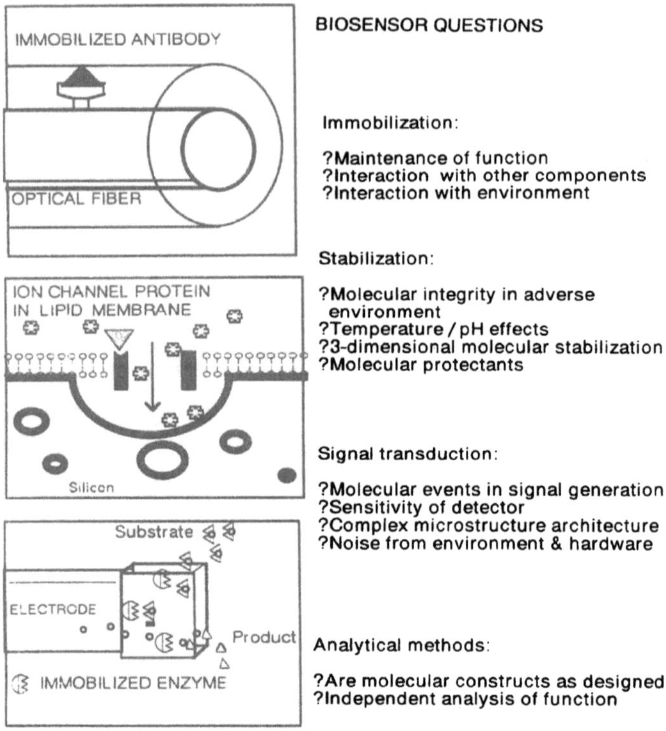

Figure 1. Issues pertinent to understanding the function of molecular structures at the solid-liquid interface.

are used as the detection component of the biosensor described here in order to exemplify the complexity of a comparatively simple system.

Antibodies belong to the family of glycoproteins known as immunoglobulins, produced by the mammalian immune system. They are synthesized as part of the response to the presence of a foreign substance, and are subsequently capable of binding directly to the eliciting substance. Molecules called antigens are responsible for such responses. Antibodies, which consist of two kinds of polypeptide chains bound together by disulfide bridges, have a Y-shape structure. The primary advantage of using antibodies as the recognition element in a biosensor is their extreme specificity (i.e., their ability to distinguish a certain molecule from the millions of other molecules in the sample) and the ability to produce large quantities of identical molecules through monoclonal antibody technology.

Mechanisms of signal transduction employed in biosensors include fluorescence, luminescence, electron transfer, color and mass effects. Fluorescence and luminescence are particularly useful for discrimination of signal from natural background noise from samples when fluorescent labels of appropriate wavelengths are selected. Measurements of electron transfer using standard electrodes are usually only sensitive to the micromolar level, but sensitivities are improving with the development of new microelectrode techniques. Instruments that measure mass changes are easy to use, but are the least able to discriminate signal from noise in samples containing complex mixtures of components (i.e., body fluids, ground water).

Table I

Detection Element	Immobilization	Devices
Binding: Antibodies Receptors Chelators Nucleic acid probes	Thin Films	Optrodes Optical waveguides Surface acoustic waveguides Piezoelectric Capacitance Surface plasmon resonance
Substrate Modification: Enzymes Abzymes Whole cells	Thick Films Chambers	Optrodes Electrodes Light addressable poteniometer
Channel Formation: Ionophores Ion Channel Proteins	Membranes Bilayer Multilayer	Capacitance Electrodes

Devices used to read the signal generated by the biomolecular recognition event include spectrometers, electrodes, piezoelectric crystals, and surface acoustic wave systems. Factors that influence the choice of device include the sensitivity and selectivity of signal recognition required, conditions of use, sophistication of the user, and cost. Biosensors must be adapted for use in particular environments such as research laboratories, clinical settings, or the great outdoors.

COMPONENTS OF THE MICROSTRUCTURE AT THE BIOSENSOR SURFACE

The components of the microstructure at the interface between the sensor surface and the sample solution is depicted in Figure 2. The complexity of the molecular interactions at the interface of the biosensor surface with the surrounding media is yet to be carefully investigated. Key components of the microstructure include the detection molecules, molecules responsible for immobilization, blocking agents, assay reagents, and other solvent components.

Much attention has been focussed on the method of immobilization of the biomolecule and the effect the immobilization has on its function. Table II cites some of the most often used immobilization techniques. In general, the binding elements are more likely to be incorporated in a thin film mode. The most desirable methods of immobilization maintain both the function of the biomolecule and stabilize the biomolecule during storage. At least one year of shelf-life in storage is usually considered desirable.

Extensive studies on protein adsorption demonstrated that many of the biomolecules used for detection interact vigorously with the surface. Hydrophobic and

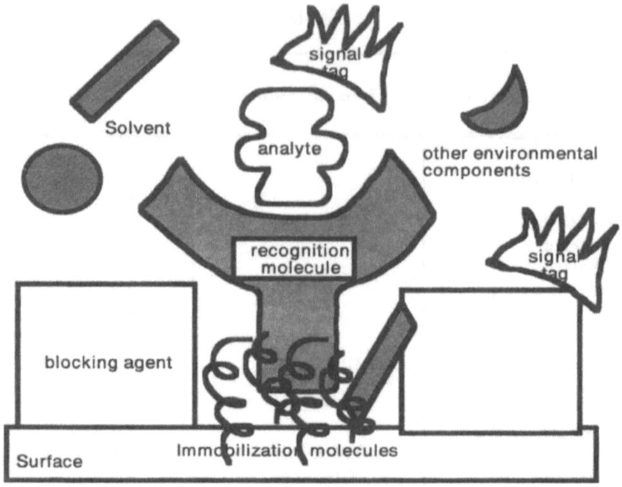

Figure 2. Components of the microstructure at the solid-fluid interface.

electrostatic interactions between the biomolecule and the surface can lead not only to strong adsorption, but also to partial denaturation and loss of function. Such interactions can still occur when biomolecules are covalently attached to surfaces. Furthermore, the ability of the immobilized biomolecule to recognize the analyte can be affected by factors in the solution such as pH, ionic strength, polarity, and density.

Table II

Thin Films	**Covalent:** **Direct binding to activated support** **Silane + Glutaraldehyde** **Silane + Heterobifunctional crosslinkers** **Noncovalent:** **Adsorption** **Through Protein A, Biotin/Avidin**
Thick Films	**Polymers (i.e., PVC, PHEMA, PVP)** **Hydrogels**
Chambers	**Dialysis Membranes** **Silicon**
Membranes **Bilayer** **Multilayer**	**Lipids** **Fatty acids** **Silanes**

Methods to minimize signals from nonspecific binding and complications due to background noise have been developed. Such methods include the use of blocking agents, detergents, and controls for subtraction of the background noise from the measurement signal. Blocking agents are molecules such as bovine serum albumin, casein, calf thymus DNA, and dextran that bind to areas of the surface that are not taken up by the biomolecule and prevent interferants in the solution from binding to the surface. Detergents, such as Triton® and Tween®, primarily serve the latter function. Despite the addition of blocking agents and detergents, other molecules from the solution may localize on the surface adjacent to the detector molecule and affect its interaction with the surrounding medium.

A variety of additional factors influence the recognition function of the complex microstructure at the biosensor surface. The biochemistry of the detection event is further affected by the interaction of the analyte nonspecifically with the surface itself and with the solvent. The availability of the secondary reactants in the solvent may affect signal generation. For instance, many enzyme substrates hydrolyze with time to produce nonspecific signal (i.e, "sample noise"). Physical parameters such as diffusion limitations and flow rate are also important. Heat, acid, and proteolytic degradation can destroy either the functional biomolecule or the analyte, resulting in signal reduction. The detection molecule is only one of many elements that effect recognition and signal transduction.

KINETICS OF ANTIBODY BINDING

In order to exemplify the effects of the microenvironment of a detector molecule on its function, the kinetics of antibody binding are analyzed free in solution, immobilized in a static system, and immobilized in flow. Several kinetic differences are apparent between free versus immobilized antibody and between antibody in equilibrium versus nonequilibrium conditions.

Antibody-antigen interactions in solutions are well understood and described in classical immunology textbooks.[1] The basic thermodynamic principles of antigen-antibody interactions, may be expressed as:

$$Ab + Ag \underset{k_d}{\overset{k_a}{\rightleftharpoons}} AbAg \tag{1}$$

where Ab represents free antibody; Ag, free antigen; AbAg, the antibody-antigen complex; k_a, the association rate constant; and k_d, the dissociation rate constant. Applying the law of mass action the equilibrium constant or the affinity of the reaction as follows:

$$K = \frac{k_a}{k_d} = \frac{[AbAg]}{[Ab][Ag]} \tag{2}$$

For antibody binding in solution, well-known equilibrium kinetics suggest that both association and dissociation are relatively rapid. For monoclonal antibodies the association constant K is typically 10^5-10^9 M^{-1}.

The two best known antibody assays, the radioimmunoassay (RIA) and the Enzyme Linked ImmunoSorbent Assay (ELISA) generally employ immobilized antibodies (or antigen). Both the ELISA and RIA are governed by the same thermodynamic constraints. The widespread use of the ELISA has triggered several investigations into the kinetics of binding of immobilized antibodies.[1] The association of the antigen with immobilized antibody is rapid but the dissociation appears to be extremely slow, probably due to mass transport limitations and rapid reassociation (Figure 3). In the time frame of the assay, binding is functionally irreversible, requiring several hours before measurable dissociation occurs.

A recently described flow immunoassay describes a mechanism for rapid detection of low molecular weight analytes in picomole quantities.[2] In contrast to many existing approaches, this assay relies on the displacement of antigen from antibody for the detection of the compound of interest. It further differs from the ELISA system in that the antigen and antibody are placed under continuous flow, nonequilibrium conditions.

THE NONEQUILIBRIUM DISPLACEMENT REACTION

The flow immunosensor is capable of detecting low molecular weight antigens in the pmole range in less than a minute by monitoring the displacement of a labeled antigen.[2] In flow, immobilized antibodies are first saturated with fluorophore-labeled antigen and placed in continuous buffer flow. Upon introduction of unlabeled antigens, a proportional amount of fluorophore-labeled antigens are released within seconds of exposure. The concentration of the released antigens is subsequently detected downstream by a fluorimeter. This rapid response time is unlike what has been described previously as functionally irreversible antigen/antibody binding in the ELISA systems, requiring several hours before measurable dissociation occurs. A theoretical model to describe the displacement of bound labeled antigens from immobilized antibodies under nonequilibrium conditions is presented herein.

A monoclonal antibody specific for the major metabolite of cocaine (benzoylecgonine) was covalently immobilized by a tresyl-chloride linkage onto Sepharose beads. The antibody binding sites were saturated with fluorophore-labeled benzoylecgonine, which is similar in structure to cocaine. The beads were then placed in a 200 μl column. When unlabeled cocaine was injected into the continuous buffer stream flowing through the column, it displaced and released the labeled bound antigen into the flow stream (Figure 4). Fluorescence in the effluent, caused by the presence of the displaced antigen was detected downstream by a fluorimeter. The intensity of the signal is directly proportional to the amount of cocaine in the sample.

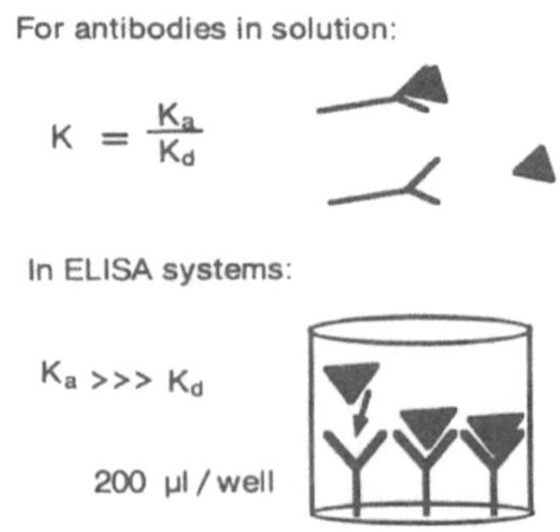

Figure 3. Association(k_a) and dissociation (k_d) rate constant relations both in solutions and in Enzyme Linked ImmunoSorbent Assays (ELISA).

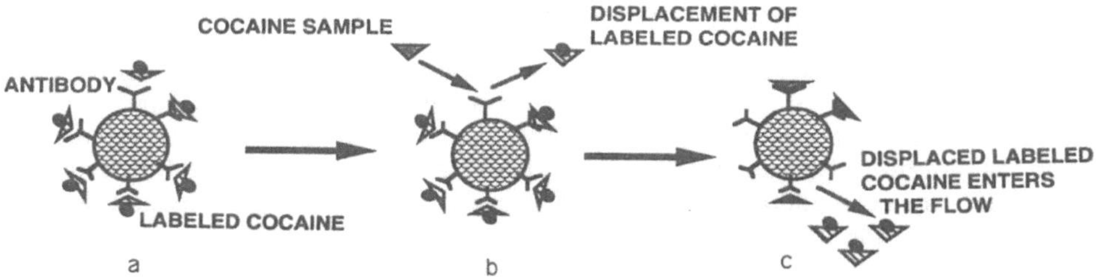

Figure 4. Flow immunosensor operation: (a) anticocaine antibodies immobilized on the Sepharose® beads are saturated with fluorescein cadaverine benzylecgonine antigen molecules; (b) a sample of cocaine is introduced resulting in a proportional amount of labeled antigen to be displaced; and (c) the labeled antigen displaced is moved by the aqueous flow to the fluorimeter, and a signal is generated.

To determine the kinetics of labeled antigen displacement, a repetitive displacement technique was employed. Identical amounts of unlabeled antigens were injected repeatedly until the column was depleted of fluorophore-labeled antigens. Using this procedure, the relatively low number of unoccupied antibody binding sites which may occur following the establishment of the buffer flow, were filled upon the introduction of antigen, and the contribution of these sites to the subsequent displacement data was minimized. Figure 5 depicts the displacement of the labeled antigens at a constant flow rate of 0.75 ml/min. The "releasable" labeled antigens were obtained by measuring the amount of the fluorophore-labeled antigen displaced during multiple additions of antigens at high excess (100-fold over available antibody binding sites). An exponential decrease for the amount of labeled antigens released as a function of time was observed. Such relation indicates a first order process. The

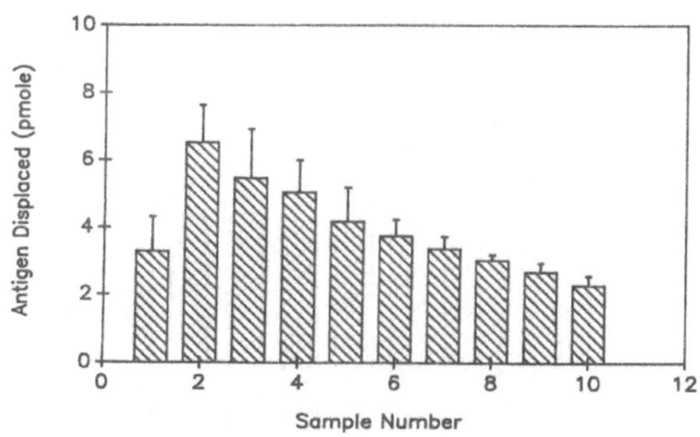

Figure 5. Repetitive displacement assay. Cocaine was loaded repeatedly onto a column of anti-benzylecgonine antibody saturated with fluorescein cadaverine benzylecgonine antigen at a ratio of 1:1 cocaine to antibody. Values represent the mean ± S.D. of two experiments conducted with a flow rate of 0.75 ml/min buffer flow through the column.

rate of depletion of labeled antigens from the column (slope) were observed to be flow-rate dependent.

The data obtained from the repetitive displacement experiments was used to investigate the effects of flow on the displacement kinetics. The undissociated fraction of labeled antigen at each step of the repetitive displacement procedure was calculated from the difference between total bound (releasable) labeled antigen and the amount displaced after each addition of unlabeled antigen. The following expression depicts the relation for the undissociated fraction (θ):

$$\theta = \frac{[bound\ Ag^*]_{t=0} - [displaced\ Ag^*]_{total}}{[bound\ Ag^*]_{t=0}} \tag{3}$$

where Ag^* represents the amount of labeled antigen, and the denominator represents the concentration of Ag^* initially bound to the column (i.e., t=0). To determine the time period available for displacement, the volume of the column containing the antibody-bound labeled antigen was divided by the flow rate (0.75 ml/min). Figure 6 illustrates the calculated values for the undissociated fraction plotted versus the displacement time. The plot illustrates that the dissociation is a first order process.

The efficiency of each displacement event in the flow immunoassay was determined utilizing the following relationship for the displacement efficiency (D_e) :

$$D_e = \frac{[displaced\ Ag^*]}{[loaded\ Ag]} \cdot \frac{1}{\theta} \tag{4}$$

The displacement at the flow rate of 0.75 ml/min was then plotted as a function of sample number (Figure 7). The displacement efficiency has also been calculated for 0.5 and 1.0 ml/min and demonstrated to increase with decreasing flow rates (data not shown).

The classical concepts for describing antibody-antigen interactions developed for fluid phase systems under equilibrium conditions cannot predict the behavior of the flow immunoassay since antibody binding and release occurs at a solid-liquid

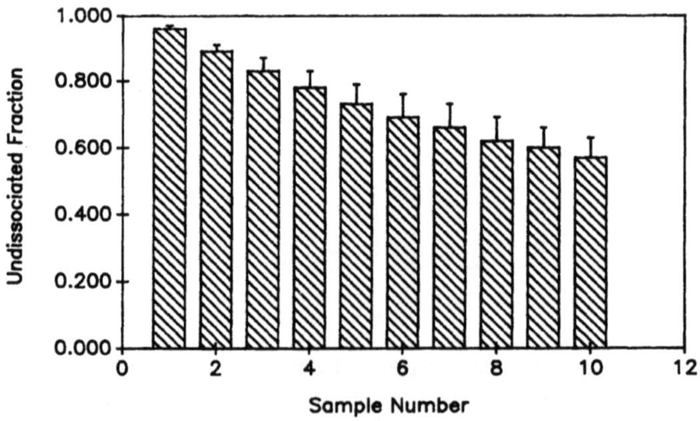

Figure 6. The undissociated fraction was calculated as described in the text for data obtained at 0.75 ml/min and plotted as a function of the cumulative time of exposure of the column to cocaine.

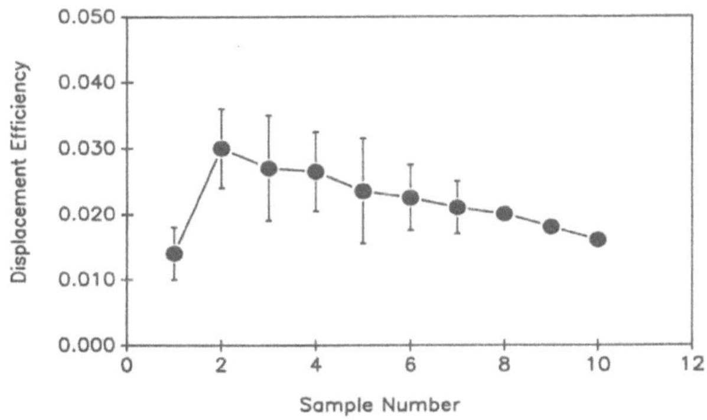

Figure 7. The calculated displacement efficiency plotted versus the sample number at the flow rate of 0.75 ml/min.

interface under nonequilibrium conditions. However, the equations developed for equilibrium systems can be modified to describe the nonequilibrium displacement event.

In addition to the physical variables such as flow rate or diffusion, biochemical variables introduced by immobilization include antibody density, steric hindrance, and spatial heterogeneity. Characterization of these parameters will enable manipulation of both the sensors's sensitivity and response time.

CONCLUSIONS

A need for faster and more effective means of (bio)chemical monitoring has resulted in the development of biosensors for application in clinical, environmental, and agricultural markets. Viewing a detection molecule in isolation on a sensor surface is an oversimplification. If additional molecules are not added to the assembly at the surface by design, they will self-assemble with the detector molecule when the sensor is exposed to a clinical or environmental sample. Insight into the molecular microstructures which form at the interface between the device and the fluid sample can be used to improve the biosensors. For example, by understanding the nonequilibrium kinetics of the immobilized antibody-antigen interaction, the sensitivity of antibody-based biosensors can be enhanced. The physical and chemical variables inherent in the detection event are just beginning to be understood at the molecular level. Yet, biosensors are very primitive compared to the complexity of the recognition and signal transduction mechanisms in biological systems.

REFERENCES

1. J. A. Berzofsky, S. L. Epstein, and I. J. Berkower, Antigen-antibody interactions and monoclonal antibodies, in: "Fundamental Immunology," W. E. Paul, ed., Raven Press Ltd., New York, pp. 315-355, (1989).

2. A. W. Kusterbeck, G. A. Wemhoff, P. Charles, R. Bredehorst, and F. S. Ligler, A continuous flow immunoassay for rapid and sensitive detection of small molecules, *J. Immunol. Meth.* 135:191-197 (1990).

CHARACTERIZATION

INTRODUCTION

The ability to fabricate microstructures is inevitably accompanied by the need to analyze those structures with ever increasing spatial sensitivity. Ultrahigh resolution characterization methods have often been the province of physicists, chemists, and engineers working with model systems. But new emphasis on biological interfaces has amplified cross-disciplinary efforts to apply such methods to synthetic microstructures. This section encompasses papers which emphasize development of new analytical tools with utility for biological applications, as well as articles which directly illustrate the capabilities of such techniques.

Hallmark et al. address the abilities and limitations of the popular new technique of scanning tunneling microscopy through a model study of closely related molecules. The article by **Lösche** communicates recent advances in neutron reflectivity measurements of molecular interfaces, with particular relevance to characterization of protein/membrane interactions. The developmental status of scanning transmission x-ray microscopy as an imaging tool for wet biological specimens is described by **Williams et al.** Fabrication and characterization are combined in the paper by **Furuno et al.** which details structural studies of protein monolayer arrays by high resolution scanning electron microscopy. **Garcia et al.** demonstrate the utility of both scanning tunneling and atomic force microscopes to evaluate particle size and structure and to assess induced morphological changes.

HOW WELL CAN THE SCANNING TUNNELING MICROSCOPE DISTINGUISH BETWEEN TWO VERY SIMILAR MOLECULES?

V. M. Hallmark, S. Chiang, J. K. Brown[a] and Ch. Wöll[b]

IBM Research Division, Almaden Research Center
650 Harry Road, San Jose, CA 95120-6099

INTRODUCTION

A decade has passed since the concept of the scanning tunneling microscope (STM) was first demonstrated.[1] After early use primarily by surface scientists seeking to characterize semiconductor[2] and metal[3] surfaces in the well controlled confines of ultrahigh vacuum (UHV), the STM has evolved into a broad range of related scanning probe microscopies[4] (SPMs) which are increasingly employed to investigate complex molecular systems in relevant ambient environments such as air and water.[5] Unfortunately, the simplicity of the method of acquiring ultrahigh resolution images belies the difficulties inherent in the interpretation of those images. The increasing commercial availability of scanning probe microscopes engenders the notion that these techniques are well understood when, in fact, there remain a large number of crucial questions about the imaging process.

Many of the unresolved issues are common to all the scanning probe microscopies: interactions of the probe tip with adsorbate and substrate, choice of appropriate substrate for molecular immobilization, and the characterization of contamination effects. Tip-sample interactions include van der Waals forces (attractive or repulsive), molecular responses to electric field gradients induced by a tunneling bias, and anomalous effects such as multiple tip images. Substrate selection poses a particularly serious problem for non-vacuum applications. The most widely employed substrate, highly oriented pyrolytic graphite (HOPG), is considered by a growing number of scientists to be the most perplexing substrate in terms of interpretation. Its wide use stems from the ease of cleaving graphite to produce large, atomically flat surfaces. This advantage is more than counteracted by observations of "giant" corrugations arising from elastic deformations of the graphite surface during imaging[6]

[a]Present address: UCSF Physics Research Laboratory, University of California at San Francisco, 389 Oyster Point Blvd., South San Francisco, CA 94080

[b]Present address: Institut für Angewandte Physikalische Chemie, Universität Heidelberg, Im Neuenheimer Feld 253, D-6900 Heidelberg, Germany.

("tunneling" on graphite may be a misnomer), peculiar electronic effects leading to preferential imaging of half of the graphite surface atoms,[7] unusual multiple tip effects,[8] and novel surface defects which can be confused with molecular adsorbates.[9] The low reactivity of the graphite surface allows imaging in gas or liquid ambient,[10] but also prevents strong adsorbate interactions which are needed to prevent molecular motion during imaging. Gold surfaces formed by epitaxial growth on mica[11] or by melting gold wire into faceted balls,[12] have also served as unreactive substrates for non-vacuum applications. But gold has its own peculiarities: gold surface mobility is high at room temperature and impurities seem to enhance mobility;[13] gold is a soft substrate and tip crashes cause long range restructuring of the surface;[14] and fresh gold contaminates quickly in air, going from hydrophilic to hydrophobic within ten minutes.[15] Contamination is a universal issue for scanning probe microscopies, and characterization and prevention of surface impurities is mandatory.

While ultrahigh vacuum tunneling experiments are handicapped by loss of "real" environmental conditions, they benefit from the ability to control many of the factors which complicate image interpretation. Thus, they provide an opportunity to explore the open questions in organic imaging in a methodical and precise manner. We describe here the early stages of a systematic investigation of a series of related molecules. The specific questions to be addressed in our experiments include:

(1) Adsorbate-surface interactions. Molecular binding site and orientation with respect to the substrate atomic lattice; molecular organization within domain structures; defects and domain boundary characteristics; order versus disorder; and molecular motion in the surface plane.
(2) Adsorbate-tip interactions. Tip induced molecular motion; multiple tip anomalies; and high resolution imaging.
(3) Impurities. Intrinsic and extrinsic contamination and their effect on substrate morphology, molecular adsorption, and imaging.
(4) Molecular structure. Aromaticity versus saturation; internal structure; and distinguishing closely related molecules and various substituent groups.

The initial studies in this series involve the two isomers naphthalene and azulene, shown schematically in Figure 1. Both molecules have the chemical formula C_8H_{10} and are aromatic, planar ring systems of quite similar physical dimensions (see Figure 1). Naphthalene is approximately 30 kcal/mol more stable than azulene. Whereas naphthalene is nonpolar, azulene is quite unusual for a hydrocarbon in that

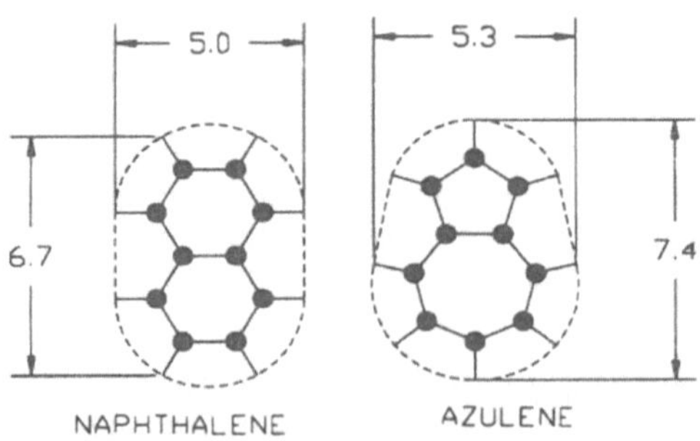

Figure 1. Scaled drawings of the isomers naphthalene and azulene. The perimeter marks the positions of the hydrogen atom centers, while the filled circles denote carbon atoms. Van der Waals radii are not included.

it possesses a significant dipole moment (0.8 Db).[16] Both molecules have been examined previously by low energy electron diffraction (LEED), while adsorbed onto the (111) face of platinum.[17,18,19] Both naphthalene and azulene are irreversibly bound to the Pt(111) surface: molecular desorption is precluded by thermal decomposition, which is marked by hydrogen evolution at ~200°C, leaving residual surface carbon.[18,19] These two adsorbate systems provide an ideal test of the capability of the STM to distinguish between very similar molecules. The experiments to be described here include submonolayer and full monolayer coverages of naphthalene and azulene individually, as well as mixed monolayers. The mixed monolayers offer the opportunity to image two similar molecules simultaneously with the same tip, stabilizing the most variable and uncontrollable parameter of the STM technique.

BACKGROUND

Naphthalene adsorption onto Pt(111) at room temperature produces a disordered overlayer which generates a 1/3 order ring diffraction pattern. Subsequent annealing between 100 and 200°C, or deposition with the substrate temperature in this range, results in an ordered overlayer characterized by a (6x3) diffraction pattern, with certain missing beams ascribed to glide plane symmetry.[18] Auger electron spectroscopy (AES) indicates that the surface coverage is consistent with that required to form an overlayer of naphthalene with the ring system parallel to the platinum substrate and the molecular centers located on a (3x3) superlattice. Disorder in this system is inferred to arise from misalignment of the molecular superlattice with the underlying platinum lattice. The ordered (6x3) structure is then generated by alignment of the two lattices and molecular packing involving two molecules per unit cell, presumably by variation in the orientations of two neighboring naphthalenes. Perhaps the simplest packing scheme which is consistent with the (6x3) unit cell is the so-called herringbone organization,[18] shown in Figure 2. Alternative structures such as pinwheels also fulfill the symmetry requirements for a (6x3) structure, but naphthalene cannot form a true pinwheel structure.[20]

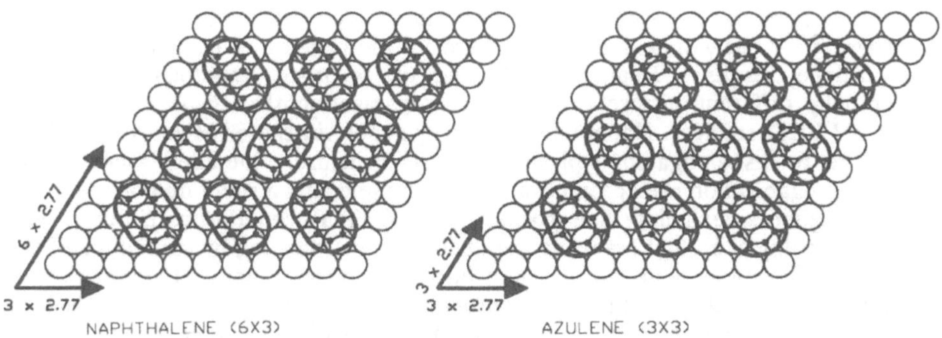

Figure 2. Schematics of the proposed "herringbone" (6x3) unit cell for naphthalene on Pt(111)[18] and the simple (3x3) unit cell for azulene on Pt(111). Two other domains of the (6x3) unit cell, rotated by 60° and 120° from the one shown, are required to generate the observed LEED pattern.

Azulene adsorption onto Pt(111) differs from that of naphthalene in that there appears to be a variety of possible molecular organizational structures depending upon both coverage and substrate temperature.[19] When adsorbed warm at a coverage near that needed for saturation, azulene forms a simple (3x3) monolayer, with the ring system parallel to the Pt(111) surface. A schematic of a possible (3x3) structure is shown in Figure 2, where all azulene molecules have been assigned the same arbitrary orientation and binding site. Higher coverage overlayers are possible, such as an observed (10x10) structure produced by tipping the azulene ring system out of the surface plane. The absence of the further restriction of two molecules per unit cell needed to promote the organization from a simple (3x3) packing to the more complex (6x3) structure seen for naphthalene implies some dissimilarity in the molecule-molecule and/or molecule-substrate interactions of the two isomers.

EXPERIMENTAL DETAILS

STM experiments were performed in a multichamber UHV system with base pressure $<10^{-10}$ torr.[21] The Pt(111) single crystal was cleaned by repeated cycles of 500-1000V Ar^+ bombardment followed by annealing at 850°C. Although carbon contamination was below AES detection levels, carbon islands were observed during STM imaging at local concentrations ranging from 0-4%. Exposures to naphthalene and azulene were made by backfilling the preparation chamber to a pressure of 5×10^{-9} torr from a dosing manifold maintained at 60-80°C. Ordered samples were prepared by dosing when the sample temperature had dropped after annealing to between 100 and 200°C (precise temperature control was prevented by the constraint of sample mobility throughout the UHV system). Mixed monolayers were prepared either by dosing sequentially or by dosing simultaneously from a naphthalene/azulene mixture. The previously described microscope was operated in constant current mode with ~0.4Hz line scan rate. Unless noted, data processing has been limited to low pass filtering.

RESULTS AND DISCUSSION

Naphthalene/Pt(111)

Our extensive STM observations for naphthalene/Pt(111) have been detailed elsewhere,[22,23,24] but will be summarized here for the purpose of comparing with the new azulene results. Real-space imaging of adsorbed naphthalene yields a conceptually different picture of molecular organization than expected from the previous LEED studies. Naphthalene is recognizable in the STM images, such as Figure 3a, by its elongated, typically bi-lobed structure. Detailed mappings of molecular organization are accomplished easily by selecting a small number of close-packed neighbors, assigning their positions to a (3x3) Pt superlattice (initially with arbitrary binding site), and assessing the quality of the mapping by coherence of the naphthalene positions with the platinum lattice at longer ranges. Examination of mappings, such as that of Figure 3b, provides conclusive evidence that the long molecular axis of naphthalene is aligned along the platinum lattice near-neighbor directions, naturally leading to three equivalent orientations separated by 120° due to the three-fold symmetry of the (111) face. The combination of three-fold binding site symmetry with two-fold molecular symmetry along the close-packed metal directions clearly restricts the binding site to a six-fold symmetric on-top site. Within close-

(a) (b)

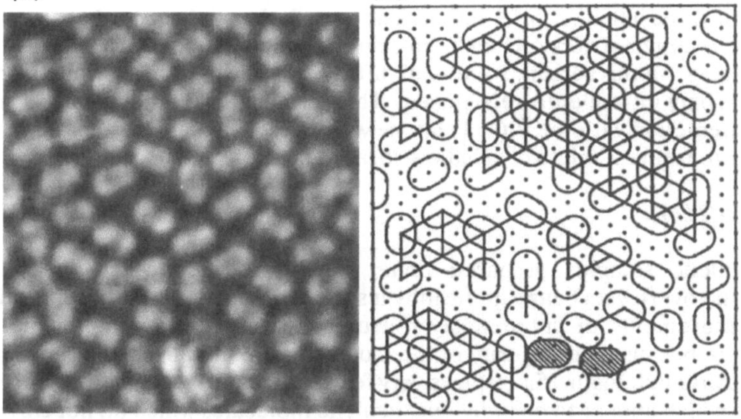

Figure 3. (a) STM image of an (6x3) ordered sample of naphthalene on Pt(111). The grey scale represents observed height with white highest. The naphthalene molecules are typically 1Å tall. Sample bias $V_s = +1.0V$ and tunneling current $I_t = 0.7nA$. (b) Detailed mapping of the molecular positions of (a) to the Pt(111) lattice. Points represent Pt atoms with 2.77Å spacing, ovals represent naphthalene molecules, and the lines indicate molecules belonging to different local (3x3) domains. The hatched molecules mark the bright features in (a) which are assigned to tilted molecules.

packed domains, the naphthalene molecules do not assume the expected herringbone structure, nor do they form true pinwheels.

Whereas the herringbone structure of Figure 2 predicts that each of three equivalent close-packed domains should consist of only two of the three possible orientations, the close-packed domains in fact contain all three orientations. Only small domains of true herringbone structure are found. Longer annealing may increase the population of herringbone packing within the overlayer, but a good (6x3) diffraction pattern is obtained with very few herringbone unit cells present. Although long range herringbone structure is not present in the STM images, the data do reveal a significant population of rows of molecules with alternating directions. This type of organizational structure is characteristic of the herringbone structure as well, and this "pseudo-symmetry" accounts for the observation of a (6x3) diffraction pattern. Domain boundaries are formed where different (3x3) superlattices abut, and usually involve a shift of the superlattice by only a single platinum lattice unit vector. Defects other than domain boundaries are also conspicuous, such as the brighter, higher features in Figure 3a which have structure consistent with the typical naphthalene molecules. Careful inspection establishes that these molecular defects occupy sites where neighboring molecules are closer than the usual naphthalene nearest neighbor distance. Because the concentration of these defects is the primary dissimilarity between ordered and disordered structures, we have interpreted these defects to be naphthalene molecules tilted on edge. Accommodation of these tilted molecules into the flat (6x3) layer is thus the source of ordering induced via annealing.

An interesting observation for adsorbed naphthalene is that both rotational and translational energy barriers are surmountable at room temperature. Both naphthalene and azulene are strongly chemisorbed to the Pt(111) surface, with decomposition at 200°C precluding molecular desorption. In Figure 4, a sequence of

STM images are marked to facilitate location of molecules undergoing both rotation among the three allowed orientations and translation between adjacent adsorption sites. The energy barriers are such that, on the timescale of these measurements (10 min. per image in this series), rotational diffusion is fairly common while translational diffusion is only infrequently recognized. Thus, although naphthalene has a strong and specific interaction with the Pt(111) lattice which results in a preferred adsorption site, this site represents a relatively shallow local minimum.

Azulene/Pt(111)

Azulene samples prepared under conditions identical to near-saturation naphthalene overlayers bear significantly less than monolayer coverages. Quantitative STM measurements of submonolayer azulene coverages are impossible at room temperature because the majority of the molecules are mobile on the timescale of the imaging. Although Figure 5a displays very few fixed features, single line traces include structures which appear to be individual azulene molecules which do not correlate well line-to-line. Increased coverage prevents molecular translation, as shown in Figure 5b for a sample near full monolayer (3x3) coverage. Near saturation coverage allows imaging of discrete molecules, but without any marked asymmetry to indicate fixed orientations with respect to the substrate lattice. The molecular spacings in the STM images are consistent with the acquisition of a (3x3) LEED pattern from this sample, and the images do not indicate the presence of any compressed molecular packing domains formed of tilted molecules. Knowing that azulene is less stable than naphthalene and that high coverage overlayers of azulene involve tipping molecules out of the surface plane, it is not surprising to find that the rotational and translational barriers for azulene are notably lower than those of naphthalene. The decreased sticking coefficient, along with out-of-plane molecular adsorption at high coverages, also implies a decreased energy barrier normal to the metal surface.

Mixed Naphthalene and Azulene Overlayers

Mixed overlayers of naphthalene and azulene give an excellent impression of the relative mobility of the two isomers. Since naphthalene is only moderately mobile, it permits some localization of azulene molecules. Figure 6 shows two consecutive images for a submonolayer sample prepared by simultaneously dosing azulene and naphthalene. While the partial vapor pressures of the two species at 70°C give a relative concentration of $P_{naph}/P_{azul} \approx 4$, the STM measurements indicate that the sticking probabilities of the two molecules are not equivalent. Estimates from a variety of dosing experiments suggest that the relative sticking probabilities are $\sigma_{naph}/\sigma_{azul} \approx 4$. Thus the azulene coverage in Figure 6 is only 5-10% of the approximately 0.2 monolayer coverage. At such low coverages azulene can occasionally be trapped within naphthalene pockets but, unless the packing is high density, azulene appears delocalized within the available area. When the partial naphthalene and overall coverage is high enough to prevent delocalization, as for the sequentially dosed $\theta_{naph}/\theta_{azul} \approx 4$, near monolayer coverage sample imaged in Figure 7a, the most notable distinction between azulene and naphthalene is the degree of elongation of the molecules. While naphthalene has an observed aspect ratio of ~2, the observed asymmetry for azulene is nearly unity, although the schematics of Figure 1 would suggest quite comparable asymmetry based on physical extent. Very high resolution can be occasionally achieved for a particular tip, at bias voltages as low as 10 mV and tunneling currents of up to 2 nA. Under these conditions, naphthalene appears as a double ring structure, with no segmentation within the two rings while

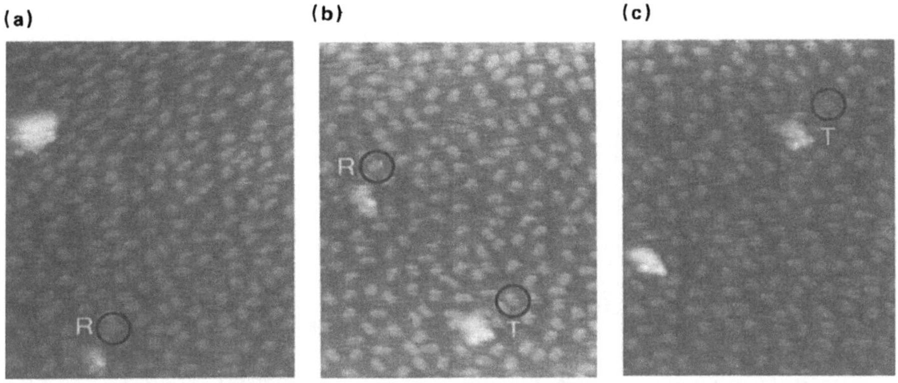

Figure 4. Sequential images of submonolayer naphthalene, proceeding from (a) to (c) at 10 min. intervals. The large defects are used as markers to track motions of individual molecules (shifts between images are due to piezo relaxation after movement to a new area). Molecule *R* undergoes a rotation of +120° between images (a) and (b), while molecule *T* translates by one Pt lattice unit vector between images (b) and (c). ~150Å x 150Å, V_s = +0.4V and I_t = 0.8nA.

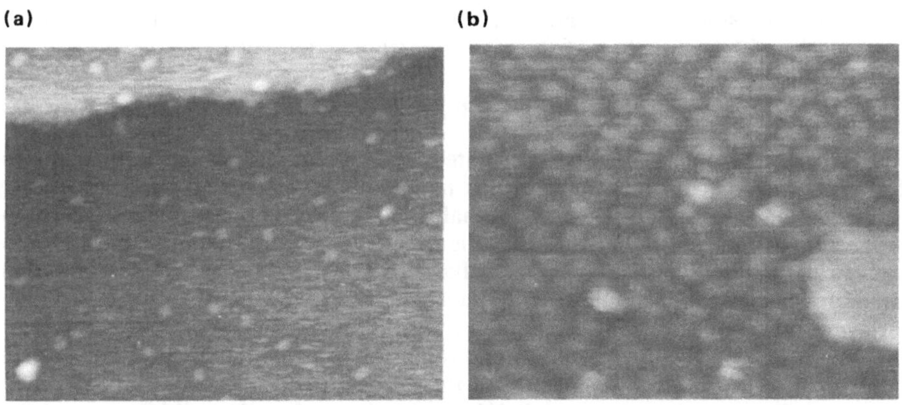

Figure 5. (a) Azulene on Pt(111) at substantially less than monolayer coverage. The line running through the upper left of the image is a Pt monatomic step. A few molecules are localized in the images, while the majority appear as background noise where molecules are observed on individual scan lines but are not well correlated line-to-line. ~300Å x 300Å, V_s = +0.18V, I_t = 0.3nA. (b) Higher coverage, producing a (3x3) LEED pattern, localizes azulene molecules, but no marked asymmetry or internal structure is evident. The smooth area in the lower right area is an impurity carbon island. ~150Å x 150Å, V_s = +0.88V, I_t = 0.2nA.

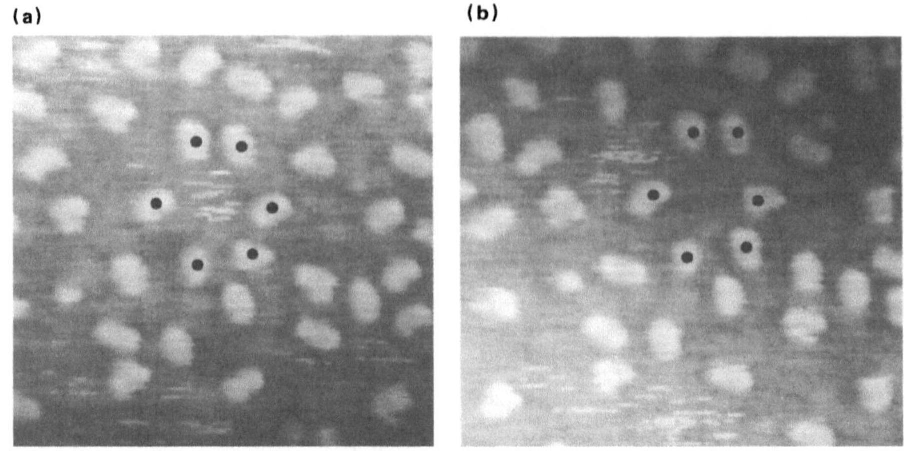

Figure 6. Sequential images of low coverage mixed naphthalene and azulene ($\theta_{naph}\approx$0.2mL, $\theta_{azul}\sim$0.02mL), separated by 10 min. The filled circles mark a ring of naphthalenee molecules. (a) An azulene molecule is trapped within the ring. (b) The azulene has escaped into the vacancy to the upper left of the ring. ~90Å x 90Å, V_s= +0.05V, I_t=0.3nA.

azulene is imaged as a single ring structure, with perhaps some slight asymmetry (Figure 7b). Although this was unexpected, electron and X-ray diffraction data have previously demonstrated that the C-C bond bridging the five- and seven-member rings of azulene is of single-bond character, with a bond length of 1.5Å as compared to bond lengths of 1.4Å for the C-C bonds of the conjugated perimeter.

Comparison of Naphthalene versus Azulene

It is quite interesting to speculate upon the forces which determine the differences in molecular organization for naphthalene versus azulene. Statistically, near neighbors within close-packed naphthalene domains do not assume an end-to-end configuration, implying either that there is an attractive inter-naphthalene potential which prefers other configurations, or that there is a repulsive inter-naphthalene potential which prohibits end-to-end packing. Since low coverage naphthalene overlayers do not exhibit aggregation over time periods of days or in response to annealing, there must exist an end-to-end repulsion for naphthalene, perhaps arising from simple steric constraints since this is the longest axis of the molecule. However, azulene is of comparable or even slightly larger dimension along this axis, and thus should encounter similar or more severe steric interactions than naphthalene. The absence of orientational effects in azulene domain packing may be due to the weak molecule-substrate interaction; fast rotation of the azulene molecules would account for the lack of molecular asymmetry in the STM images.

Further details of the interaction of azulene with the Pt(111) surface require low temperature STM capabilities which we do not possess at the present time. We can, however, explore the effects of substitution upon azulene adsorption. Preliminary experiments with a mixture of 65% 1-methylazulene and 30% 2-methylazulene indicate that single methyl substitution is not sufficient to eliminate diffusion of azulene within the surface plane. However, in the case of 4,6,8-trimethylazulene, localization of the

(a) (b)

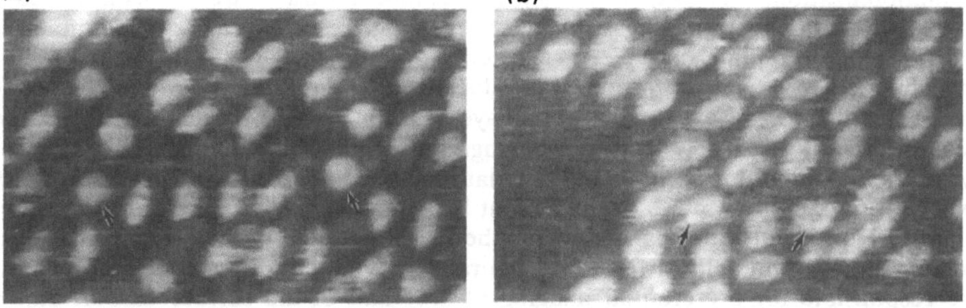

Figure 7. (a) Low resolution image of a mixed naphthalene and azulene monolayer near saturation coverage. The arrows mark azulene molecules which are distinctly round compared with the elongated naphthalene molecules. ~90Å x 60Å, $V_s = +0.25V$, $I_t = 2nA$. (b) High resolution image for a similar sample. The elongated naphthalene molecules are composed of double rings, while the rounder azulene molecules have single holes. ~90Å x 60Å, $V_s = +0.04V$, $I_t = 1nA$.

substituted azulene molecules is achieved, and clear orientational effects are evident. It is quite intriguing to note that in the latter experiments, resolution of the three methyl substituent groups is realized. Further investigations of a larger number of related molecules by STM as well as comparisons with molecular orbital distributions are in progress to extend our understanding of the STM's ability to distinguish among similar molecules.

CONCLUSIONS

In summary, we have initiated a systematic STM study of closely related molecules to elucidate how details of molecular and substrate structure determine the observed STM images. Utilizing standard UHV surface science techniques permits preparation and detailed characterization of controlled organic overlayers. The first experiment in this project has compared naphthalene and azulene adsorbed onto Pt(111). Although these adsorption systems represent well-characterized model systems, the STM results have radically altered our conceptions of molecular organization and there are several lessons which should be heeded as the STM is used to image more complex molecules.

By previous standards, both naphthalene and azulene are strongly chemisorbed to the Pt(111) surface. The STM investigations have raised awareness that molecules which are tightly bound to a surface should be assumed only to have high barriers to desorption, i.e. diffusion normal to the substrate. Expectations of rotational and translational motion within the plane of the surface should not be naively correlated with desorption behavior. Thus, in all the probe microscopies, the search for an appropriate substrate must consider that irreversible or strong binding to a substrate may not be sufficient to immobilize the molecules of interest. In many studies, such as those of biomolecules, a particular advantage of the probe microscopies over other imaging techniques is the hope that molecules can be imaged in more relevant environments, without changing the molecules. While molecules imaged on inert substrates may meet this condition, they may prove difficult to image because of a lack of immobilization. STM experiments with graphite substrates are particularly subject to questions about the imaging mechanism. Although results for a variety of molecules adsorbed onto graphite have been reported, we are unable to image naphthalene or the azulene derivatives on carbon islands localized on the Pt(111) surface. Either the molecules do not adsorb to this monolayer carbon or adsorbates on such carbon islands do not produce electronic corrugations which can be imaged under our experimental conditions. Of course, monatomic height carbon islands should not be expected to exhibit graphite behavior, but it is interesting that corrugation on these islands is consistent with imaging every other carbon atom. In the current theoretical understanding of tunneling to graphite, interactions with the second layer carbon atoms are responsible for preferential imaging of only half of the surface carbon atoms.[7] Perhaps some further theoretical investigations are warranted to understand this similar effect when platinum forms the second layer. In our experience, tunneling tips which achieve atomic resolution on the platinum surface do not yield recognizable molecular images, while tips which produce stable molecular images do not provide simultaneous imaging of substrate atoms. Substrate and tip development probably constitute the areas of most needed investigation for future STM applications.

A new picture of adsorbate and substrate order has also emerged from the last few years of STM research. Although long range order was never a realistic view, the degree of disorder and the quantity of surface defects as imaged in real space with the STM are still sobering. Notions of order and disorder can go far astray when characterization is not sensitive to very localized structures. The packing of naphthalene on the Pt(111) surface serves as a good reminder that inference of unit cells from diffraction methods and visualization of long range molecular organization as involving repeating unit cells are deceptive.

Distinguishing between closely related molecules is of great importance to applications where molecular recognition or sequencing is desired. Although we have succeeded in differentiating among naphthalene and various substituted azulenes, the degree of detail that may be acquired for individual molecules may not be as high as desired. The thermal motion which prevents localization of molecules can be eliminated by reducing temperature during the imaging experiments. The ability to control temperature poses obvious advantages for exploring kinetics. Beyond thermal motion however, another source of molecular motion involves interactions between the probe tip and the adsorbed species. This interaction can be taken advantage of to position atoms and molecules at desired locations,[25] but since the highest resolution of internal molecular structure requires close proximity of tip and molecule, induced motion may hinder detailed imaging of internal structure.

REFERENCES

1. G. Binnig, H. Rohrer, Ch. Gerber, and E. Weibel, Surface studies by scanning tunneling microscopy, *Phys. Rev. Lett.* 49:57 (1982); G. Binnig and H. Rohrer, Scanning tunneling microscopy, *Helv. Phys. Acta* 55:726 (1982); G. Binnig and H. Rohrer, The scanning tunneling microscope, *Sci. Am.* 253:50 (1985).
2. G. Binnig, H. Rohrer, Ch. Gerber, and E. Weibel, 7x7 Reconstruction on Si(111) resolved in real space, *Phys. Rev. Lett.* 50:120 (1983); J. A. Golovchenko, The tunneling microscope: a new look at the atomic world, *Science* 232:48 (1986); R. M. Tromp, R. J. Hamers, and J. E. Demuth, Quantum states and atomic structure of silicon surfaces, *Science* 234:304 (1986).

3. G. Binnig, H. Rohrer, Ch. Gerber, and E. Weibel, (111) Facets as the origin of reconstructed Au(110) surfaces, *Surf. Sci.* 131:L379 (1983); G. Binnig, H. Rohrer, Ch. Gerber, and E. Stoll, Real space observation of the reconstruction of Au(100), *Surf. Sci.* 144:321 (1984).
4. H. K. Wickramasinghe, Scanned probe microscopies, *Sci. Am.* 261:74 (1989).
5. P. K. Hansma and J. Tersoff, Scanning tunneling microscopy, *J. Appl. Phys.* 61:R1 (1987).
6. J. M. Soler, A. M. Baro, N. Garcia, and H. Rohrer, Interatomic forces in scanning tunneling microscopy: imaging of individual states, *Phys. Rev. Lett.* 57:444 (1986).
7. I. P. Batra, N. Garcia, H. Rohrer, H. Salemink, E. Stoll, and S. Ciraci, A study of the graphite surface with scanning tunneling microscopy and electronic structure calculations, *Surf. Sci.* 181:126 (1987); D. P. E. Smith, G. Binnig, and C. F. Quate, Atomic point-contact imaging, *Appl. Phys. Lett.* 49:1166 (1986).
8. H. A. Mizes, S.-i. Park, and W. A. Harrison, Multiple-tip interpretation of anomalous scanning tunneling microscopy images of layered materials, *Phys. Rev.* B36:4491 (1987); T. R. Albrecht, H. A. Mizes, J. Nogami, S.-i. Park, and C. F. Quate, Observation of tilt boundaries in graphite by scanning tunneling microscopy and associated multiple tip effects, *Appl. Phys. Lett.* 52:362 (1988).
9. C. R. Clemmer and T. P. Beebe, Jr., Graphite: a mimic for DNA and other biomolecules in scanning tunneling microscope studies, *Science* 251:640 (1991).
10. S.-I. Park and C. F. Quate, Tunneling microscopy of graphite in air, *Appl. Phys. Lett.* 48:112 (1986); R. Sonnenfeld and P. K. Hansma, Atomic-resolution microscopy in water, *Science* 232:211 (1986); J. Schneir, R. Sonnenfeld, P. K. Hansma, and J. Tersoff, Tunneling microscopy study of the graphite surface in air and water, *Phys. Rev.* B34:4979 (1986).
11. V. M. Hallmark, S. Chiang, J. F. Rabolt, J. D. Swalen, and R. J. Wilson, Observation of atomic corrugation on Au(111) by scanning tunneling microscopy, *Phys. Rev. Lett.* 59:2879 (1987).
12. J. Schneir, R. Sonnenfeld, O. Marti, P. K. Hansma, J. E. Demuth, and R. J. Hamers, Tunneling microscopy, lithography, and surface diffusion on an easily prepared, atomically flat gold surface, *J. Appl. Phys.* 63:717 (1988).
13. E. Holland-Moritz, J. Gordon II, G. Borges, R. Sonnenfeld, Motion of atomic steps of Au(111) films on mica, *Langmuir* 7:301 (1991).
14. D. D. Chambliss and R. J. Wilson, Relaxed diffusion limited aggregation of Ag on Au(111) observed by scanning tunneling microscopy, *J. Vac. Sci. Technol.* B9:928 (1991).
15. T. Smith, The hydrophilic nature of a clean gold surface, *J. Coll. Interface Sci.* 75:51 (1980).
16. H. J. Tobler, A. Bauder, and Hs.H. Gunthard, The microwave spectrum and dipole moment of azulene, *J. Mol. Spectrosc.* 18:239 (1965).
17. J. L. Gland and G. A. Somorjai, Low energy electron diffraction and work function studies of benzene, naphthalene, and pyridine adsorbed on Pt(111) and Pt(100) single crystal surfaces, *Surf. Sci.* 38:157 (1973).
18. D. Dahlgren and J. C. Hemminger, Symmetry extinction of LEED beams for naphthalene adsorbed on Pt(111), *Surf. Sci.* 109:L513 (1981).
19. D. Dahlgren and J. C. Hemminger, Chemisorption and thermal chemistry of azulene and naphthalene adsorbed on Pt(111), *Surf. Sci.* 114:459 (1982).

20. S. C. Fain, Jr. and H. You, Diffraction from pinwheel and herringbone structures of nitrogen and carbon monoxide on graphite, in: "The Structure of Surfaces: Selected Papers," M. A. Van Hove and S. Y. Tong, eds., Springer-Verlag, Berlin (1984).
21. S. Chiang, R. J. Wilson, Ch. Gerber, and V. M. Hallmark, An ultrahigh vacuum scanning tunneling microscope with interchangeable samples and tips, *J. Vac. Sci. Technol.* A6:386 (1988).
22. V. M. Hallmark, S. Chiang, J. K. Brown, and Ch. Wöll, Real-space imaging of the molecular organization of naphthalene on Pt(111), *Phys. Rev. Lett.* 4 66:48 (1991).
23. V. M. Hallmark, S. Chiang, and Ch. Wöll, Molecular imaging of ordered and disordered naphthalene on Pt(111), *J. Vac. Sci. Technol.* B9:1111 (1991).
24. S. Chiang, D. D. Chambliss, V. M. Hallmark, R. J. Wilson, and Ch. Wöll, Imaging adsorbates on metals by scanning tunneling microscopy, in: "The Structure of Surfaces III," S. Y. Tong, M. A. Van Hove, X. Xide, and K. Takayanagi, eds., Springer-Verlag, Berlin, p. 204 (1991).
25. D. M. Eigler and E. K. Schweizer, Positioning single atoms with a scanning tunneling microscope, *Nature* 344:524 (1990).

ADDITIONAL READING

1. G. Binning and H. Rohrer, The scanning tunneling microscope, *Scientific American* 253:50 (1985).
2. P.K. Hansma and J. Tersoff, Scanning tunneling microscopy, *Applied Physics* 61:R1 (1987).
3. H.K. Wickramasighe, Scanned probe microscopies, *Scientific American* 261:74 (1989).

STRUCTURAL CHARACTERIZATION OF MOLECULAR INTERFACE LAYERS USING NEUTRON AND X-RAY REFLECTIVITY TECHNIQUES

Mathias Lösche

Institute of Physical Chemistry, Mainz University, D-6500 Mainz, FRG

INTRODUCTION

Synthetic microstructures used in biological research today include a variety of different systems that serve very diverse purposes and require very different fabrication and characterization techniques. One class of microstructures that is at the boundary between life and materials sciences, and certainly has been associated closer to the latter in the past, is that of molecularly structured planar interface films. In fact, Langmuir-Blodgett (LB) films,[1] i.e. molecularly layered films comprised of amphiphatic organic molecules on solid substrates, have been used for many years as model systems for the investigation of molecular interactions between chromophores or donor/ acceptor couples, and for the development of molecular electronic, optical and opto-electronic devices.[2]

In the context of life sciences, planar structured functionalized interfaces have been introduced into research more recently[3,4] and, so far, have not earned as much attention as have LB films in materials science. This may be due in part to the fact that biologists and physiologists tend to get intimidated by the necessarily sophisticated characterization techniques utilized to resolve the structure of such interfaces, and, vice versa, that most physicists and engineers may have a hard time understanding that biomolecular systems are not that ill-defined after all. This paper is an attempt to bring these two disparate views closer together and as such spares much of the formal physics underlying the characterization techniques presented. These can be looked up in the references cited. It should rather serve to communicate recent developments in the characterization of functionalized bio/organic interfaces *at a molecular length scale*, to summarize results from some of our recent work, and to suggest future experiments.

What is so exciting about functionalized surfaces? Due to the reduced dimensionality of an interface and to its ability to orient adjacent molecules, heterogeneous reactions may occur with much higher efficiency than reactions in isotropic solutions. In artificial systems this enables very sensitive tracking of molecular events, and has been amply utilized to implement analytical techniques in bio/medical sciences and in biotechnology. This advantage, however, poses a serious problem when it comes

Synthetic Microstructures in Biological Research, Edited by
J.M. Schnur and M. Peckerar, Plenum Press, New York, 1992

to the characterization of such interfaces: due to the small number of molecules present at a surface as compared with the bulk volume, interrogating techniques have to be very sensitive and have to discriminate between signals arising at the surface and noise emanating from the bulk. Classical surface sensitive characterization techniques have concentrated on solid state interfaces which can be analyzed under high vacuum conditions. Clearly, such experimental conditions are undesirable when it comes to the investigation of systems of biological relevance. One approach to overcome this inherent difficulty is to use *planar* interfaces for the characterization of large sample areas, and this has revived efforts to develop new investigation methods for such systems.[5] J. Georger contributed to this conference an in-depth account of the in-plane patterning of planar functionalized interfaces, with their structure parallel to the surface normal remaining unresolved.

Specular x-ray reflectivity measurements have been utilized for a number of years to probe the structure of liquid interfaces and surface films.[6-8] Based on very similar principles, the measurement of the specular neutron reflectivity of interfaces is an emerging tool for the investigation of stratified samples, such as polymer/polymer interfaces and polymers interacting with external interfaces. It has, for example, been used[9] to characterize block copolymer[10] films. As we will demonstrate, this method is equally well suited to characterize protein/membrane interactions. Whereas the advent of highly intense Synchrotron x-ray sources enabled the characterization of ultrathin organic surface films, such as lipid monolayers,[11,12] neutron radiation draws its appeal from the fact that contrast variation is viable, e.g. by D/H (^{2}H/^{1}H) exchange. This is a powerful feature, since one can characterize an object by deuterating different molecules, or masking particular parts of a molecule, under the *assumption* that the structure of that object is not affected by the manipulation.

We have begun our efforts to characterize protein/membrane interactions by re-investigating the structure of a phospholipid monolayer on an aqueous subphase,[13] where we have combined older x-ray reflectivity data[14] with our neutron reflectivity results. This provided a very detailed picture of the molecular lipid film which would not have been disclosed with either one of the methods by itself. We proceeded by looking at a specific recognition reaction of a model protein with a monolayer-bound receptor which we characterized using neutron reflectivity in connection with fluorescence microscopy, with an emphasis on checking the assumption of structural identity of isotopically substituted samples.[15] This report is a synopsis of experiments conducted in our laboratories so far and indicates a powerful technique evolving to tackle some of the unresolved problems in structural biology.

METHODS

Reflectivity Measurements

The physics of specular reflection of x-ray or neutron beams from a surface is not very different from reflecting photons of visible light from an interface between media having different indices of refraction. The chromatic or angular effects of the deposition of uniform transparent films with thickness values in the order of the wavelength of light on the reflectivity are due to interference effects and are quantitatively described in text books.[16] The reflected light, in turn, contains information about the index of refraction and the thickness of the interface layer. If one sets out to probe *molecular* layers at interfaces one has to resort to other sources of radiation with suitable wavelengths for which the structured interface presents a difference in its scattering length density and hence, refractive index.

In molecular interface films, where absorption and anomalous dispersion are usually negligible, working out the optical properties of the media involved is straightforward.[9,17,18] The refractive index of a medium is,

$$n = 1 - \delta;\ \delta \sim \lambda^2 * N_A * \sum_i \rho_i \qquad (1)$$

where ρ_i denotes the scattering length density of the ith atomic species ($\rho_i = \nu_i Z_i/V$ is the *electron density* in the case of x-ray optics or $\rho_i = \nu_i b_i/V$ the *neutron scattering length density* in the other case; ν_i is the number of scatterers i within a unit volume, V). The summation extends over the unit volume, and λ and N_A denote wavelength and Avogadro's number. At this point it is important to recognize that it is possible in the case of x-ray and neutron diffraction to *calculate* the expected refractive index of a homogeneous medium from first principles once a model for its molecular content has been proposed, since the values of the neutron scattering lengths, b_i, for different atoms i are known, as are the number of electrons Z_i per atom i.

Figure 1 depicts the geometric setup of a specular reflection experiment. An impinging beam (wave vector, $\vec{k}_o$) is partially reflected from the interface between two media with wave vector $\vec{k}_1$. The transmitted beam is not shown. The momentum transfer on reflection is,

$$\vec{Q} = \vec{k}_1 - \vec{k}_o;\ |\vec{Q}| = Q_z = 2\,k \sin\alpha = \frac{4\pi}{\lambda} \sin\alpha \qquad (2)$$

where $\vec{z}$ denotes the direction of the interface normal. The reflectivity, $R = R(Q_z)$, is the ratio of the intensities I_1 and I_o of the two beams. It can be experimentally determined as a function of momentum transfer either by measuring both intensities as a function of λ at constant incident angle, α, or as a function of α at constant λ. The reflectivity from a perfectly smooth interface with a jump in scattering length density from ρ_1 to ρ_2 is called the Fresnel reflectivity, R_F, and has a characteristic Q_z dependence:[17]

$$R_F \sim \left(\frac{Q_c}{2\,Q_z}\right)^4 \qquad (3)$$

$Q_c = 2\,k * \alpha_c$ and α_c is the critical angle for total reflection. Equation (3) is valid for $Q_z \gg Q_c$.

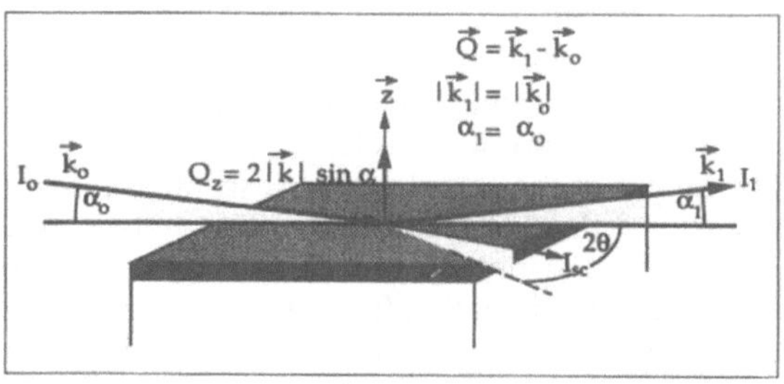

Figure 1. Schematic depiction of the geometry of a reflection experiment from a structured interface. An incoming beam (intensity, I_o) is specularly reflected (I_1) with a resulting momentum transfer, Q_z. Neutron or x-ray reflectivity experiments determine the ratio I_1/I_o as a function of Q_z. Periodic in-plane structure within the surface film may give rise to a scattered intensity, I_{sc}, at a horizontal scattering angle, 2Θ.

The reflectivity from a stratified interface with a more complicated scattering length density profile, $\rho = \rho(z)$, along the $\vec{z}$ direction is given by[9,17,18]

$$R(Q_z) = R_F(Q_z) \left| \frac{1}{\rho_2} \int \frac{d\rho(z)}{dz} * e^{iQ_z z} \, dz \right|^2 \tag{4}$$

Here, ρ_2 is the scattering length density of the bulk medium. According to equation (4), the reflectivity from a complex interface is connected with the Fresnel reflectivity via the Fourier transform of the scattering length density gradient profile across that interface. This implies that there is no direct way of determining a unique function $\rho(z)$ from $R(Q_z)$. Rather, equation (4) can be used to verify or reject models of ρ that describe the interface structure. Furthermore, this result directly reflects the *phase problem* that afflicts any diffraction technique: the phases of the reflected amplitudes remain indeterminate, which may prevent the discrimination between a valid and a wrong model that both give rise to *identical* reflectivity curves.

Instrumentation

Details of the liquid surface reflectometer located at Risø National Laboratory, Denmark, have been described earlier.[13] A monochromatic neutron beam with a wavelength in the range, $\lambda = 3.7...4.8$ Å, is directed downward to impinge on the sample at a small angle, α, that can be varied between 0 and 7°. Beyond the monochromator crystal (Highly Oriented Pyrolytic Graphite) the incident beam passes through an entrance slit that controls its vertical height between 0.01 and 3 mm, an 80' Soller collimator for horizontal collimation, and a second height control slit. Together the two slits determine the incident angle and divergence of the beam. A counter (0.05 atm ^3He) monitors a small and constant percentage of its intensity. The footprint of the beam on the sample, i.e. the investigated sample area, is typically kept constant (at ~ $3 * 10$ cm^2) during the measurement of the reflectivity when α is varied. Consequently the incident intensity is comparably small near grazing incidence (α ~ 0°), where reflectivity is high, but the full beam intensity (with the height control slits wide open) is utilized for large α (large Q_z), where reflectivity is tiny. The maximum intensity of the impinging beam is about 40,000 counts/sec. After reflection, the beam passes through a second slit/collimator/slit system arranged symmetrically to the entrance side. A high efficiency counter (5 atm ^3He) measures the reflected intensity at an equivalent position to the monochromator on the incidence arm of the instrument. Optionally, a Be filter operated at liquid nitrogen temperature can be inserted between the last exit slit and the detector to discriminate against higher harmonic components of the operating wavelength passing through the monochromator. As the height of the monochromator is fixed (determined by the incoming beam from the reactor), the height of the sample table is scanned together with α in order to maintain the specular reflection geometry. The entire detector arm of the instrument may be rotated horizontally out of the specular position in order to measure horizontal diffraction originating from in-plane structure within the sample film or to determine background intensities. The latter facilitates the experimental determination of the background for each momentum transfer value, Q_z. Due to the low neutron source intensity (compared to Synchrotron x-ray experiments) the determination of a reflectivity curve in the range between $Q_z = 0...0.3$ Å^{-1} on D_2O takes typically 12 hours.

The sample is prepared on a Langmuir film balance encapsulated in a gas-tight and temperature-controlled Al container. In order to damp surface waves on the aqueous subphase, a glass block underneath the beam's footprint serves to keep the depth there at a small value (300 μm). Both film balance and reflectometer are computer controlled.

Fluorescence microscopic measurements were carried out in Mainz on a dedicated microscope integrated into a film balance.[19] In these experiments, the lateral distribution of dye molecules at the interface is determined using a highly sensitive video detector. The fluorophores may be attached to lipid molecules within a lipid surface film or to protein molecules that bind to functionalized lipids. With standard fluorophores (e.g., FITC or NBD⁺), the instrument is sensitive enough to generate micrographs from label densities as low as $10^{12}/cm^2$ (1/10,000 $Å^2$). Standard image processing procedures were used for contrast enhancement and quantitative evaluation.[20]

Materials and Sample Preparation

For the aqueous subphases, water was used that was either Millipore filtered (H_2O) or distilled at least five times (four times from $KMnO_4$) in an all quartz/Teflon (PTFE) apparatus (Normag, Hofheim, FRG) in the case of D_2O. Resistance values of these samples were greater than 18 MΩcm (H_2O) or 10 MΩcm (D_2O). The water was stored in Teflon (FEP) bottles (Berghof, Eningen, FRG) prior to using it on the film balance. The isotopic purity of the D_2O was greater than 99.9% in the lipid experiments and between 98% and 99% in the protein experiments. For the determination of the structure of a lecithin monolayer, pure water was used. For protein experiments, 0.5 mol/l NaCl was added to the subphase. Salts were of p.a. grade (Merck, Darmstadt, FRG).

L-α-perdeutero-dipalmitoyl-phosphatidylcholine (DPPC-d_{62}, purity > 99 %; Avanti Polar Lipids, Birmingham, AL) and L-α-dipalmitoyl-phosphatidylcholine (DPPC, > 99%, Sigma, München, FRG) were used as supplied. The phospholipids were spread from a $CHCl_3/CH_3OH$ (3:1) solution (both solvents: p.a.; Merck). A biotinylated lipid, N-biotinyl-S-[1,2-bis[(octadecyloxy)carbonyl]ethyl]cysteine[21] (B-BOCEC, see insert in Figure 7), was a generous gift from H. Ringsdorf, Mainz, and was spread from $CHCl_3$ to give a low surface coverage in the range of 1 molecule per 150 to 1250 $Å^2$ in different experiments.

Streptavidin[22] (SA) was used as a model protein for receptor/ligand interactions due to its exceptionally high binding affinity and specificity for biotin.[23] Its three-dimensional structure has been solved.[24,25] Recent studies have demonstrated that SA binds to biotinylated lipid monolayers where it forms macroscopic two-dimensional (2D) domain structures with a pronounced optical anisotropy.[26] These structures have been transferred to electron microscopy grids and investigated by electron microscopy after uranyl staining.[27] After these procedures 2D protein crystals were observed and an electron density map at 15 Å resolution has been determined with image processing techniques.[27]

SA used in our protein experiments was from ICN Biomedical (Meckenheim, FRG) and from Boehringer Mannheim (Penzberg, FRG). For a stock solution, streptavidin was diluted in D_2O at a concentration of 0.5 mg/ml. 1 ml aliquots were added to the electrolytic subphase (400 ml) contained within the film balance for neutron reflectivity experiments. Protein was usually injected underneath a lipid monolayer which had been pre-spread on a NaCl solution and characterized by reflectivity measurements; results were identical if a lipid monolayer was spread on a subphase that already contained the protein.

⁺FITC, fluorescein isothiocyanine; NBD, nitrobenzoxadiazol

RESULTS

The Structure of a DPPC Monolayer at High Lateral Pressure

Room temperature pressure-area isotherms (lateral film pressure, π, vs. area per molecule in the film, A) of DPPC or DPPC-d_{62} on H_2O or D_2O show small but significant differences at low π where the films are in the 2D liquid and gel phases (data not shown). These differences have tentatively been attributed to different kink densities in the lipid's chains depending on the isotopic constitution of the methylene groups and to an increase in hydrogen bond strength between the lipid's head groups and adjacent water molecules.[13] At high lateral pressure, however, the isotherms are reasonably close to assume that the structure of the molecules constituting the molecular films on H_2O and on D_2O is identical.[13]

Figure 2 shows the neutron reflectivity data R vs. Q_z (diamond symbols) from a DPPC-d_{62} monolayer on D_2O at $\pi = 42$ mN/m, T = 18°C. Included are the Fresnel reflectivity (dashed line) and the reflectivity computed from a model of the interface structure (see below), optimized to describe the experimental data (solid line). Both Fresnel and model reflectivities have values, R = 1, below the critical momentum transfer value, $Q_c \sim 0.017$ Å^{-1}, and decrease sharply at higher values of Q_z. Whereas the Fresnel reflectivity falls steadily according to equation (3), some structure is seen in the reflectivity from the stratified interface indicative of interference effects due to the film. At Q_z larger than 0.25 Å^{-1}, R has decreased to about 10^{-6} and continues to fall according to the Q_z^{-4} law. This prevents a determination of R at even higher Q_z values which would increase the amount of information about the film structure revealed from the measurement. With lipid films on an H_2O subphase, data collection is confined to smaller Q_z regimes since the difference in the scattering length densities of water and air is an order of magnitude smaller and the drop in reflectivity is therefore even steeper (still obeying the Q_z^{-4} law). Deposition of one monolayer of the perdeuterated DPPC-d_{62} in this case leads to an increase of R by a factor of ~1,000 over a large regime of the experimentally accessible reflectivity curve[13] (data not shown).

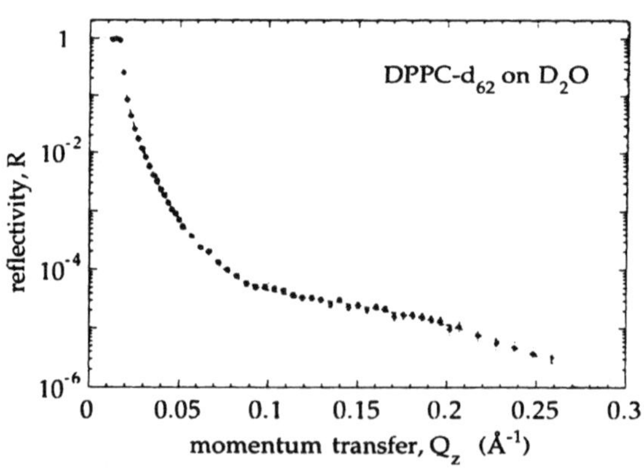

Figure 2. Neutron reflectivity from a perdeuterated DPPC monolayer at $\pi = 42$ mN/m, T = 18°C on D_2O. Experimental data are shown as diamond symbols. The reflectivity calculated for a model describing the scattering length density, ρ_n, parallel to the interface normal is included as a solid line. The dashed line indicates the Fresnel reflectivity of pure D_2O.

In our evaluation of the data we took advantage of the fact that the x-ray reflectivity of a (non-deuterated) DPPC monolayer on H_2O at π = 42 mN/m had been determined previously.[14] Figure 3 shows the neutron data (DPPC-d_{62} on D_2O and H_2O) as well as the x-ray data (DPPC on H_2O), displayed as R/R_F vs. Q_z. It is clearly seen that the x-ray data extend over roughly twice the Q_z regime covered by the neutron data, due to the much greater source intensity. A pronounced minimum in the x-ray reflectivity data occurs at $Q_z^{min} \sim 2\pi/d_{film}$ because of destructive interference due to the total film thickness. For the system studied, the corresponding neutron reflectivity minimum was not observable.

We used the combined neutron *and* x-ray data sets[28] for the definition and refinement of a general model for the interface structure in a composition-space refinement procedure.[29] In such a procedure, different parts of the molecules constituting the film were assigned to regions with different constant values of the scattering length density.[13] In the case of DPPC it was natural to divide the lipid into head group and chain compartments such that a hydrophobic film of thickness, d_{chain}, lies on top of a hydrophilic film with d_{head} which incorporates the lipid head group *and* an unknown number of water molecules. As far as the lipid is concerned, the atomic contents of these compartments are precisely known from the structure, and scattering lengths for both neutrons and x-rays can be assigned to these regions where the contribution of head group water molecules is a free parameter. Variation of the geometric model parameters now changes neutron and x-ray scattering length densities in a predictable way, and a least-squares fitting procedure optimizes the model to fit all data sets simultaneously. In addition to these film-dependent model parameters a film-independent surface roughness, conceivably due to capillary waves,[30] was incorporated as a Gaussian smearing σ convoluted into the two-slab model of the surface film. Therefore, five adjustable parameters defined the model of the stratified interface. The scattering density profiles along the z axis are included as insets in the panels of Figure 3 for the different experimental situations. Note that the neutron scattering length densities of the head group regions differ slightly due to the contributions of the different water isotopes.

The results of our evaluation are listed in Table 1 and schematically displayed in Figure 4. Confidence limits were determined by screening the sensitivity of χ^2 to variations of a model parameter while readjusting the other parameters to a new local minimum.[13] A similar procedure revealed that a determination of parameters based on either neutron or x-ray data alone was far from being able to define a model with comparable accuracy to the combined data sets.[13]

Specific Binding of Streptavidin to a Biotin Functionalized Lipid Monolayer

For a first characterization of the structural organization of a protein specifically interacting with a functionalized lipid monolayer *in-situ* and with *molecular resolution* we chose the biotin/streptavidin receptor/ligand couple as a model system. In excitation polarized fluorescence microscopy we observed that FITC-labeled[31] streptavidin readily forms macroscopic aggregates which show pronounced optical anisotropies if injected underneath dilute monolayers of B-BOCEC into aqueous subphases of high ionic strengths. Figure 5 presents as an example a fluorescence micrograph of labeled SA (~ 1 fluorophore per protein) interacting with a lipid monolayer (~120 Å^2 per B-BOCEC) on 0.5 M NaCl solution. The excitation is polarized parallel to the short edge of the image. The length of the bar is 50 μm. A number of regions with different but homogeneous brightness values are observed. On rotating the polarizer, darker and brighter regions switch emission intensities. The

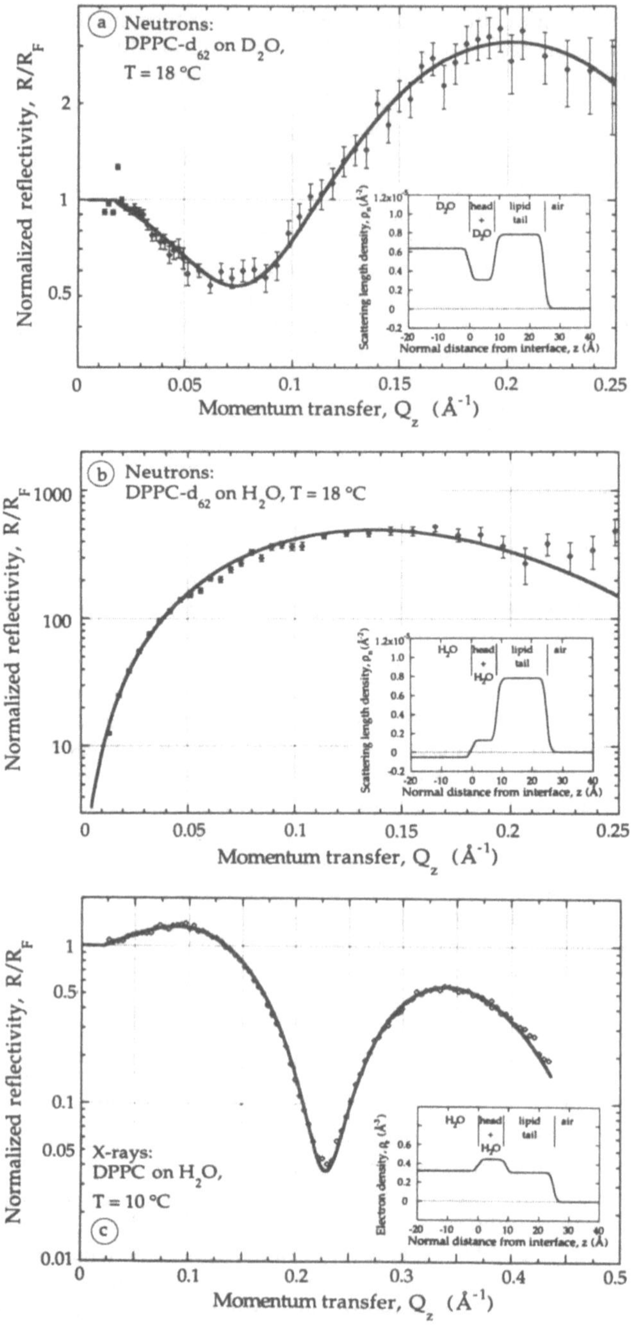

Figure 3. Fresnel normalized neutron and x-ray reflectivities, R/R_F, of DPPC on pure water at $\pi = 42$ mN/m. The insets show the scattering length density profiles parallel to the interface normal that were used to generate the computed reflectivities from a universal model (see text) included with the data as solid lines.

Table 1

Structure of a DPPC monolayer on the surface of pure water at π = 42 mN/m. Upper part: best fit parameters from a simultaneous fit to the experimental data from neutron and x-ray reflectivity measurements as described in the text; lower part: resulting values for dependent quantities.

A_o	41.2 ± 0.5 Å^2	(Cross-section of a lipid molecule in the film)
β	$33 \pm 3°$	(Average chain tilt angle from surface normal)
d_{head}	9.3 ± 1 Å	(Thickness of lipid head group layer)
N_W	4 ± 2.5	(Number of water molecules in head group)
A $(=A_o/\cos\beta)$	49.1 ± 2 Å^2	(Area per lipid molecule in surface film)
d_{chain} $(=l_{chain} \cdot \cos\beta)$	16.0 ± 0.6 Å	(Thickness of lipid chain group layer)
d_{total} $(=d_{head}+d_{chain})$	25.3 ± 0.4 Å	(Total interface film thickness)
V_{head} $(=A \cdot d_{head}-N_W \cdot V_W)$	335 ± 8 Å^3	(Volume of the "dry" lipid head group)
V_{total} $(=V_{head}+A \cdot d_{chain})$	1124 ± 4 Å^3	(Volume of lipid molecule in film)

(constants: l_{chain} = 19.15 Å; V_W = 30 Å^3)

narrow stripes separating different regions show the highest emission intensities and are not affected by the polarizer. This suggests that in the large areas the fluorophores show hindered rotational mobility and that each one is attached to SA in a similar geometric orientation which is not optimal for excitation. In contrast, those boundary regions show higher rotational mobility and their average transition moment direction is closer to the electric field vector of the excitation beam. It is therefore evident that the large protein areas are of higher order than the boundary regions. As judged from the isotherm (cf. Figure 3 in reference 21), the lipid monolayer is in a homogeneous fluid state ($\pi <$ 10 mN/m) and no differences in its

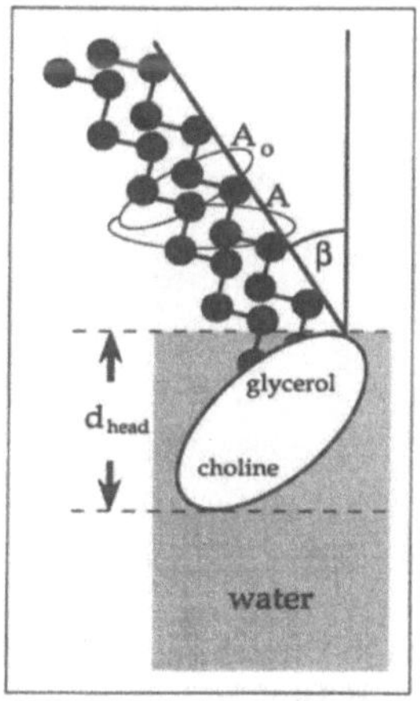

Figure 4. Schematic illustration of a DPPC molecule at the air/water interface with some of the structural parameters of Table 1 depicted.

structural organization are expected on H_2O and on D_2O subphases. However, it cannot be taken for granted that the protein interacts with the monolayer exactly the same way in both media. We therefore took the macroscopic morphological properties of SA aggregates as reported by fluorescence microscopy as a criterion for the structural identity of samples with different isotopic constitution. In fact, we found that nucleation of the optically anisotropic regions shown in Figure 5 occurs more efficiently in D_2O than in H_2O, leading to a greater number of ordered areas (which grow to a smaller final size) under otherwise identical experimental conditions.

The neutron reflectivities of SA bound to functionalized lipid monolayers are shown in Figure 6 as diamond symbols. The experimental conditions under which the data shown in two panels have been obtained were identical (120 Å^2/B-BOCEC, 0.5 M NaCl), except that Figure 6a displays the data set taken on D_2O whereas 6b shows the data set taken on H_2O. Again, dashed lines show the Fresnel reflectivities (which are identical to the reflectivities of the lipid monolayers, since the hydrogenated lipid chains at low densities give virtually no contrast against air) and solid lines indicate the reflectivities calculated for an optimized model simultaneously describing both experimental data sets. Due to the increased total interface thickness, interference extrema are now obviously accessible in the neutron experiments, and a maximum is observed at Q_z - 0.13 Å^{-1} in Figure 6a. As before, the model's scattering length density profiles along the interface normal are included as inserts. A pronounced difference in the scattering length densities of the protein compartment due to intervening water (D_2O or H_2O) is observed.

As a model independent result from these experiments we found the reflectivity data unequivocally indicative of a single protein monolayer immediately underneath the interface. We find that the scattering length density due to the protein is strongly confined at the interface, and is, in particular, definitively different from what would be expected if the driving force for protein association to the interface were electrostatic in nature. This suggests that strong and specific binding of SA to the biotinylated lipid occurs. We were not able to resolve any water between the protein

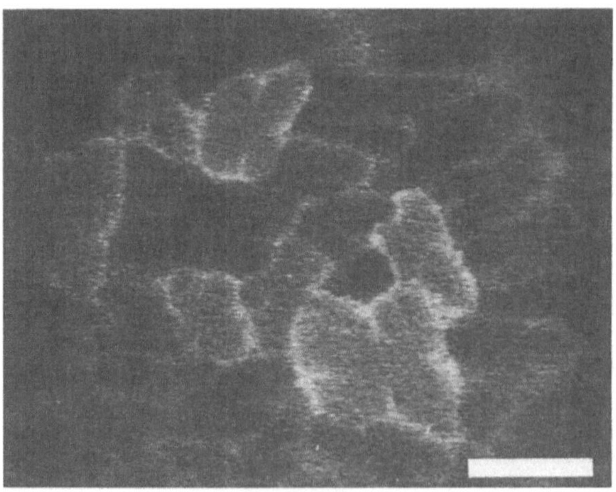

Figure 5. Fluorescence micrograph showing the lateral arrangement of FITC-labeled SA interacting with a B-BOCEC monolayer (120 Å^2 per lipid) on H_2O (0.5 M NaCl) at T = 18°C. The excitation is polarized parallel to the short edge of the picture. The bar is 50 μm.

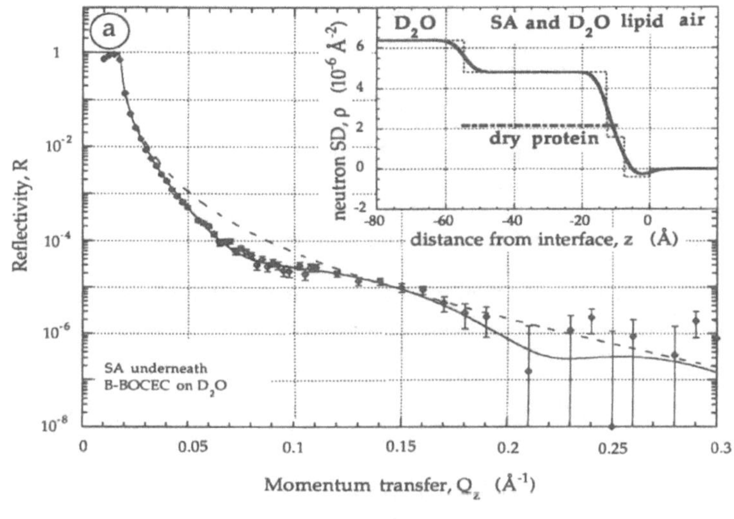

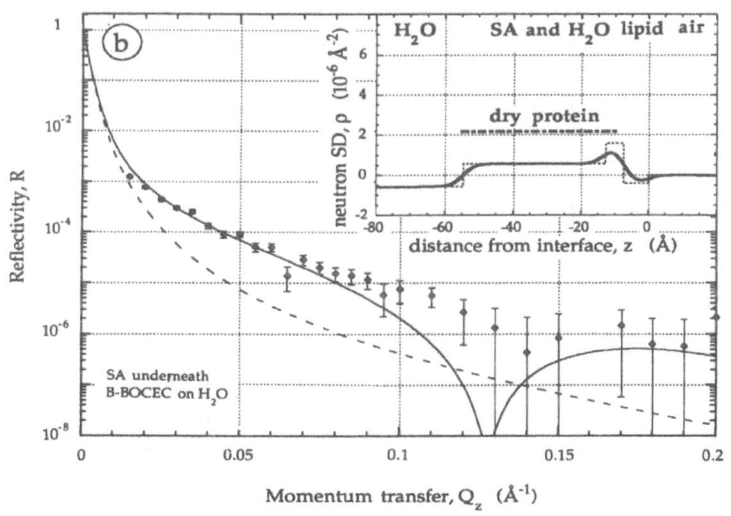

Figure 6. Neutron reflectivities of monomolecular layers of SA specifically bound to B-BOCEC in lipid surface monolayers on D_2O and on H_2O. Further details of the presentation as in Figures 2 and 3.

and the lipid, consistent with the fact that the ligand's spacer group between its hydrophobic anchor and the biotin's thiolane cycle is rather short, ~10 Å.

Table 2 is a compilation of the numerical results for the best fit parameter values found from the simultaneous refinement of the two data sets. The qualitative results are schematically depicted in Figure 7. The atomic content of the SA molecule was taken from Green.[23] In our first evaluation,[15] the model-dependent interpretation of the scattering length density was based on an estimate of the protein's water excluded volume, V_{SA}^{dry}, from the general properties of 3D SA crystals.[25] In connection with the atomic content of the molecule this quantity determines its scattering length density and thus the perception of the molecular content of the protein layer. Since then we have experimentally determined V_{SA}^{dry} with measurements of the x-ray reflectivity of SA/lipid monolayers and by simultaneously refining x-ray and neutron data sets.[32] We find a value of V_{SA}^{dry} ~ 65,000 Å^3, which is much smaller than the estimated number.[15] Similarly, we have determined the volume of the biotin head group of the lipid from the experimental data.

Since fluorescence microscopic data suggested that interactions between SA molecules may differ in H_2O and D_2O we have allowed for different *lateral* organizations of the protein layer in our global model of the two data sets. The result, however, shows that the optimized structural parameters agree within experimental error.

The area per lipid, A_b, is known from the amount of lipid applied to the interface. The lipid monolayer is accounted for in the model with the (small) scattering length it contributes due to its presence at the interface. Its structural parameters, however, are not refined with respect to the experimental data set. The model is in fact very insensitive to their actual values.

In our re-interpretation of the scattering length density of the protein layer an ambiguity in its possible molecular structure, which was due to the phase problem (see discussion in reference 15), has been removed, and it has been concluded that the coverages of the surface with the protein on H_2O and on D_2O are close to 100%.

Table 2

Structural parameters derived from the data of Figure 6. Upper part: independent parameters determined from the best fit to the data; lower part: dependent quantities.

d_p	41.6 ± 5 Å	(protein layer thickness)
$A_o(H_2O/D_2O)$	3650/4200 + 1000/- 800 Å^2	(area occupied in the film by the unit volume)
σ	4 Å	(surface roughness)
n_w (= $(A_o * d_p - V_{SA}^{dry} - n_b * V_b)/V_w$)	2300/2950 ± 1000	(number of water molecules per SA in film)
n_b (= A_o/A_b)	30/35	(number of ligand molecules per SA)

(constants: V_{SA}^{dry} = 65,000 Å^3; V_w = 30 Å^3; V_b = 680 Å^3; A_b = 120 Å^2)

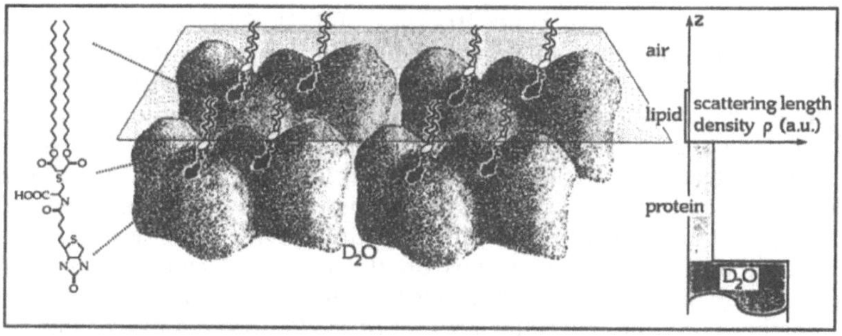

Figure 7. Schematic representation of a monomolecular layer of SA specifically bound to B-BOCEC (see inset) in a monolayer on 0.5 M NaCl.

DISCUSSION

The implications of our results on the structural properties of the lipid, DPPC, in a monomolecular surface layer at high lateral pressure have been discussed in detail elsewhere.[13] A striking resemblance exists between our structural model and the structure of DPPC in bilayers as revealed from x-ray scattering from vesicle dispersions.[33-35] Structural parameters have been determined with comparable accuracy as in the vesicle investigations, and are in very close agreement with their numerical results. In turn, this provides a strong argument that the lipid *in bilayers* experiences a rather high lateral pressure, of the order of 40 mN/m or higher. The correspondence between bilayers and monolayers in this respect has been discussed vividly and controversially for a long time.[36-39] Looking into details of the comparison, the head group conformation seems slightly more extended in the monolayer than in bilayers. The number of water molecules associated with the head group is very small but corresponds to results for example found in DSC investigations.[40]

While comparing these different systems, it has to be kept in mind that the results may be *slightly* model dependent. In fact, it has been shown explicitly for the case of bilayer diffraction studies[41] that this dependence is small. Another important caveat is that the assumption of the insensitivity of the structure to isotopic exchange may be critical. We have addressed this problem carefully in our original work[13] and we are confident that it is not an issue in the situation that we have studied. But for a generalization of such experiments to surfactant and to lipid surface films one has to keep in mind that chain order parameters are affected by deuteration[42] as is the interaction between lipid head groups and adjacent water molecules.[43-45]

Addressing the question if the structural model based on data can be refined even further we have learned that the picture presented here defines the limit of what can be deduced with confidence. Specifying more complicated models by pursing the molecules into smaller sections taught us that the conclusions drawn from these efforts were not definite if tested by mapping confidence limits.[13]

Our work on the streptavidin/biotin model system may be the first *in-situ* structural characterization of a molecular recognition process on a nanometer length scale. The neutron reflectivity measurements demonstrated unequivocally the formation of a single monomolecular protein layer underneath a dilute lipid monolayer obtained from a SA bulk concentration of 10^{-8} M. We have indeed observed SA monolayers forming at surface concentrations of B-BOCEC as low as 1 lipid per 1,250 $Å^2$(-2.5 biotin moieties per SA in the protein film). The determination of the lateral protein density in the 2D aggregates does not depend on crystallinity of the sample. In fact, we cannot directly decide whether or not a sample *is* in a crystalline state without detecting horizontally scattered intensity (cf. Figure 1). Due to the weak neutron source intensity it was not too surprising that our efforts to find scattering signals did not meet with success. The experimental situation presented here was, however, most favorable for reflectivity investigations: Due to the short spacer moiety between anchor and recognition groups on the lipid molecule, the protein binds *immediately* to the lipid monolayer, rendering the structural model as simple as possible. Furthermore, the low water content of the 2D protein aggregates results in a compact protein layer which provides high contrast against the bulk subphase. This facilitated a rather precise determination of the average area, A_o, of the protein in the film and of its water content. The hydration of the protein in the film is low, and the value of A_o is close to that reported for transferred uranyl stained structures.[27] Together with our fluorescence microscopic observations this leads indirectly to the conclusion that our protein sample is probably in a 2D crystalline state in the *in-situ* situation.

As in the case of lipid monolayers, the isotopic influence on the structure is an important issue. In both reflectivity and fluorescence microscopy experiments we have consistently found a small dependence of the structure on the water isotopic content. Differences in the strength of hydrogen (or deuterium) bonds affecting the protein-protein interactions are held responsible for these observations.[15] However, the microscopic structural parameters determined from the neutron reflectivity data sets presented here are identical for the systems in H_2O and in D_2O within experimental error.

We have continued investigations of the SA/biotin system in x-ray[32] and in further neutron[46] reflectivity experiments with an emphasis on testing the isotope/structure relationship and the influence of the lipid monolayer on structure formation in the adjacent protein compartment. We have been able to resolve a water layer intervening between the protein and lipid monolayers if a lipid with a longer spacer was used. The structure formation of SA binding to binary lipid monolayers including species with perdeuterated and hydrogenated methylene groups in their chains showed a pronounced dependence on the isotopic content of the lipids as well as the subphase water. A quantitative analysis of these data is under way.

CONCLUSIONS

The main objective of this paper is to demonstrate the potential of reflectivity measurements from planar liquid surfaces as a new emerging tool for structural investigations of molecular bio/organic interface layers *in-situ*. Although the methods (and in particular, neutron sources) are not readily available for routine use, they may provide invaluable information on selected systems where the association of proteins with interfaces, or partial or entire interpenetration of proteins into lipid membranes is of concern, or where the lateral packing of proteins at interfaces has to be controlled. F. S. Ligler in her account on the challenges of the development of protein based sensor devices[47] has posed such problems. In that case, it is necessary to arrange functional proteins in an optimal density and in an optimal orientation within 2D structures to achieve the highest possible sensitivity. Similar requirements are met in the optimization of surfaces for biochemical and clinical analytical applications. Iteration of the formation of different functionalized molecular films at interfaces may lead to very sophisticated configurations under structurally well-defined conditions.[26]

The mutual structural arrangement of proteins at membranes is of immediate interest for biological function, e.g. in the case of the GTPase-based amplification system universally used for signal processing in living cells.[48,49] The investigation of biomineralization processes, where surface associated protein species control the structure formation of inorganic molecules, may be another potential field of application of the methods presented. Finally, these techniques may be valuable to establish and control in a general way the formation of 2D crystals of proteins, mediated by functionalized lipid monolayers,[50] for their investigation by electron diffraction.[51] In this field, efforts are made to develop a third method of collecting high-resolution structural information, in addition to x-ray diffraction from 3D crystals and to NMR, on systems with the interesting property that ligands are inevitably bound to their receptor molecules in the crystalline state.

ACKNOWLEDGMENTS

I am most grateful to my colleagues D. Vaknin (Ames), K. Kjaer and J. Als-Nielsen (Risø), M. Piepenstock and A. Diederich (Mainz) who were involved in the construction of the neutron reflectometer and/or the experiments, and to W. Knoll, H. Möhwald and H. Ringsdorf (Mainz) for valuable discussions. H. Ringsdorf also provided the biotinylated lipid. Financial support came from from the NOVO and the Carlsberg foundations, the Danish Science Research Council and the German Bundesministerium f. Forschung und Technologie under contract no. 05 453 FA I9.

REFERENCES

1. H. Kuhn, D. Möbius, and H. Bücher, Spectroscopy of monolayer assemblies, in: "Physical Methods of Chemistry," A. Weissberger and P. Rossiter, eds., Wiley Interscience, New York (1972).
2. J. D. Swalen, D. L. Allara, J. D. Andrade, E. A. Chandross, S. Garoff, J. Israelachvili, T. J. McCarthy, R. Murray, R. F. Pease, J. F. Rabolt, K. J. Wynne, and H. Yu, Molecular monolayers and films, *Langmuir* 3:932 (1987).
3. D. G. Hafeman, V. von Tscharner, and H. M. McConnell, Specific antibody-dependent interactions between macrophages and lipid haptens in planar lipid monolayers, *Proc. Natl. Acad. Sci. USA* 78:4552 (1981).
4. L. K. Tamm, Lateral diffusion and fluorescence microscope studies on a monoclonal antibody specifically bound to supported phospholipid bilayers, *Biochemistry* 27:1450 (1988).
5. H. Möhwald, Phospholipid and phospholipid-protein monolayers at the air/water interface, *Annu. Rev. Phys. Chem.* 41:441 (1990).
6. K. Kjaer, J. Als-Nielsen, R. M. Kenn, C. Böhm, P. Tippmann-Krayer, C. A. Helm, H. Möhwald, F. Leveiller, D. Jacquemain, M. Lahav, L. Leiserowitz, and M. Deutsch, X-ray scattering studies of organic monolayers on electrolytic solutions: arachidic acid on $CdCl_2$, in: "Surface X-ray and Neutron Scattering," H. Zabel and I. K. Robinson, eds., Springer, Berlin (1992).
7. J. Als-Nielsen and H. Möhwald, Synchrotron x-ray scattering studies of Langmuir films, in: "Handbook of Synchrotron Radiation, Vol. 4," S. Ebashi, E. Rubinstein, and M. Koch, eds., Elsevier North-Holland, Amsterdam (1990).

8. D. Jacquemain, S. Grayer Wolf, F. Leveiller, M. Deutsch, K. Kjaer, J. Als-Nielsen, M. Lahav, and L. Leiserowitz, Two-dimensional crystallography on amphiphilic molecules at the air-water interface, *Angew. Chem., Int. Ed. Engl.* 31:130 (1992).
9. S. H. Anastasiadis, T. P. Russell, S. K. Satija, and C. F. Majkrzak, The morphology of symmetric diblock copolymers as revealed by neutron reflectivity, *J. Chem. Phys.* 92:5677 (1990).
10. F. S. Bates, Polymer-polymer phase behavior, *Science* 251:898 (1991).
11. K. Kjaer, J. Als-Nielsen, C. A. Helm, L. A. Laxhuber, and H. Möhwald, Ordering in lipid monolayers studied by Synchrotron x-ray diffraction and fluorescence microscopy, *Phys. Rev. Letters* 58:2224 (1987).
12. P. Dutta, J. B. Peng, B. Lin, J. B. Ketterson, M. Prakash, P. Georgopoulos, and S. Ehrlich, X-ray diffraction studies of organic monolayers on the surface of water, *Phys. Rev. Lett.* 58:2228 (1987).
13. D. Vaknin, K. Kjaer, J. Als-Nielsen, and M. Lösche, Structural properties of phosphatidylcholine in a monolayer at the air/water interface. Neutron reflection study and reexamination of x-ray reflection experiments, *Biophys. J.* 59:1325 (1991).
14. C. A. Helm, H. Möhwald, K. Kjaer, and J. Als-Nielsen, Phospholipid monolayer density distribution perpendicular to the water surface. A Synchrotron x-ray reflectivity study, *Europhys. Lett.* 4:697 (1987).
15. D. Vaknin, J. Als-Nielsen, M. Piepenstock, and M. Lösche, Recognition processes at a functionalized lipid surface observed with molecular resolution, *Biophys. J.* 60:1545 (1991).
16. M. Born and E. Wolf, "Principles of Optics," Pergamon Press, London (1967).
17. J. Als-Nielsen and K. Kjaer, X-ray reflectivity and diffraction studies of liquid surfaces and surfactant monolayers, in: "Phase Transitions in Soft Condensed Matter," T. Riste and D. Sherrington, eds., Plenum Press, New York (1989).
18. T. P. Russell, X-ray and neutron reflectivity for the investigation of polymers, *Mater. Sci. Rep.* 5:171 (1990).
19. M. Lösche and H. Möhwald, Fluorescence microscope to observe dynamical processes in monomolecular layers at the air/water interface, *Rev. Sci. Instrum.* 55:1968 (1984).
20. M. Lösche, H. Duwe, and H. Möhwald, Quantitative analysis of surface textures in phospholipid phase transitions, *J. Colloid Interf. Sci.* 126:432 (1988).
21. R. Blankenburg, P. Meller, H. Ringsdorf, and C. Salesse, Interaction between biotin lipids and streptavidin monolayers: Formation of oriented two-dimensional protein domains induced by surface recognition, *Biochemistry* 28:8214 (1989).
22. L. Chaiet and F. J. Wolf, The properties of streptavidin, a biotin-binding protein produced by streptomycetes, *Arch. Biochem. Biophys.* 106:1 (1964).
23. N. M. Green, Avidin, in: "Advances in Protein Chemistry," M. L. Anson and J. T. Edsell, eds., Academic Press, New York (1975).
24. P. C. Weber, D. H. Ohlendorf, J. J. Wendolowski, and F. R. Salemme, Structural origins of high-affinity biotin binding to streptavidin, *Science* 243:85 (1989).
25. W. A. Hendrickson, A. Pähler, J. L. Smith, Y. Satow, E. A. Merritt, and R. P. Phizackerley, Crystal structure of core streptavidin determined from multiwavelength anomalous diffraction of synchrotron radiation, *Proc. Natl. Acad. Sci. USA* 86:2190 (1989).

26. M. Ahlers, R. Blankenburg, D. W. Grainger, P. Meller, H. Ringsdorf, and C. Salesse, Specific recognition and formation of two-dimensional streptavidin domains in monolayers: Applications to molecular devices, *Thin Solid Films* 180:93 (1989).
27. S. A. Darst, M. Ahlers, P. H. Meller, E. W. Kubalek, R. Blankenburg, H. O. Ribi, H. Ringsdorf, and R. D. Kornberg, Two-dimensional crystals of streptavidin on biotinylated lipid layers and their interactions with biotinylated macromolecules, *Biophys. J.* 59:387 (1991).
28. D. Vaknin, K. Kjaer, J. Als-Nielsen, and M. Lösche, A new liquid surface neutron reflectometer and its application to the study of DPPC in a monolayer at the air/water interface, *Makromol Chem., Macromol. Symp.* 46:383 (1991).
29. M. C. Wiener and S. H. White, Fluid bilayer structure determination by the combined use of x-ray and neutron diffraction II. "Composition-space" refinement method, *Biophys. J.* 59:174 (1991).
30. M. K. Sanyal, S. K. Sinha, K. G. Huang, and B. M. Ocko, X-ray-scattering study of capillary-wave fluctuations at a liquid surface, *Phys. Rev. Letters* 66:628 (1991).
31. R. D. Nargessi and D. S. Smith, Fluorometric assays for avidin and biotin, *Methods Enzymol.* 122:67 (1986).
32. D. Vaknin, K. Kjaer, H. Ringsdorf, R. Blankenburg, M. Piepenstock, A. Diederich, and M. Lösche, manuscript in preparation.
33. G. Büldt, H. U. Gally, J. Seelig, and G. Zaccai, Neutron diffraction studies on phosphatidylcholine model membranes. I. Head group formation, *J. Mol. Biol.* 134:673 (1979).
34. G. Zaccai, G. Büldt, A. Seelig, and J. Seelig, Neutron diffraction studies on phosphatidylcholine model membranes. II. Chain conformation and segmental order, *J. Mol. Biol.* 134: 693 (1979).
35. M. C. Wiener, R. M. Suter, and J. F. Nagle, Structure of the fully hydrated gel phase of dipalmitoylphosphatidylcholine, *Biophys. J.* 55:315 (1989).
36. D. Oldani, H. Hauser, B. W. Nichols, and M. C. Phillips, Monolayer characteristics of some glycolipids at the air-water interface, *Biochim. Biophys. Acta* 382:1 (1975).
37. S. W. Hui, M. Cowden, D. Papahadjopoulos, and D. F. Parsons, Electron diffraction study of hydrated phospholipid single bilayers. Effects of temperature, hydration and surface pressure of the "precursor" monolayer, *Biochim. Biophys. Acta* 382:265 (1975).
38. A. Blume, A comparative study of the phase transitions of phospholipid bilayers and monolayers, *Biochim. Biophys. Acta* 557:32 (1979).
39. J. F. Nagle, Theory of the main lipid bilayer transition, *Ann. Rev. Phys. Chem.* 31:157 (1980).
40. H. Dörfler and G. Brezesinski, Phasenumwandlungserscheinungen in Lecithin/Wasser-Systemen I. Einfluβ des Wassers auf die Phasenumwandlungen homologer Lecithin/Wasser-Monohydrate, *Coll. Polym. Sci.* 261:286 (1983) [In German].
41. G. I. King and S. H. White, Determining bilayer hydrocarbon thickness from neutron diffraction measurements using strip-function models, *Biophys. J.* 49:1047 (1986).
42. G. Schmidt and W. Knoll, Densitometric characterization of aqueous lipid dispersions, *Ber. Bunsenges. Phys. Chem.* 89:36 (1985).
43. K. D. Dreher and D. F. Sears, Stearic acid monolayers on heavy water, *Trans. Faraday Soc.* 62:741 (1966).

44. G. Lipka, B. Z. Chowdhry, and J. M. Sturtevant, A comparison of the phase transition properties of 1,2-diacylphosphatidylcholines and 1,2-diacylphosphatidylethanolamines in H_2O and D_2O, *J. Phys. Chem.* 88:5401 (1984).
45. M. C. Wiener, S. Tristram-Nagle, D. A. Wilkinson, L. E. Campbell, and J. F. Nagle, Specific volumes of lipids in fully hydrated bilayer dispersions, *Biochim. Biophys. Acta* 938:135 (1988).
46. D. Vaknin, M. Piepenstock, and M. Lösche, manuscript in preparation.
47. F. S. Ligler and S. Rabbany, Biological microstructures in biosensors, this volume.
48. G. L. Johnson and N. Dhanasekaran, The G-protein family and their interaction with receptors, *Endocrine Rev.* 10:317 (1989).
49. H. R. Bourne, D. A. Sanders, and F. McCormick, The GTPase superfamily: conserved structure and molecular mechanism, *Nature* 349:117 (1991).
50. E. E. Uzgiris and R. D. Kornberg, Two-dimensional crystallization technique for imaging macromolecules, with an application to antigen-antibody-complement complexes, *Nature* 301:125 (1983).
51. G. Mosser and A. Brisson, Conditions of two-dimensional crystallization of cholera toxin B-subunit on lipid films containing ganglioside GM1, *J. Struct. Biol.* 106:191 (1991).

IMAGING WITH THE BROOKHAVEN SCANNING TRANSMISSION X-RAY MICROSCOPE

Shawn Williams, Chris Jacobsen, Janos Kirz, and Xiaodong Zhang

Department of Physics, State University of New York at Stony Brook
Stony Brook NY 11794

ABSTRACT

Current x-ray microscopes offer the possibility of imaging unstained, unfixed, unsectioned, micron thick, wet biological specimens with a resolution approaching 30nm. Because the absorption of x-rays depends primarily on the elemental composition of the specimen, the images lend themselves to quantitative analysis of sample structure. The Scanning Transmission X-ray Microscope at Brookhaven National Lab uses a zone plate to form a focused probe through which the sample is scanned. The modulation transfer function of the microscope has been measured and agrees well with calculations based on the zone plate construction. This knowledge permits recovery of the intrinsic contrast of object features using image deconvolution. Current zone plates provide a Rayleigh resolution of 50 nm, and the ability to see 30 nm objects. This resolution has been achieved in practice for images of test specimens and dried biological specimens, where high doses can be tolerated. Single images of wet *V. faba* chromosomes show 80 nm resolution. Subsequent images show that specimens were degraded by radiation damage following the initial exposure. Many researchers have used the Brookhaven STXM for problems in neurobiology, protein transport, elemental imaging, and chromosome structure.

INTRODUCTION

Useful resolution involves both instrument resolution and specimen contrast. The ultimate resolution of imaging systems depends critically on the wavelength of the radiation used to image the sample. The wavelength of x-rays makes them attractive candidates for imaging beyond the limit of visible light microscopy, and with elemental sensitivity.[1] Microscopes capable of imaging specimens with soft x-rays[2] use either reflection or diffraction based optics to focus the x-rays. In the scanning transmission x-ray microscope (STXM) at Brookhaven National Laboratory, a sample is scanned through x-rays focused by a zone plate, while the transmitted radiation is measured with a proportional counter. The STXM offers: 1) the ability to image areas up to

Synthetic Microstructures in Biological Research, Edited by
J.M. Schnur and M. Peckerar, Plenum Press, New York, 1992

several millimeters square, 2) direct digitization of the signal, and 3) the smallest specimen radiation dose for a given signal. The resolution obtained with the microscope is found to be diffraction limited by the zone plate used to focus the x-rays.[3] X-rays with wavelengths between the carbon and oxygen K edges (4.3 - 2.2 nm) interact more strongly with organic materials than with water, so that imaging in this wavelength range permits one to examine unstained, wet biological specimens[3,4] with a superior resolution than is possible with visible light. Higher energies are suitable for studying the chemical environment of surfaces by analyzing spatial distributions of photoelectrons and Auger electrons[5] (SPEM). The high contrast of biological specimens promises a reduction of radiation damage compared to imaging with electrons.[6] Additionally, the attenuation length of the x-rays is suitable for examining micron thick specimens without sectioning. Fourier transforms of images of test objects and dried specimens are above the noise level to the resolution limits of the microscope. Images of wet *V. faba* chromosomes have been obtained with 80 nm resolution. Subsequent images show that the chromosomes were degraded by radiation damage following the initial exposure. Studies are currently under way to define and extend the limits to resolution caused by radiation damage. Because the absorption of x-rays is determined primarily by the energy of x-ray photons and the elemental composition of the sample, quantitative densitometry of images can be converted directly to mass mapping. Many researchers have used the Brookhaven STXM for problems in neurobiology, protein transport, elemental imaging, and chromosome structure.

THE BROOKHAVEN SCANNING TRANSMISSION X-RAY MICROSCOPE

X-rays cannot be refractively focused by conventional lenses since the attenuation is comparable to the phase shift produced. X-ray microscopes use reflection or diffraction optics to focus the x-rays. Many microscopes using reflecting optics are based on total reflection at angles less than the critical angle of a particular material. Others make use of the higher reflectivity afforded by multi-layer mirrors at larger angles. Image resolutions of 3.8 x 1.7 μm have been obtained with total reflection optics,[7] while 100 nm resolution images have been recorded with multi-layer Schwarzschild objectives.[8] To date, the highest resolution x-ray images have been recorded using zone plates to focus x-rays by diffraction. The STXM and SPEM at the National Synchrotron Light Source are zone plate based x-ray microscopes. Other zone plate based microscopes are the scanning microscopes located in Daresbury[9] and Gottingen,[10] and the Gottingen imaging microscope.[11]

The x-ray source for the Brookhaven microscopes is a 35 period undulator, which is located 23 meters from the experimental end station, accelerates the relativistic electrons radially to the orbital plane of the synchrotron, producing a narrow beam of partially coherent radiation. The energy distribution of the radiation produced is concentrated to integer multiples of a fundamental energy, and can be tuned by varying the size of the gap between the rows of magnets. The X1 soft x-ray undulator at the National Synchrotron Light Source can be tuned to produce a fundamental between 207 and 775 eV, and is currently the brightest continuous source of soft x-rays in the world. The horizontal divergence (0.25 mr) of the x-ray beam is dominated by the divergence of the electron beam in the straight section of the synchrotron. Consequences of the finite electron beam divergence are the widening of the spectral lines produced by the undulator towards longer wavelengths, and the presence of even as well as odd order harmonics on axis. The electron beam divergence is actually beneficial, as both the fundamental and second orders are used in separate end stations simultaneously. The spherical grating in the bichromator horizontally focuses the x-rays of the fundamental and second harmonic to the exit slits of the "long" and "short" wavelength branches, respectively. The fundamental and second harmonic are relayed to the two branches of the beam line via mirrors which increase the separation of the two branches and provide vertical refocusing if desired (Figure 1). Because of the horizontal focusing performed by the bichromator, the exit slits become the horizontal sources (3 meters from the experiment), while the vertical source is the undulator. For experiments that use zone plates to demagnify the source, the horizontal and vertical source separation produces a slight, but measurable, astigmatism. During the spring of 1992 a vertical refocusing mirror has replaced the primary mirror upstream of the bichrometer, which decreases the astigmatism and increases the x-ray flux delivered to the experiment. The Scanning Transmission X-ray Microscope is the primary user of the long wavelength x-rays of the undulator fundamental, while the shorter x-rays of the first harmonic are shared by the Scanning Photo Electron Microscope,[4] the Gabor holographic microscope,[12] and high resolution diffraction experiments.[13] The two beam line branches operate simultaneously 75% of the time. The high vacuum of the beam line is separated from the atmospheric pressure of the microscope by a 250 μm wide, 100 nm thick silicon nitride exit window, which transmits 70% of the incident soft x-rays.

Zone plates are circular diffraction gratings with alternating open and opaque zones constructed so that light passing through the zone plate will add coherently at the focal point.[14] If the zone plate is coherently illuminated, light is focused on the axis of the zone plate with focal lengths $f_n = 2r\partial r/n\lambda$ where r is the radius of the zone plate, ∂r is the thickness of the finest zone, n is the order of the focus, and λ is the wavelength. A perfect zone plate with a 1:1 mark-to-space ratio has no even order foci, and has diffraction (focusing) efficiencies proportional to $1/(n\pi)^2$. A variation on this theme is to construct alternate zones of a material that shifts the phase of the light by $\approx\pi$, thereby quadrupling the diffraction efficiency. The size of the focal spot formed by the zone plate is determined by the finest zone width, and has a radius of approximately $1.22^* \partial r$ at the first minimum of the Airy pattern. The zone plate in use by the NSLS STXM is 90 μm in diameter, and has alternate zones made of 100 nm thick nickel, which shifts the phase of the light by 0.54π for 3.64 nm x-rays. The diffraction efficiency at the first order focus was measured to be 12.9% compared to a theoretical value of 14.5%. This zone plate has an opaque central stop of 40 μm diameter. An "order sorting aperture" with a diameter smaller than the central stop is aligned on the zone plate axis to block the unfocused radiation, as well as the higher order foci (Figure 2). For 3.64 nm wavelength x-rays there is generally ≈210 μm between the order sorting aperture and the sample. The modulation transfer function of this zone plate was determined experimentally by comparing STXM images of a test object with a smallest feature size of 36 nm to STEM images of the same object.[3] Both the efficiency and modulation transfer function of the zone plate agree well with predicted values, allowing the point spread function for the microscope to be calculated. Because this characterization has been performed, image deconvolution can be employed to recover the original contrast of the sample features (Figure 3). The flux can be increased at the expense of increasing the width of the point spread function (decreasing the resolution) by opening the exit slit of the bichromator. Both the zone plate and order sorting aperture are mounted on a computer controlled stepping motor driven stage for semi-automatic focusing. The microscope can be completely disassembled and reassembled for a different experiment in 90 - 120 minutes.

Samples are magnetically mounted on a phosphor-bronze flexure stage[15] which is driven by two piezoelectric actuators with capacitance feedback control. These

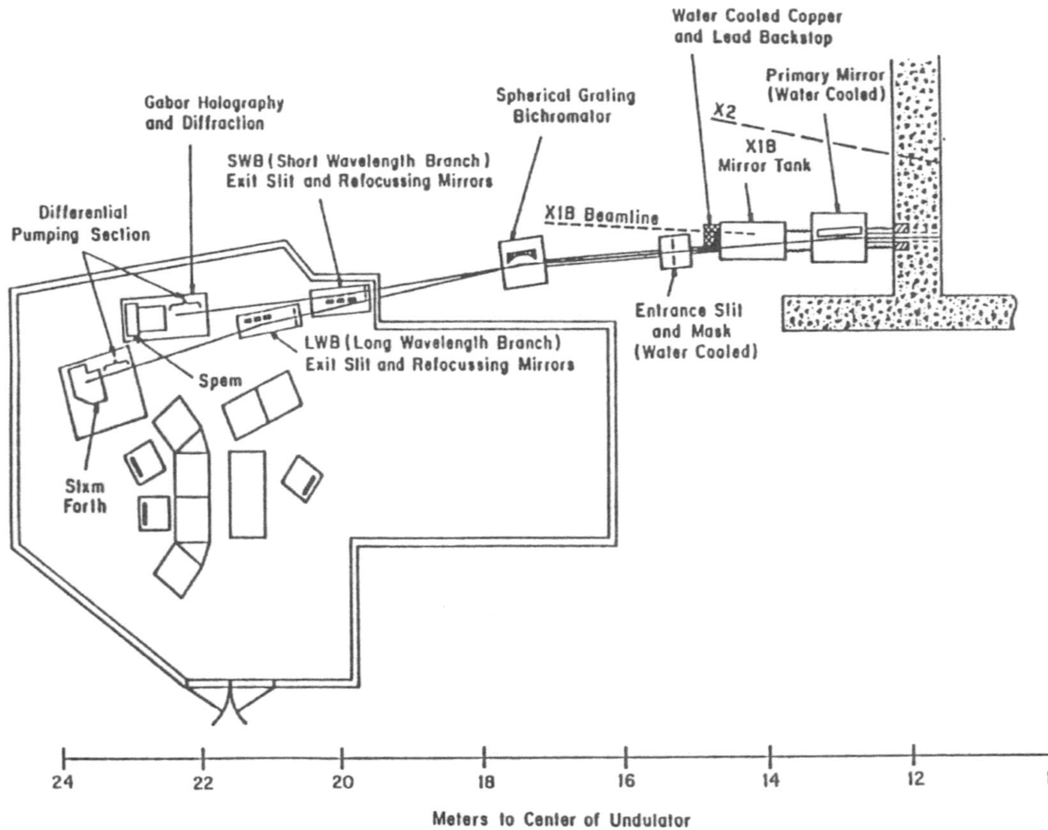

Figure 1. Beam line optics deliver soft X-rays from the undulator to the experiments. X-rays are steered by the water-cooled primary mirror to the entrance slit of the bichromator. The spherical grating in the bichromator focuses the undulator fundamental and first harmonic on the exit slits of the two branches. The SWB and LWB refocusing mirrors increase the separation and provide vertical refocusing if desired.

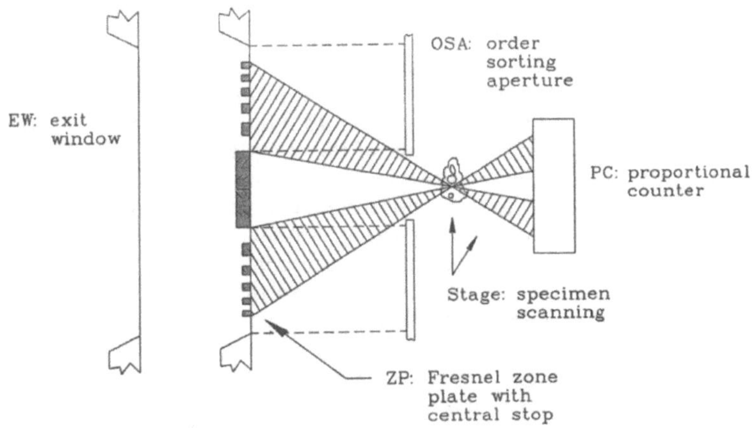

Figure 2. Schematic diagram of the Brookhaven STXM: X-rays pass from the ultra-high vacuum of the beam line to atmospheric pressure through the 100 nm thick silicon nitride exit window (EW). Light is diffractively focused by the zone plate (ZP). Only light reaching the primary focus of the zone plate will avoid the central stop and the order sorting aperture (OSA). After passing through the sample, unabsorbed x-rays are counted by the proportional counter (PC).

piezoelectric actuators provide digital positioning capability and are linear over a 75 μm range. With a step size of 4.6 nm, the fine scanning stage has the accuracy required to take advantage of the zone plate resolution. The entire fine stage is mounted on a stepper motor driven coarse stage which delivers 1 μm accuracy over a 25 mm range in X and Y. This two stage arrangement permits the operator to make large area coarse scans at low dose to locate specimens of interest, and then to sample more finely for high resolution imaging. Past the sample is a high efficiency proportional counter using a 1:9 mixture of CO_2 in argon. The proportional counters

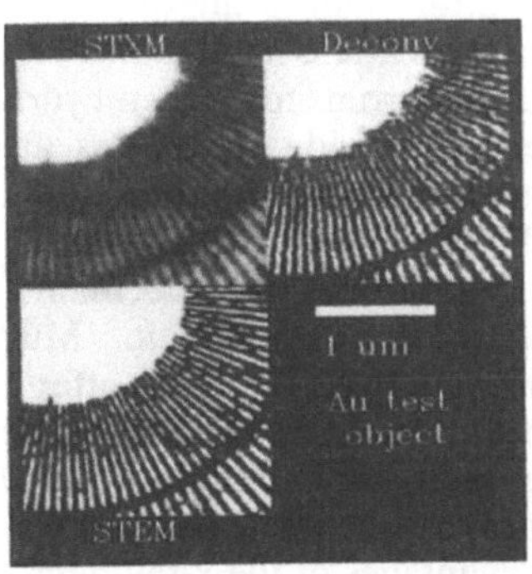

Figure 3. Once the microscope modulation transfer function is known, the original object contrast can be recovered from the image by deconvolution. STEM, STXM and deconvolved STXM (Deconv.) images of a spoked test object. The finest features in the pattern are 36 nm.

use hybrid electronics and provide a linear range of 0.1 - 2000 KHz, with a maximum count rate of 4000 KHz. Although these counting rates are presently sufficient, updates in beam line optics will necessitate large changes in counter design or alternative strategies for data acquisition. Typical pixel dwell times are 1 - 10 msec, which means that a 300 x 300 image takes 1.5 - 15 minutes to record. By keeping the relatively inefficient x-ray optics upstream of the sample stage, the radiation dose to the specimen is minimized. Helium flows into a Plexiglass box enclosing the area from the silicon nitride exit window to the proportional counter, maximizing the number of photons which reach the counter after passing through the sample.

All beam line optics and stage motion are controlled by interactive software which runs on a VAXstation 3200 computer. Even the wavelength can be changed interactively, though large changes require resetting the undulator gap. Most new users can pilot the microscope independently after spending a few hours with experienced operators. Images are stored on the computer hard disk and can be transferred to magnetic tape for archiving.

RADIATION DAMAGE

All techniques producing structural information make use of the fact that some form of radiation interacts differently with the specimen than with the surrounding material. For the STXM, the absorption cross-section for soft x-rays having energies between the carbon and oxygen K edges (284 - 531 eV) for biological macromolecules is much larger than the cross-section for water (Figure 4). This large difference in cross-section yields large specimen contrast, so that on a per dose basis, far fewer soft x-rays than electrons are required to image wet biological specimens at a given resolution.[2,4] Unfortunately, the same soft x-rays that produce the image are known to be very effective in damaging biological specimens. A 25 Rad dose of carbon Kα x-rays has been shown to kill 50% of human diploid fibroblasts,[16] and a dose of 275 Rad causes 1 chromosome aberration per human lymphocyte.[17] Foster et al.[18] have found that a soft x-ray dose of 0.3 MRad will stop contraction in active myofibrils. It is clearly necessary to relate the dose needed to produce the soft x-ray images with the specimen damage produced.

For microscopists the crucial parameter is the resolution attainable before radiation damage compromises interpretation of the images obtained. Though the electron microscope offers better than 2 angstrom resolution for some material specimens, studies of biological specimens have shown that radiation damage limits the available resolution to several angstroms for frozen hydrated crystals at liquid helium temperatures,[19] while images of individual particles at room temperature show a resolution limitation closer to 2 nm.[20] The diffraction limited resolution of 50 nm for the STXM was obtained with test specimens, but for conventional samples the resolution will be determined largely by the specimen's ability to endure the dose needed to obtain a significant signal to noise ratio. Multiple images of freeze-dried *V. faba* chromosomes show no significant mass loss after 150 MRad, so are insensitive to soft x-ray irradiation. Fourier transforms of some of these images show features out to the cutoff frequency of the microscope. Studies to quantitatively compare electron micrographs with STXM images of the same sample are currently underway.

One of the principal advantages of the x-ray microscope is the ability to image wet specimens, but under these conditions radiation damage plays a more visible role. Gilbert et al.[21] have shown that initially live fibroblasts isolated from chick dorsal root ganglia appear structurally intact for more than 0.5, but less than 3 hours following a 95 nm resolution soft x-ray image, with a dose of 2 MRad. Experiments

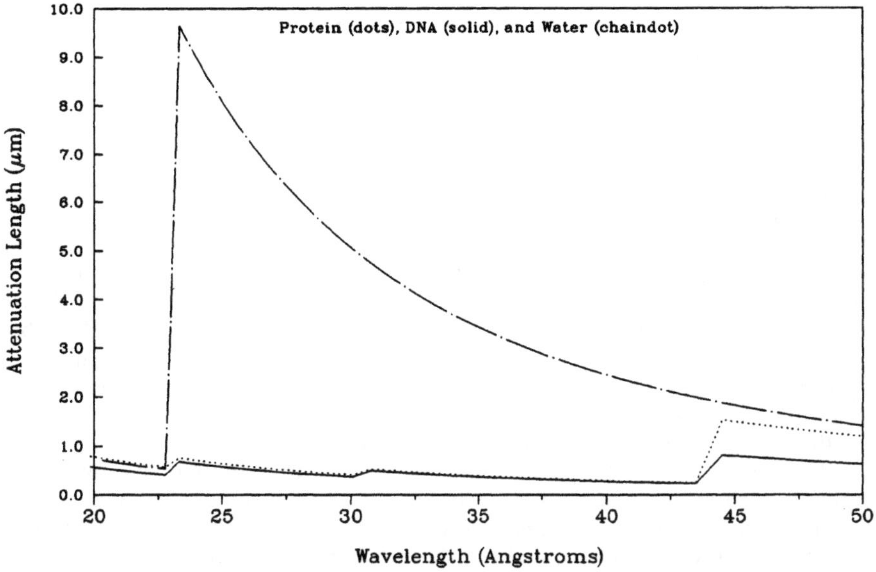

Figure 4. Attenuation length of protein, DNA, and water. Attenuation lengths of biological macromolecules and water in the water window. The difference in attenuation lengths between organic material and water provides good specimen contrast.

performed on fixed hydrated *V. faba* chromosomes imaged at 50 nm resolution have shown that damage occurs slowly compared to the time required to image neighboring pixels for dwell times of 1 - 10 msec. Assuming first order kinetics for mass loss, first images of chromosomes show a rate constant of 0.001 $MRad^{-1}$ for doses of 10 - 10^3 MRad, under the current imaging conditions of the STXM. Subsequent images reveal the chromosomes to be severely shrunken, and have degraded following their initial exposure. To retain 90% of the initial mass in the second image, the first image dose must be less than 34 MRad. By inspection, chromosomes are among the most sensitive of specimens, but it is not known to what extent the damage rate is influenced by the sample buffer. Currently under study are the effects of alcohols and sulfhydryl compounds on the measured mass loss and resolution. Buffers containing sulfhydryls are known inhibitors of DNA breakage[23] because they are efficient free radical scavengers.[23,24]

The results for chromosomes agree with the postulate that the time scale for biological damage resulting from the diffusion of radiolysis products will be in the msec range.[25] A 1 - 10 msec per pixel dwell time may be acceptable at the current resolution level of scanning x-ray microscopes if radiation damage was limited to these short range effects. But, longer range damage events and cell movement as a result of exposure to the beam were clearly observed in studies of initially live neurons isolated from chick and dorsal root ganglia and rat pancreatic acinar cells. The desire to study these more sensitive living specimens, and to push to higher resolution, prompts the need for higher data acquisition rates. Flash data collection using imaging or holographic microscopes powered by x-ray lasers or plasma sources[26] is an attractive option for getting the highest data acquisition rates. As the popularity of x-ray microscopy increases, undoubtedly methods will be found for minimizing the effects of radiation damage during the imaging process.

CURRENT STUDIES USING THE MICROSCOPE

Imaging with soft x-rays between the carbon and oxygen K edges allows the examination of wet, unsectioned, unstained biological specimens. The challenge is to design a wet cell which keeps the sample fully hydrated, extends less than 200 μm in front of the sample plane, and has a total thickness of less than 1 millimeter. To date there have been two designs which have successfully incorporated all of these features. The first involves keeping the specimen totally immersed in a 1 - 5 μm thick buffer layer which is sandwiched between two silicon nitride windows[27] or organic foils.[28] The second involves sealing the specimen in a humidity chamber which can be filled with buffer, then drained and filled with humid air or helium.[29] Both designs have been utilized by different groups at the Brookhaven STXM.

A major subject of interest to us has been the folding and dynamics of metaphase chromosome structure. Studies of chromosomes isolated from the meristematic root tissue of *V. faba* show that internal structure is best visualized under ionic conditions which expand the chromosome, yet maintain chromatin higher order structure.[30] Low ionic strength expansion of the chromosomes, a common preparation procedure for electron microscopy, eliminates secondary structure. Comparisons of dried and hydrated samples have shown that freeze-dried and freeze-substituted chromosomes have a morphology most closely approximating wet samples. Chromosomes which have been dehydrated at room temperature are severely shrunken with decreased visibility of internal structure (Figure 5). The severe condensation of the room-temperature-dried chromosomes reduces the already low signal in the conventional transmission electron microscope, due to multiple scattering in thick specimens. Freeze-drying, and freeze-substitution should be employed for examination of dry chromosome specimens. Current studies are aimed at finding conditions minimizing radiation damage, and maximizing the contrast of chromosome internal structure in both wet and dry specimens.

A growing number of researchers are taking advantage of the high resolution and native specimen environment accorded by the Brookhaven x-ray microscope (for a good review see "X-ray Microscopy III", e.g., reference 14). The mechanism of protein transport is being examined in studies of isolated wet unfixed zymogen granules,[27] and rat pancreatic acinar cells. Cultures of fibroblasts and neurons from chick dorsal root ganglia have been studied by Pine[29] and Gilbert.[21] Using pairs of images taken on either side of the calcium L edge, calcium deposits are being localized and measured in sections of cartilage and tendon.[31] Exploratory work on imaging of collagen and myofibrils have also been undertaken by visitors from King's College, London.

The microscope is being further developed to extend the current capabilities to provide additional information. X-ray-induced visible light fluorescence has already provided images of phosphor grains with better than visible light resolution, and some common fluorescent labels have demonstrated significant visible light fluorescence with soft x-ray stimulation.[32] Simultaneous acquisition of the x-ray absorbance and fluorescence signals will permit chemical specificity to accompany quantitative mass mapping. Design and testing of detectors which will measure the absorption and phase change of x-rays passing through the sample has also begun.

Figure 5. Comparison of wet to chromosomes dried by several common procedures. *V. faba* chromosomes that were (counter clockwise from upper right): critical point dried after being put through a graded acetone series at room temperature; critical point dried after being put through a graded ethanol series at room temperature; critical point dried after being snap frozen in liquid nitrogen slush and ethanol substituted at -70°C (freeze-substitution); and wet mounted. All chromosomes started out in 65 mM ionic strength phosphate buffer. All images are displayed so that grey levels scale linearly from 0% (black) to 85% (white) transmission.

FUTURE IMPROVEMENTS

Improvements in beam line optics will provide an eight-fold increase in useful flux within a year, which will further minimize the loss of resolution caused by sample movement and radiation damage during imaging.[25] Increasing the flux should then provide images closer to representing a snapshot of the original sample. Photon counter technology will have to keep up with this trend. Design and testing of new proportional counters and alternative methods of data acquisition have already begun. Currently the zone plate used to focus the x-rays determines the resolution limit of the microscope. Decreasing the finest zone width not only provides the ability to resolve smaller objects, but also results in increased contrast of larger object features. With more facilities for providing soft x-rays being constructed, and improvements in data acquisition rate and resolution, the soft x-ray microscope will become an increasingly integral part of the structural biologist's toolbox.

ACKNOWLEDGEMENTS

This work is supported in part by the Department of Energy under grant #DE-FG02-89ER60858 and the National Science Foundation under grant #DIR-9005893. Work was performed at the NSLS, which is supported by the Department of Energy Office of Basic Energy Sciences under contract DE-AC02-76CH00016. The zone plate used in the microscope was made by Dr. Erik Anderson under a collaborative agreement between the Lawrence Berkeley Laboratory Center for X-ray Optics and the Nanostructure Fabrication Group at IBM Research.

REFERENCES

1. A. Engstrom, Quantitative micro and histochemical elementary analysis by roentgen absorption spectography, *Acta Radiol. Suppl.* 63:1 (1946).
2. H. Wolter, Spiegelsysteme streifenden einfalls als abbildende Optiken fur Roentgenstrahlen, *Ann. Phys.* 10:94 (1952).
3. C. Jacobsen, S. Williams, E. Anderson, M. T. Browne, C. J. Buckley, D. Kern, J. Kirz, M. Rivers, and X. Zhang, Diffraction-limited imaging in a scanning transmission x-ray microscope, *Optics Comm.* 86:351 (1991).
4. D. Sayre, J. Kirz, R. Feder, D. Kim, and E. Spiller, Potential operating region for ultrasoft x-ray microscopy of biological samples, *Science* 196:1339 (1970).
5. H. Ade, J. Kirz, S. Hulbert, E. Johnson, E. Anderson, and D. Kern, Scanning photoelectron microscope with a zone plate generated microprobe, *Nucl. Inst. Meth. Phys. Res.* A291:126 (1990).
6. D. Sayre, J. Kirz, R. Feder, D. M. Kim, and E. Spiller, Transmission microscopy of unmodified biological materials: comparative radiation dosages with electrons and ultrasoft x-ray photons, *Ultramicroscopy* 2:337 (1977).
7. Y. Suzuki and F. Uchida, Hard x-ray microprobe with total reflection mirrors, *Rev. Sci. Instr.* 63:578 (1992).
8. R. R. Freeman, J. E. BJorjkholm, J. Bokor, L. Eichner, J. Gregus, T. E. Jewell, W. M. Mansfield, A. A. MacDowell, E. L. Raab, W. T. Silfvast, L. H. Szeto, D. M. Tennant, W. K. Waskiewicz, D. L. White, and O. R. Wood II, X-ray reduction imaging at the NSLS, *Synchrotron Radiation News* 4:33 (1991).
9. D. Morris, C. J. Buckley, G. R. Morrison, A. G. Michette, P. A. F. Anastasi, M. T. Browne, R. E. Burges, P. S. Charalambous, G. F. Foster, J. R. Palmer, P. J. Duke, Recent advances in scanning x-ray microscopy, *Scanning* 13:7 (1991).
10. B. Niemann, The Gottingen scanning x-ray microscope, in: "Soft X-ray Optics and Technology," E. E. Koch and G. Schmahl, eds., *SPIE proc.*, 733:423 (1987).
11. D. Rudolph, B. Niemann, G. Schmahl, and O. Christ, The Gottingen x-ray microscope, and x-ray microscopy experiments at the Bessy storage ring, in: "X-ray Microscopy," G. Schmahl and D. Rudolph, eds., Springer, Berlin, (1984).
12. M. R. Howells, C. Jacobsen, J. Kirz, K. McQuaid, and S. S. Rothman, Progress and prospects in soft x-ray holographic microscopy, in: "Modern Microscopies," P. J. Duke, and A. G. Michette, eds., Plenum Publishing, New York, (1990).
13. W. Yun, J. Kirz, and D. Sayre, Observation of the soft x-ray diffraction pattern of a single diatom, *Acta Cryst.*, A43:131 (1987).
14. E. H. Anderson and D. Kern, Nanofabrication of zone plate lenses for high resolution x-ray microscopy, in: "X-ray Microscopy III," A. Michette, G. Morrison, and C. J. Buckley, eds., Springer, Berlin (1992).
15. M. Browne, Aspects of nanopositioning in stage design for high resolution x-ray microscopy, in: "X-ray Microscopy III", A. Michette, G. Morrison, and C. J. Buckley, eds., Springer, Berlin (1992).
16. D. T. Goodhead, J. Thacker, and R. Cox, Effectiveness of 0.3 keV carbon ultrasoft x-rays for the inactivation and mutation of cultured mammalian cells, *Int. J. Rad. Biol.* 36:101 (1979).
17. R. P. Virsik, Ch. Schafer, D. Harder, D. T. Goodhead, R. Cox, and J. Thacker, Chromosome aberrations induced in human lymphocytes by ultrasoft Alk and Ck X-rays, *Int. J. Rad. Biol.* 38:545 (1980).
18. G. F. Foster, P. M. Bennet, C. J. Buckley, and R. E. Burge, Investigation of radiation damage to biological specimens at water window wavelengths, *Rev. Sci. Instr.* 63:599 (1992).

19. J. M. Baldwin, R. Henderson, E. Beckman, and F. Zemlin, Images of purple-membrane at 2.8Å resolution obtained by cryo-electron microscopy, *J. Mol. Biol.* 202:585 (1988).
20. M. S. Isaacson, Specimen damage in the electron microscope, in: "Principles and Techniques of Electron Microscopy," M. A. Hayat, ed., Van-Nostrand Publishing Co., New York (1976).
21. J. R. Gilbert, J. Pine, J. Kirz, C. Jacobsen, S. Williams, C. J. Buckley, and H. Rarback, Soft x-ray imaging of whole wet tissue culture cells, in: "X-ray Microscopy III," A. Michette, G. Morrison, and C. J. Buckley, eds., Springer, Berlin (1992).
22. R. C. Fahey, B. Vojnovic, and B. D. Michael, The effects of counter-ion condensation and co-ion depletion upon the rates of chemical repair of ploy(U) radicals by thiols, *Int J. Rad. Biol.* 59:885 (1991).
23. R. Roots and S. Okada, Estimation of life times and diffusion distances of radicals involved in x-ray-induced DNA strand breaks and killing of mammalian cells, *Radiation Research* 64:306 (1975).
24. R. Roots and S. Okada, Protection of DNA molecules of cultured mammalian cells from radiation-induced single-strand scissions by various alcohols and SH compounds, *Int. J. Rad. Biol.* 21:329 (1972).
25. K. Shinohara and A. Ito, Radiation damage in soft x-ray microscopy of live mammalian cells, *J. Microsc.* 161:463 (1991).
26. J. C. Solem, Imaging biological specimens with high intensity soft x-rays, *J. Opt. Soc. Am.*, B3:1551 (1986).
27. S. S. Rothman, K. Goncz, and B. Loo Jr., Following protein transport with the high resolution x-ray microscope, in: "X-ray Microscopy III," A. Michette, G. Morrison, and C. J. Buckley, eds., Springer, Berlin (1992).
28. G. Schneider and B. Niemann, Environmental chamber for x-ray imaging of wet biological specimens, in: "X-ray Microscopy III," A. Michette, G. Morrison, and C. J. Buckley, eds., Springer, Berlin (1992).
29. J. Pine and J. R. Gilbert, Live cell specimens for x-ray microscopy, in: "X-ray Microscopy III," A. Michette, G. Morrison, and C. J. Buckley, eds., Springer, Berlin (1992).
30. S. Williams, C. Jacobsen, J. Kirz, S. Lamm, and J. Van't Hof, Scanning transmission x-ray microscopy of hydrated mitotic chromosomes, in: "X-ray Microscopy III," A. Michette, G. Morrison, and C. J. Buckley, eds., Springer, Berlin (1992).
31. C. J. Buckley, G. F. Foster, R. E. Burge, S. Y. Ali, and C. A. Scotchford, Elemental mapping of biological tissue by x-ray absorption difference imaging in the STXM, in: "X-ray Microscopy III," A. Michette, G. Morrison, and C. J. Buckley, eds., Springer, Berlin (1992).
32. C. Jacobsen, S. Lindaas, S. Williams, and X. Zhang, Scanning luminescence x-ray microscopy: imaging fluorescent dyes at sub-optical resolution, submitted to *Journal of Microscopy.*

SCANNING ELECTRON MICROSCOPY OF PROTEIN MONOLAYERS ON A SILICON WAFER

Taiji Furuno, Ayano Sato, and Hiroyuki Sasabe

Frontier Research Program, RIKEN (Institute of Physical and Chemical Research), 2-1 Hirosawa, Wako, Saitama 351-01, Japan

ABSTRACT

Two-dimensional regular arrays of ferritin and catalase were prepared by adsorption to a polypeptide film spread at the air/water interface. These arrays and purple membrane (two-dimensional crystal of bacteriorhodopsin) were negatively stained on a silicon wafer and observed with a high resolution scanning electron microscope. The resolution of an image was characterized by Fourier transformation, and the diffraction spots around 30Å spacings were observed for all of these images. The potential ability of the high resolution scanning electron microscopy (SEM) for protein monolayers on a solid support is demonstrated.

INTRODUCTION

The fabrication of densely packed monolayer arrays of protein on a solid support is significant both in basic research and technology. The technique to prepare two-dimensional (2D) crystals of protein is useful in the structural studies by transmission electron microscopy (TEM), since three-dimensional crystallization of many interesting proteins for X-ray analysis is still difficult. On one hand, 2D crystals of protein would be one of the structural entities of the future micro-devices based on biological molecules. In the device application, a uniform film, in which the protein molecules are densely packed in a desired orientation on the purposed solid plate, might be more valuable in a practical sense than patches of 2D crystal which give higher resolution electron microscopic images.

A few proteins have so far been crystallized in two dimensions by versions of Langmuir-Blodgett (LB) techniques,[1] in which protein molecules were bound to alkylated ligands[2-6] incorporated into lipid monolayers, or concentrated protein solution was directly spread on a clean mercury surface.[7] Another method for 2D crystallization of protein at the air/water interface is to utilize a binding of protein to a monolayer of charged lipid.[8,9] In the 2D crystallization of protein at the air/water interface, the preformed film must be fluid enough to facilitate the crystallization.[2]

In our studies a synthetic polypeptide, poly(1-benzyl-L-histidine) (PBLH), was used as such material.[10] The interfacial film of PBLH is supposed to have the following features feasible for binding proteins, and probably for 2D crystallization.

1) its imidazole group, positively charged at acidic conditions, is assumed to be electrostatically interactive with acidic proteins in an appropriate range of subphase pH,
2) due to its polymeric structure, some of the hydrophobic benzyl groups are facing the subphase, allowing us to assume hydrophobic interactions between PBLH and proteins, which might stabilize the binding further,
3) the film is flexible and slightly fluid, as previously discussed.[11]

The recent progresses in TEM have proved that it is possible to obtain a three-dimensional structure of protein molecule near atomic resolution.[12] In application of protein 2D crystals to devices, however, we need another tool to see packing of proteins on a bulk substrate. In these several years ultra-high resolution SEMs capable of 7Å resolution at an accelerating voltage of 30kV are commercially available. In this paper, the ability of an ultra-high resolution SEM (HRSEM) for imaging protein molecules on a bulk silicon wafer is vividly demonstrated for three kinds of negatively stained monolayer arrays of protein.

The 2D arrays of protein are negatively stained conventionally in TEM to enhance contrast and to preserve the structure. In the scanning electron microscopy the secondary electron yield which varies depending on the surface structure is converted into an image.[13] The negative staining is expected to be useful also in SEM for imaging ordered and/or densely packed protein arrays. The shape of each protein molecule would be contrasted by accumulation of heavy metal stain at the spaces between protein molecules and the specific sites due to morphological and/or chemical reasons. Several proteins have already been observed by HRSEM as bright objects with three-dimensional view on a carbon plate.[14,15] Three-dimensional imaging of protein arrays is also possible by TEM with fine metal-shadowing.[16] In this paper, however, we do not discuss the relationship between our images of negatively stained protein arrays and those on a carbon plate or metal-shadowed.

The first example we present here is ferritin.[17] The shape of the molecule is a spherical shell with outer and inner diameters of 13 and 7 nm, respectively. A core of iron oxide composed maximally of 4000 iron atoms is held inside the polypeptide shell. The SEM images of unstained ordered arrays and stained isolated molecules have already been presented in our previous paper,[10] where the protein shell was resolved from the core by negative staining. The second protein, beef liver catalase,[18] is composed of 4 identical subunits and with total molecular weight of 240k. The shape of the molecule when viewed from the direction which gives its maximum dimensions is roughly a rectangle with the size of 10x11nm.[2] The third example is purple membrane (PM) which is intrinsically two-dimensional crystal of membrane protein, bacteriorhodopsin (bR), and membrane lipids. The structure of bR has already been solved to an atomic resolution by electron crystallographic method.[12]

The protein arrays presented here are supposed to have three different kinds of contrast. Ferritin contains a large body of iron oxide which emits a lot of secondary electrons through the polypeptide shell. Then the contrast and brightness of ferritin would be the most intense among the three arrays. The negatively stained array of catalase would give an image with moderate contrast due to accumulation of stain at the spaces between neighboring molecules. An image of lowest contrast may be obtained for purple membrane (thickness about 4.5 nm), since any space between bR molecules is filled with membrane lipids, and there would not be such an accumulation of stain as outlining the molecular shape. However, a subtle surface morphology with crystalline order would be captured if the thickness of stain matches the shallow escape depth of secondary electrons.

METHODS AND RESULTS

Binding of Ferritin and Catalase to PBLH Film

PBLH (purchased from Sigma) was dissolved in chloroform containing a small amount of dichloroacetic acid (1 μl/1mg of PBLH). The film of PBLH was spread over the solution of horse spleen ferritin (purchased from Sigma) and incubated for 1-3h around pH 6 (concentration of ferritin 10-40 μg/ml, for details of experimental setup see reference 10).

The 2D array of beef liver catalase was prepared by a similar method by spreading a PBLH film over the protein solution. The subphase pH was adjusted to 4.8-6.4 with phosphate buffer. We have already reported some images of catalase taken by HRSEM, including a monolayer assembly prepared by the PBLH method.[19] In the present study, a thermal treatment of repeated heating and cooling between room temperature and 35-37°C was preliminarily applied to the catalase/PBLH film *in situ* on the water surface. The crystal growth was facilitated by this treatment and larger crystals than those formed at room temperature were obtained with ease.

Deposition of the Interfacial Films onto Silicon Wafer

The interfacial films were transferred onto a silicon wafer (~8x12 mm^2) with hydrophobic surface by a horizontal transfer method.[10] The transferred film which emerged wet was rinsed with pure water or dilute buffer of the same pH with the subphase, and then stained. The purple membrane fragments were suspended in a staining solution, and a drop of suspension was deposited and spread on a silicon wafer which had been rendered hydrophilic by irradiation with UV light from a low-pressure mercury lamp. To quickly dry up a stained film, the wafer plate wetted with the solution was placed in a microcentrifuge tube and the excess solution was blown off by centrifugation (at 6000 rpm, 2000g) for 10-20s. The specimens of catalase and purple membrane were stained and dried with this method. The surface films of ferritin and catalase could be horizontally transferred and entirely covered the surface of silicon wafer when inspected by SEM at low magnifications, although some holes and clefts were seen in the film.[10]

High Resolution Scanning Electron Microscopy and Image Processing

A high resolution scanning electron microscope (HRSEM) (S900, Hitachi Ltd.) was used at an accelerating voltage of 30kV. During observation the specimen temperature was kept at 60-80°C to reduce surface contamination. The electron dose was typically $200e^-/Å^2$ for taking a fresh image by one scan at magnification of 100,000.

The SEM images were digitized with a CCD camera and fed to an IBM PC compatible computer equipped with a frame grabber and array processor boards (both from Data Translation Inc., USA). The digitized images were Fourier transformed for characterizing the resolution and for obtaining noise-filtered images. An image processing by correlation averaging[20] was applied to the 2D array of catalase

(Figure 2). A correction for lattice distortion was applied to the images of ferritin (Figure 1(a)) and catalase (Figure 2) according to the following simple ways. The selected region was Fourier transformed first, filtered weakly and then the cross-correlation between the entire image and a cut-out small area was calculated. After surveying the correlation peaks, the ideal lattice points were determined, and the lattice was then corrected for distortion by rearranging translationally the cut-out original images with unit cell size centered on the correlation peaks. The Fourier transform of the corrected image will be a measure for the resolution of the correlation-averaged one (uncorrected for rotational distortion).

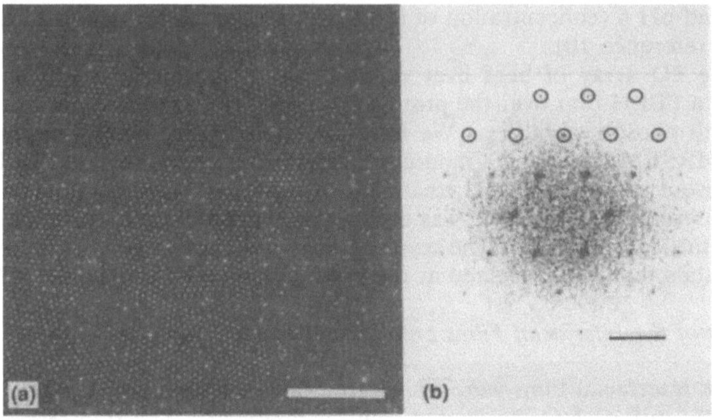

Figure 1. (a) Scanning electron micrograph of monolayer arrays of ferritin stained with ammonium molybdate containing glucose. The specimen was kept at 80°C during observation. The arrays of ferritin were covered inhomogeneously with a layer of the stain. At the right-edge region ferritin arrays are faintly seen beneath a thick stain layer. Scale bar=200 nm. (b) Computed diffraction pattern of a lightly stained well-ordered array of ferritin. The lattice unbending was applied. Diffraction bar=0.1 nm^{-1}.

Ferritin

The ordered monolayer array of ferritin was negatively stained with ammonium molybdate (0.2% w/v, pH 4.9) containing 0.2% glucose. The film was naturally dried by keeping the wet substrate vertical. In Figure 1(a) ferritin molecules are arranged in a uniform dense packing, as confirmed by observing the unstained film prepared simultaneously. However, as can be seen in the figure, there are two kinds of region: one region is the assembly of brightly seen ferritin particles, and in the other corpuscular images are almost invisible or faintly visible. The secondary electrons from iron cores of ferritin are blocked where the stain is accumulated too thick over the ferritin arrays. This result indicates clearly that in the scanning electron microscopy of stained protein monolayers the image is seriously dependent on staining, especially on the thickness. Figure 1(b) is a Fourier transform of a well ordered region in Figure 1(a). The spacing of the highest diffraction spot is 3.8 nm.

Catalase

A monolayer specimen was thermally treated between 23°C and 35°C (pH 4.9, 10mM phosphate, 10 mM NaCl concentration of catalase 10 µg/ml). The view of the catalase monolayer (Figure 2) is considerably different from that of ferritin (Figure 1(a)). We interpret that the molecular shape is visible due to a stain accumulation at the spaces between neighboring molecules. This kind of view would be common to a dense packing of as large proteins as catalase, when successfully stained.

The binding of catalase to PBLH film is not simply explained by electrostatic forces between the oppositely charged molecules, since from the pI value of beef liver catalase they should be positively charged at pH 4.9. However, it was suggested that the electrostatic interaction regulated the binding rate of catalase to PBLH film. A formation of a dense packing was slower at pH 4.9 than at pH 6.0 where catalase molecules should have net negative charge.

The computed diffraction pattern of the image corrected for lattice distortion is shown in Figure 3(a). The highest order diffraction spot corresponds to 3.4 nm spacing. Figure 3(b) is a Fourier-filtered image, and the inset is a correlation-averaged one. The stain excluding region (protein) is shown as bright. The averaged image represents the shape characteristic to a rectangular view of catalase.

Purple Membrane

Purple membranes negatively stained with 3% PTA (pH 4.8) were deposited on a silicon wafer and observed with the HRSEM. In Figure 4(a) the patch form of

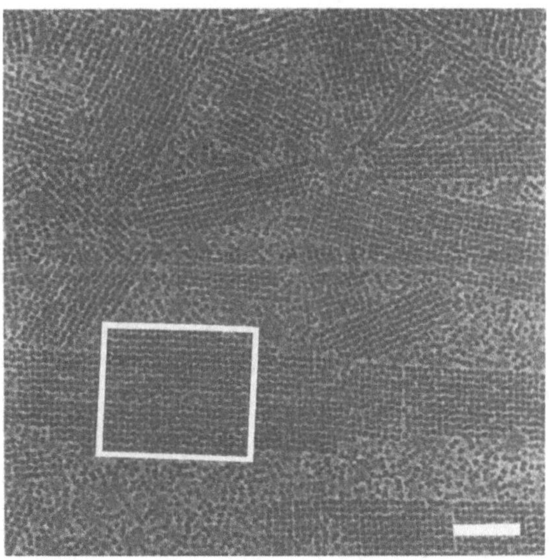

Figure 2. Scanning electron micrograph of 2D crystals of catalase bound to PBLH film. A thermal treatment was applied to the film at the air/water interface. The transferred film was stained with phosphotungstic acid (pH 5.3, 3% w/v). The specimen was heated to 80°C during the observation. The boxed region was subjected to image processing. The image was slightly contaminated on the entire area. Scale bar = 100 nm.

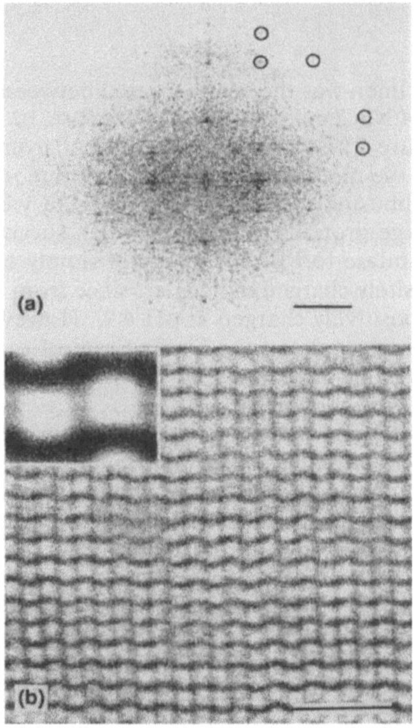

Figure 3. (a) Computed diffraction pattern of the boxed area in Figure 2 after lattice unbending. Diffraction bar = 0.1 nm^{-1}.
(b) Fourier-filtered image and correlation averaged one (inset) of the boxed area in Figure 2. Scale bar = 50 nm.

PM is distinctly seen on the silicon wafer. However, any crystal structure (unit cell dimensions, $a=b=6.2$ nm with p3 symmetry[12]) is invisible by visual inspection. This is because the surface morphology of PM is on the order of a few angstroms, so that the crystal structure of the surface of PM is too weakly contrasted by staining to ensure a direct imaging. Furthermore, it was a difficult task to take a well-focused image of PM due to this low-contrasted image. However, it was possible to demonstrate the presence of crystalline order in the PM fragment shown in Figure 4(a). In the computed Fourier transform (Figure 4(b)) there appeared diffraction spots extending to 1/3.1 nm^{-1}.

DISCUSSION

Protein molecules are considered to be more stable in a 2D crystal form than existing as a sole particle on a substrate. However, a certain process of fixation is required to avoid deformation which should occur during drying. Staining is a convenient and inevitable process also in HRSEM for the purpose of preserving the structure and for increasing the contrast. In the present study we have confirmed that the HRSEM is able to give images of negatively stained protein approximately at 30Å resolution. If a resolution of e.g. 20Å is achieved for a nicely treated 2D crystal of

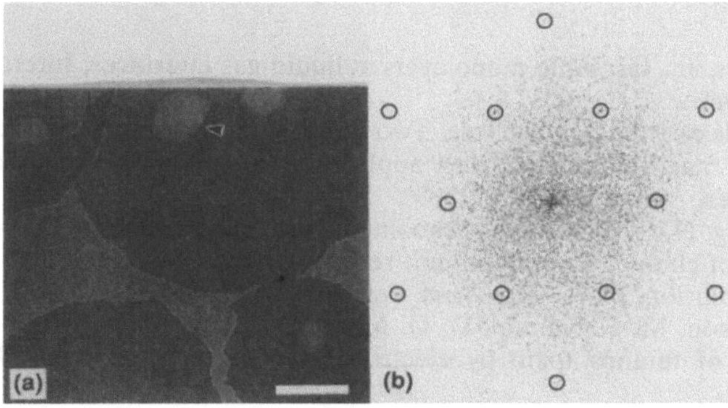

Figure 4. (a) Scanning electron micrograph of negatively stained purple membrane patches. A bright domain indicated with an arrow head is a granule of accumulated stain. Scale bar=200 nm.
(b) Computed diffraction pattern of an area in a PM patch in (a).

protein, it would be possible to draw out fine structure from the crystal which afford complementary structural information to that obtained by TEM. Unfortunately, the technique of specimen treatment is not established yet in HRSEM for protein observation. We tried to observe a negatively stained specimen using a cryoholder at -167°C, but any improvement for the PTA-stained specimen has not been obtained yet probably due to a granularization of the stain. Another trial for the uniform staining of the 2D arrays with a mixture of glucose and stain often led to an overstain which prohibited the imaging of proteins underneath, and the surface seemed to be more sensitive to contamination by electron irradiation than without glucose. Anyhow, we must further pursue the treating method of 2D arrays for higher resolution by scanning electron microscopy.

Any electroconductive substrate with smooth surface, e.g., polished semiconductor wafer, may be used as a substrate in HRSEM for monolayer arrays of protein. The usability of such bulk substrates with flat surface is an advantage of the SEM. However, we must be aware of the surface roughness. The bare silicon wafer we used in this study presented a roughness image. But, fortunately, such a surface image was hidden to some degree with the overlying protein and stain layers due to a shallow escape depth of secondary electrons. Furthermore, we must be conscious of the possibility that the staining procedure may rearrange the packing of protein molecules when the molecules are not covalently bound to the surface of the substrate.

CONCLUSION

Three kinds of negatively stained monolayer arrays with high contrast (ferritin), medium contrast (catalase) and low contrast (bacteriorhodopsin) have been examined by HRSEM. It is demonstrated that the HRSEM is able to analyze the crystal structure of such proteins on a silicon wafer at a resolution of approximately 30Å. The demonstrations in this study confirm the potential of the HRSEM as a tool to characterize protein monolayers on a solid support for device applications. Although a damage by electron dose[21] should limit the resolution, the structure analysis of 2D crystals of protein is possible by HRSEM, if specimens are treated successfully.

REFERENCES

1. G. L. Gaines, Jr., Insoluble monolayers at liquid-gas interfaces, Interscience, New York, 1966
2. E. E. Uzgiris and R. D. Kornberg, Two-dimensional crystallization technique for imaging macromolecules, with application to antigen-antibody-complement complexes, *Nature* 301:125 (1983).
3. D. S. Ludwig, H. O. Ribi, G. K. Schoolnik, and R. D. Kornberg, Two-dimensional crystals of cholera toxin B-subunit-receptor complexes: Projected structure at 17Å resolution, *Proc. Natl. Acad. Sci.* 83:8585 (1986).
4. J. P. Robinson, M. F. Schmid, D. G. Morgan, and W. Chiu, Three-dimensional analysis of tetanus toxin by electron crystallography, *J. Mol. Biol.* 200:367 (1988).
5. H. O. Ribi, P. Reichard, and R. D. Kornberg, Two-dimensional crystals of enzyme-effector complexes: Ribonucleotide reductase at 18-Å resolution, *Biochemistry* 26:7974 (1987).
6. R. Blankenburg, P. Meller, H. Ringsdorf, and C. Salesse, Interaction between biotin lipids and streptavidin in monolayers: Formation of oriented two-dimensional protein domains induced by surface recognition, *Biochemistry* 28:8214 (1989).
7. H. Yoshimura, M. Matsumoto, S. Endo, and K. Nagayama, Two-dimensional crystallization of proteins on mercury, *Ultramicroscopy* 32:265 (1990).
8. S. A. Darst, H. O. Ribi, D. W. Pierce, and R. D. Kornberg, Two-dimensional crystals of *Escherichia coli* RNA polymerase Holoenzyme on positivly charged lipid layers, *J. Mol. Biol.* 203:269 (1988).
9. P. Fromherz, Electron microscopic studies of lipid protein films, *Nature* 231:267 (1971).
10. T. Furuno, H. Sasabe, and K. M. Ulmer, Binding of ferritin molecules to a charged polypeptide layer of poly-1-benzyl-L-histidine, *Thin Solid Films* 180:23 (1989).
11. T. Furuno, H. Sasabe, R. Nagata, and T. Akaike, Studies of Langmuir-Blodgett films of poly(1-benzyl-L-histidine)-stearic acid mixtures, *Thin Solid Films* 133:141 (1985).
12. R. Henderson, J. M. Baldwin, T. A. Ceska, F. Zemlin, E. Beckmann, and K. H. Downing, Model for the structure of bacteriorhodopsin based on high-resolution electron cryo-microscopy, *J. Mol. Biol.* 213:899 (1990).
13. L. Reimer, Scanning Electron Microscopy. Physics of image formation and microanalysis, in: "Springer Series in Optical Sciences," vol. 45 Springer-Verlag (1985).
14. T. Nakadera, A. Mitsushima, and K. Tanaka, Application of high-resolution scanning electron microscopy to biological macromolecules, *J. Microsc.* 163:43 (1991).
15. S. Inaga, H. Osatake, and K. Tanaka, SEM images of DNA double helix and nucleosomes observed by ultrahigh-resolution scanning electron microscopy, *J. Electron Microsc.* 40:181 (1991).
16. H. Gross, T. Muller, I. Wildhaber, and H. Winkler, High resolution metal replication, quantified by image processing of periodic test specimens, *Ultramicroscopy* 16:287 (1985).
17. E. C. Theil, Ferritin: Structure, gene, regulation, and cellular function in animals, plants, and microorganisms, *Ann. Rev. Biochem.* 56:289 (1987).

18. M. R. N. Murthy, T. J. Reid III, A. Sicignano, N. Tanaka, and M. G. Rossmann, Structure of beef liver catalase, *J. Mol. Biol.* 152:465 (1981).
19. T. Furuno, K. M. Ulmer, and H. Sasabe, Scanning electron microscopy of negatively stained catalase on a silicon wafer, *Microsc. Res. Tech.* 21:32 (1992).
20. R. Durr, R. Hegerl, S. Volker, U. Santarius, and W. Baumeister, Three-dimensional reconstruction of the surface protein of *pyrodictium brockii*: Comparing two image processing strategies, *J. Struct. Biol.* 106:181 (1991).
21. D. L. Misell, Image analysis, enhancement and interpretation. in: "Practical Methods in Electron Microscopy," Vol.7, North-Holland (1978).

CHARACTERIZATION OF A ß-GLUCAN PARTICLE USING THE SCANNING TUNNELING AND ATOMIC FORCE MICROSCOPES

Antonio A. Garcia*, Patrick Oden**, Uwe Knipping**, Gary Ostroff***, and Roberta Druyor*
*Department of Chemical, Bio & Materials Engineering, Arizona State University, Tempe, AZ 85287-6006
**Department of Physics, Arizona State University, Tempe, AZ 85287-1504
*** Alpha-Beta Technology, Two Biotech Park, 373 Plantation Street, Worcester, MA 01605

The Scanning Tunneling (STM) and Atomic Force Microscopes (AFM) can routinely image uncoated Adjuvax™ particles immobilized on graphite under aqueous solution. Adjuvax™ is a particle derived from yeast cells which is mainly composed of ß-glucan, but also contains chitin. Particle sizes from the AFM, laser light scattering, and the light microscope are larger than particle sizes from STM. This result suggests that the tunneling probe compresses particles during imaging. Yet, the STM can provide detailed images of the Adjuvax™ surface electronic topography/morphology, substantiating theoretical predictions of an interwoven, mat-like structure for the fully hydrated particle. After treatment with ß-glucanase (gluculase, ICN Biomedicals Inc., Costa Mesa, CA), STM images of the treated particles show a disruption of the "pleated" morphology and a presence of bundles of strand-like structures.

STM AND AFM IMAGING OF BIOPOLYMERS

The Scanning Tunneling and Atomic Force Microscopes are capable of determining nanometer-scale structure of surfaces in air or in solution. There has been much interest in the use of these probe microscopes for determining the structure of adsorbates, especially uncoated, hydrated biological colloids. The STM has generated topographical images of biological molecules[1] such as DNA,[2-3] RNA,[4] and fibronectin,[5] and complexes such as DNA-RecA,[6] microtubules,[7] tobacco mosaic virus,[8] phage Φ29,[9] and a bacterial wall.[10] Some of these images were obtained by coating with a conducting film. However, recent work has shown that biological molecules can be imaged without coatings in a vacuum, in air, or under water. Angstrom resolution of uncoated DNA on a graphite substrate in a vacuum has been recently demonstrated.[11] The recently developed Atomic Force Microscope provides a better understood mechanism for image generation of biological molecules, and allows for the imaging of nonconductive substrates and adsorbates. AFM images of a DL-leucine crystal,[12] immunoglobulins,[13] blood cells,[14] bacteria,[15] bacteriorhodopsin,[16] and DNA[17] have been published.

However, one of the principle difficulties blocking the wide-spread use of both STM and AFM for uncoated adsorbate imaging is the development of a broadly applicable technique for tight binding from solution onto suitable substrates. Many researchers recognize that successful STM/AFM imaging of adsorbates under water requires relatively strong forces in order to eliminate the possibility that the adsorbate would be swept away by the probe. In STM the tunneling probe can exert a force on the substrate on the order of 10^{-7}-10^{-6} N when imaging graphite in air with a tungsten[18] or a platinum-rhodium tip.[19] The AFM can be operated with as low a force as 10^{-9} N. However, even a nanoNewton of force is substantial when it acts over a small area. Persson[20] suggests that if forces greater than 10^{-11} N are used in scanning probe microscopy, deformation of biomolecules is likely.

In order to image Adjuvax™, we utilized the inherent hydrophobicity of the material to immobilize the particles onto a graphite surface (highly oriented pyrolytic graphite: HOPG). In experiments with the AFM we have verified the need for a hydrophobic surface such as graphite since imaging with mica as a substrate proved futile due to scanning probe "sweeping" of the weakly bound particles.

HYDROPHOBIC INTERACTIONS

The forces between two crossed cylinders of DODA-coated mica have been measured[21-23] using the surface force apparatus developed by Israelachvili.[24] These studies yielded an attractive force curve with normalized force (F/R) reaching magnitudes greater than -0.1 N/m for surface separations less than about 2 nm. It can be shown that for two cylinders of equal radii, the interaction of two crossed cylinders is the same as that for a sphere approaching a flat plate where the sphere has the same radius as the cylinders. Thus for a sphere having a radius of 100 nm, the resultant hydrophobic attractive force would be greater than 10^{-5} N for surface separations less than 2 nm. Eriksson and colleagues[25] have proposed a mean-field theory describing the hydrophobic attractive forces which agrees well with the experimental results and predicts a more rapidly increasing force at small separation distances than the function proposed by Rabinovich and Derjaguin.[26] It would not be correct to take the results obtained by coating mica spheres and universally apply it to all hydrophobic sphere/plate interactions. However, these results suggest that hydrophobic forces may be quite substantial for hydrophobic biopolymers with mean radii of 10 nm or greater or for hydrophobic colloidal particles. The resultant hydrophobic forces would be quite sizable for hydrophobic particles with diameters in the range of most Adjuvax™ particles (i.e. 3-5 μm). However, weaker hydrophobic forces will result if surface charges are present. In fact, Eriksson and colleagues[25] point out that the detailed structure of the mica surfaces greatly influences the measured hydrophobic attractive forces. Thus, although we cannot calculate the hydrophobic force between an Adjuvax™ particle and a graphite surface, the above cited research suggests that hydrophobic forces may be strong enough to keep the particles "fixed" on the surface during probe microscopy scanning.

EXPERIMENTAL

Adjuvax™ is a ß-glucan microsphere manufactured by Alpha-Beta Technology

(Worcester MA) as a vaccine adjuvant.[27-28] Adjuvax™ is prepared from the cell walls of *Saccharomyces cerevisiae* (Baker's Yeast) by a process which retains the size and 3-dimensional morphology of the yeast cell. The hollow yeast "cell ghost-like" particles are composed of >95% ß-glucan and >3% chitin.

For both the STM and AFM experiments, dried Adjuvax™ particles were suspended in a pH=7.2 phosphate buffer containing sodium azide. Sonication dispersed the particles and minimized the occurrence of aggregates. A range of Adjuvax™ concentrations (0.6-1 mg/ml) were used in the STM experiments. The AFM was difficult to operate at concentrations above 30 mg/ml because of laser light scattering and particle agglomeration on the graphite surface caused by multilayer adsorption.

The Scanning Tunneling Microscope used was based on a prototype designed at Arizona State University. A simple wet chemical cell (120 µl capacity) was fashioned by cutting a polypropylene microcentrifuge tube and placing it over the HOPG Grade ZYB (Union Carbide, OH) substrate after applying apiezon wax around the tube to prevent liquid leakage. To operate the STM in aqueous solutions, a platinum iridium tip coated with apiezon wax[29] was used. Images of Adjuvax™ particles were collected at various tip bias voltages and tunneling currents, although many images were obtained at -300 mV tip bias and a tunneling current between 0.2-0.5 nanoAmperes. A scanning head capable of a maximum z extension range of 3100 nm and a maximum x-y scan range of 75 µm was primarily used.

A Nanoscope II (Digital Instruments, Santa Barbara, CA) was used for the AFM experiments. The Nanoscope II AFM glass electrochemical cell of 150 µl capacity housed the Adjuvax™ suspension; no electrochemistry was performed. As with the STM experiments, HOPG Grade ZYB was used as the substrate. A 200 µm thin leg cantilever with a silicon nitride tip (Digital Instruments, Santa Barbara, CA) was mounted on a piezo tube with a maximum x-y scan range of 160 µm. The hard-core repulsive force curve was optimized so that about 2 nanoNewtons of force was used during constant force mode scanning. Images were collected in the constant force mode.

GENERAL OBSERVATIONS ON STM/AFM IMAGING OF ADJUVAX™

STM images of scan areas from 400-4900 μm^2 typically do not show as high surface coverages as with the AFM. Moreover, using bulk liquid-phase concentrations which would yield multilayer coverages of Adjuvax™ do not result in STM images showing multi, mono or even near monolayer coverage. On the other hand, AFM experiments have yielded images with near monolayer coverage over as much as 10,000 μm^2. This discrepancy is probably due to two factors: lower tip-induced forces in AFM as compared to STM; and the physics of tunneling over an "insulator". Both factors involve forces that result in either the STM probe "sweeping" away particles from the surface or particle collapse or outright destruction due to high pressures applied by the probe to the particles. Since a ß-glucan particle is thought to be an electrical insulator, in order to maintain a constant tunneling current the probe either removes the particle from the surface or induces electron tunneling by some currently unknown method. One mechanism for tunneling through an insulator proposed by Lindsay and colleagues[30] is based on electronic modifications caused by tip-induced distortion. This tunneling theory is proposed to explain uncoated DNA imaging in solution. If insulator particles with dimensions on the order of 3-5 µm in diameter must be compressed in order to induce tunneling currents, the shear forces would be substantial and would possibly result in measurable distortions of particle morphology as well as the clearing of particles from the surface. It is evident from the work of various researchers that STM images cannot be interpreted solely as surface morphology. Because of this we will refer to STM images of Adjuvax™ particles as being images of "surface features" or "structures" with the quotes indicating that the images are really a combination of electronic topography and morphology.

Although AFM experiments produce images of a larger number of particles per unit surface area than STM (see Figures 1a, 1b and 1c) and the contrast mechanism for insulator imaging is much better understood in AFM, the STM images yield information of what appears to be fine surface features on the Adjuvax™ particles. The AFM images are generally featureless (see Figure 1b) whereas the STM images show fine particle "structure" even at relatively low resolution scans (see Figure 2). Resolution limitation on the AFM may well be due to particle elasticity.

We are currently running AFM experiments at higher contact forces to see if surface features can be imaged. STM imaging of Adjuvax™ surface features may be facilitated by particle collapse due to high pressures induced by the probe.

We have run experiments with both the STM and AFM where particles are adhering to the substrate in real-time as evidenced by an increase in the number of particles on the surface observed by reviewing repeat scans of the same area. Also, during AFM experiments the particles adhere more strongly with time since the ability to image particle fields improves if an incubation period precedes scanning. A possible scenario that could explain these observations is as follows. First, when the probe achieves contact or tunneling with the surface, the adsorbed layer (without an incubation period) is disturbed by the approach. After a period of time whereby stable scanning over an area is achieved, particles diffuse to the surface and adsorb via hydrophobic interactions. Spreading of the Adjuvax™ particles on graphite follows adsorption, just as proteins have been found to change their tertiary structure and spread onto a surface upon hydrophobic adsorption. In using an incubation period before contact of the probe with the surface, particle spreading allows a strong attachment to develop that is not disturbed by the probe approach to the surface. The main advantage for utilizing hydrophobic adsorption in a real-time mode over imaging a coated or fixed sample is that (in either STM or AFM) particles can migrate back to the surface if the probe disturbs the deposition layer. Moreover, the dynamics of hydrophobic adsorption/desorption can be followed with this method.

STATISTICAL EVALUATION OF PARTICLE SIZES FROM STM/AFM IMAGES

The hydrophobic adsorption immobilization method allows reproducible STM and AFM imaging of Adjuvax™ particles. However, there was evidence of tip induced particle movement in STM since near monolayer, monolayer, or multilayer surface coverages were not observed with the STM even though sufficiently concentrated solutions were used. Particle "sweeping" is more easily minimized with the AFM allowing for the imaging of higher surface concentrations. From a total of ten separate STM imaging experiments, the x-y size information of 91 particles is summarized in Figure 3. Also shown in Figure 3 are AFM x-y data based on 113 particles and size distribution data from laser light scattering experiments. Using a public domain software package,[31] the 2-dimensional particle areas from the STM and AFM scans were compiled. The size distribution is given in Figure 3 in terms of a normal probability plot based on equivalent spherical diameters. According to the STM data, 90% of the particles have an equivalent diameter of less than 4 µm, while

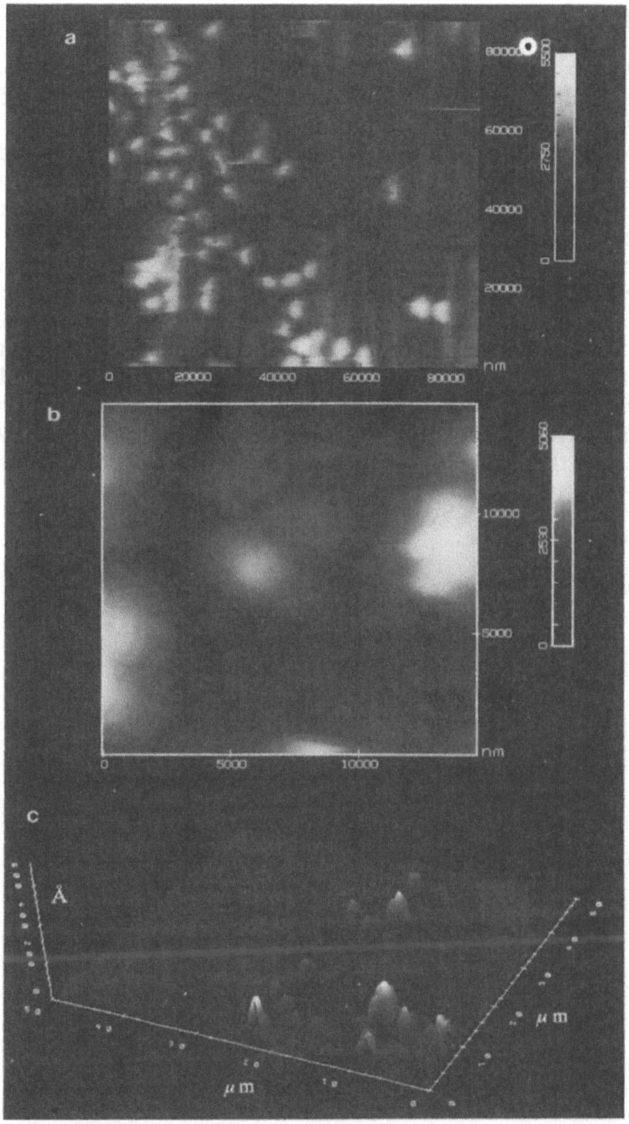

Figure 1a. AFM Image of Adjuvax™ particles showing high surface coverage. The verticle streaks are due to the cantilever "crashing" into the particles because the tip is offset from the end of the cantilever.

Figure 1b. AFM image of Adjuvax™ particles showing featureless surface morphology.

Figure 1c. STM image of field of Adjuvax™ particles.

both the AFM and light scattering data give much larger particle sizes. Although the discrepancy is not great, the AFM gives larger particle sizes than light scattering probably due to either probe distortion causing the particles to be flattened or to errors in measuring equivalent diameters due to image streaking. Several factors may contribute to the smaller particle sizes from STM: (1) the maximum z-extension away from a surface on the STM PZT tube was 1.5 μm while the AFM PZT tube has a maximum z-extension away from a surface of 2.5 μm; (2) low tunneling contrast on the particle periphery may lead to lower particle x-y area; (3) STM tip-induced particle shrinkage. Due to the poor understanding of electron tunneling with so-called insulators, the relative importance of these effects on the STM size distribution is difficult to ascertain.

Evidence of particle image distortion by STM is more pronounced when analyzing the z-data or contrast information. A normal probability plot of the STM data from 91 particles is given in Figure 4. The average particle thickness by STM is 0.24 μm with a standard deviation of 0.23 μm. This is much lower than the typical 1.5 μm thickness from AFM. The relatively large standard deviation in STM maximum particle contrast suggests that the tunneling contrast mechanism produces quite variable results. This observation corroborates theory proposed by Garcia and Garcia[32] who argue that tunneling with insulators such as biological structures will

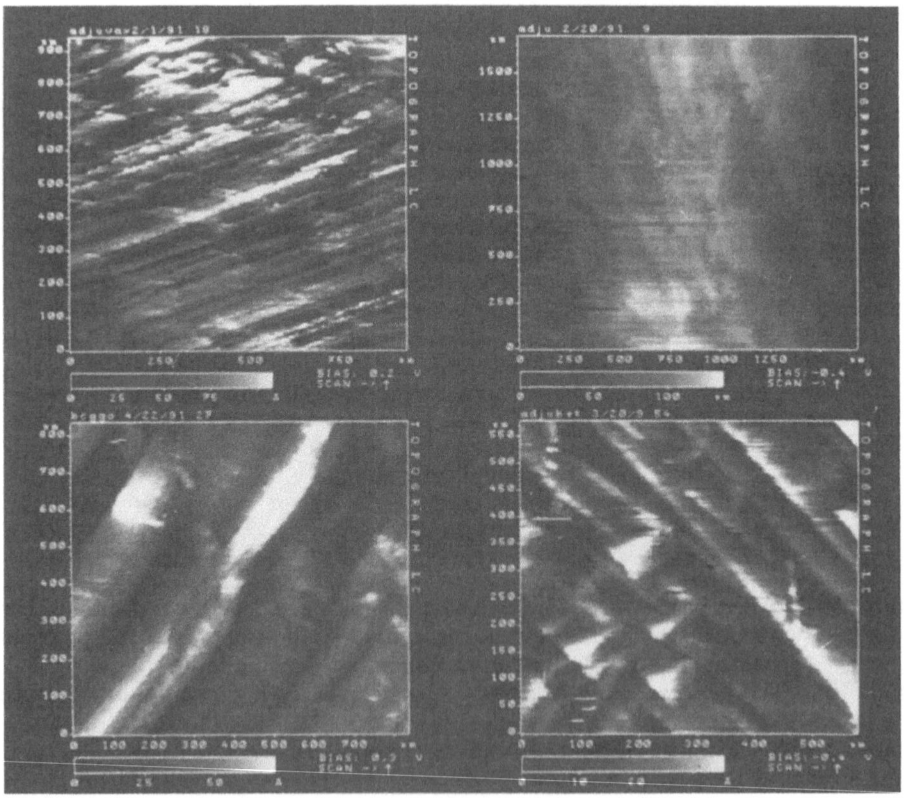

Figure 2. STM images of Adjuvax^{TM} surface features.

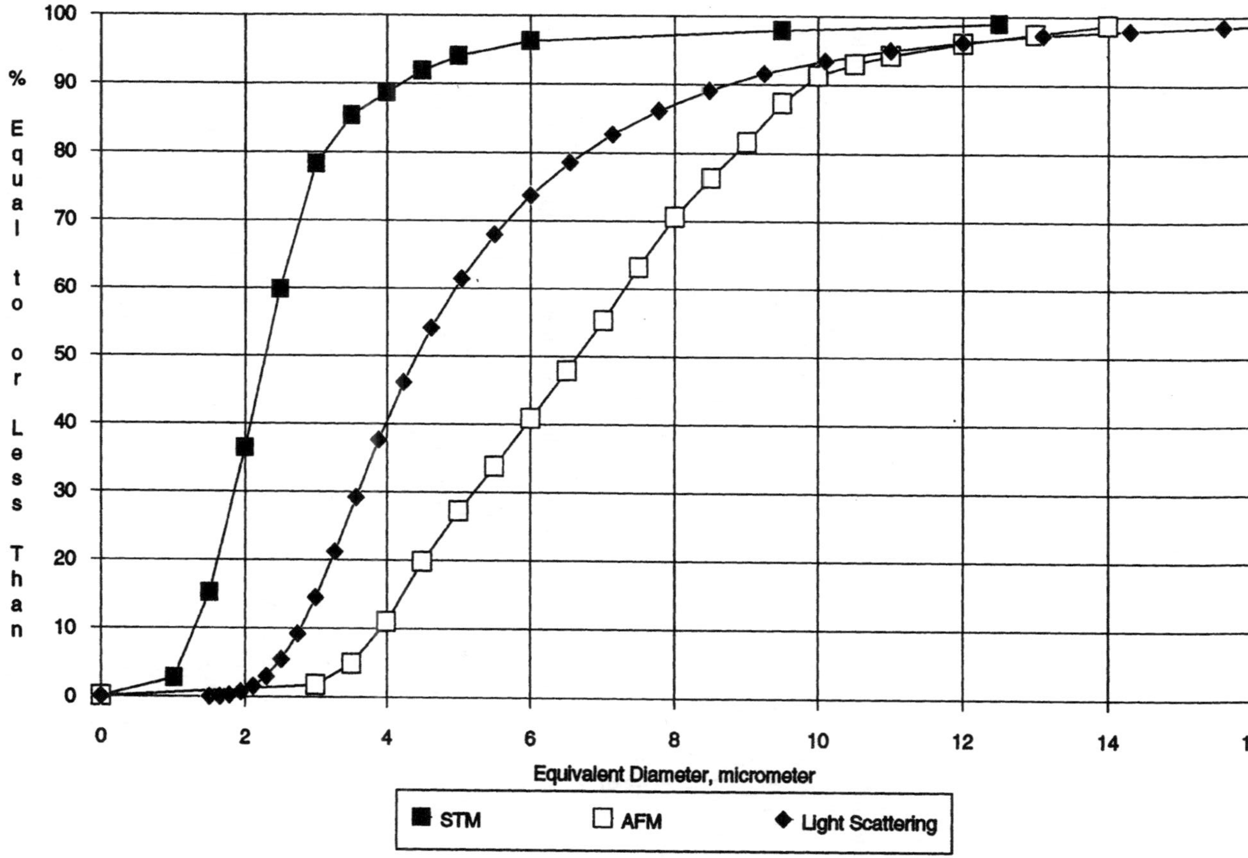

Figure 3. Adjuvax™ particle size distributions from STM and AFM x-y data and light scattering.

produce widely varying contrast information. Although tip-induced "flattening" of the Adjuvax™ particles appears to be the norm in STM imaging, the particles can be rescanned many times without radically altering "surface features" or without overall particle degradation. In fact, tip-induced pressure on the particles may help bring out "surface features" as discussed in the next section. However, in order for STM imaging of uncoated, hydrated biopolymers to realize its potential for providing true morphological information, an understanding of "insulator" tunneling physics is needed. We are currently involved in exploring a wide range of tunneling conditions such as tip/substrate bias voltage, tip forces, buffer composition, tunneling current, and tip geometry in order to assess the importance of these parameters with respect to the maximum tunneling contrast of Adjuvax™.

INTERPRETATION OF DETAILED "SURFACE FEATURES" FROM STM IMAGES

Figure 2 summarizes results from four separate STM imaging experiments showing distinct "surface features" on Adjuvax™ particles. The scan on the upper left hand side was obtained with a scan head of 12 μm max x-y deflection while the other images were acquired with a 75 μm maximum x-y scan range head. All the images show a pleated electronic topography/morphology with distances between pleats ranging from about 50 - 150 nm. The region between the main or "mother" particle and a daughter yeast cell bud scar (Figure 5) shows a marked difference in morphology from the rest of the particle surface. Over a x-y scanning range from 500-2000 nm, this boundary region exhibits a criss-cross pattern with strands of approximately 30-50 nm in diameter. The strands in Figure 5 when compared to the "pleats" of Figure 2 suggest that the STM is imaging chitin within the chitin-rich bud scar in the ß-glucan cell wall microfibrils. The average length of the microfibrils cannot be determined from the STM images probably because the microfibrils form a "woven" network making it difficult to follow the strands by topographical imaging of an intact network. Because of the introduction of image "noise", higher resolution imaging of the individual microfibril strands (which could resolve the underlying molecular structure) is not currently possible. In the next section, the utility of STM imaging of the uncoated, hydrated topography of Adjuvax™ is extended to help understand the effect of enzyme treatment on particle morphology.

MORPHOLOGY OF ß-GLUCANASE TREATED ADJUVAX™

Limited digestion of Adjuvax™ with ß-glucanase (ß-1,3 and ß-1,6 glucanases) appears to disrupt some of the pleat "features" leaving a surface electronic topography/morphology which can contain bundles of strand-like subunits or an overall amorphous morphology. The images shown in Figure 6 are scans of some gluculase-treated Adjuvax™ particle surfaces. Individual fiber bundles are most prevalent in the lower left-hand image of Figure 6. Figure 7 contains a STM image of a gluculase-treated Adjuvax™ particle showing a tile-like pattern towards the right-hand-side. The individual tile-like subunits are approximately 10-20 nm in length. An interpretation of the "structural" changes shown in these images as compared to the images in Figure 2 is that the ß-glucan is largely responsible for the regular pleated structure. Since gluculase disrupts ß-1,3 and ß-1,6 glycosidic bonds thereby depolymerizing ß-glucan, the surface of gluculase-treated Adjuvax™ exhibits enrichment in chitin. Similar strand-like features were observed in the chitin-rich bud

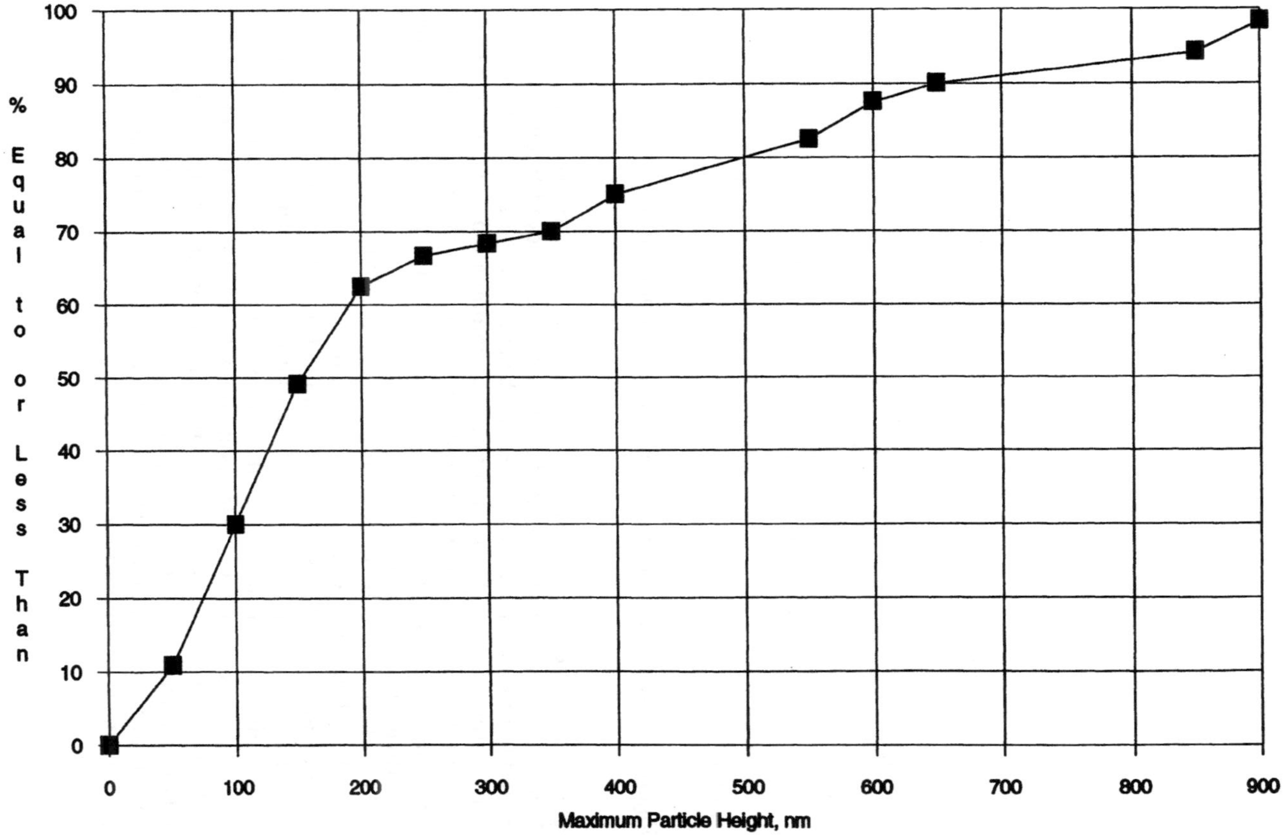

Figure 4. Adjuvax™ maximum z-data (contrast) from STM images.

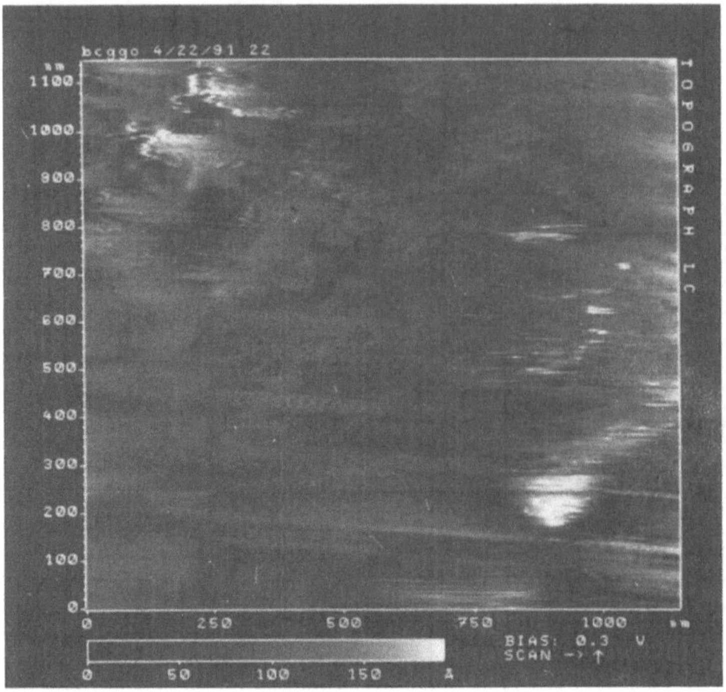

Figure 5. STM image of interface between main particle and bud.

scar regions of untreated Adjuvax™. Thus, the electronic topographical/morphological differences in the treated Adjuvax™ are believed to be attributed to the inherent differences between a surface enriched in chitin versus one dominated by a ß-glucan structure.[33] Light microscopic examination of untreated and gluculase-digested Adjuvax™ particles confirms the enhancement in the prominence of the chitin bud scars, but light microscopy cannot detect any differences in surface structure, particle size, or shape.

Even with tip-induced particle distortion, the STM can be used to follow changes in the electronic topography/morphology due to surface treatment of Adjuvax™ and to aid in the study of cell wall structure. STM images of BCG cells[34] also show that biological surfaces do not all yield the same "features", but that in fact imaging due to tunneling is capable of giving distinct morphology depending mostly on the surface structure. However, tip-induced distortions and the poorly understood "insulator" tunneling mechanism do not allow for straight-forward image interpretation. Further studies of Adjuvax™ samples which have had carefully controlled surface modifications will provide heuristics for uncoated biological structure STM image interpretation.

CONCLUSIONS

Scanning Tunneling and Atomic Force Microscopy can image uncoated, ß-glucan particles which are immobilized onto a hydrophobic surface from an aqueous solution. The AFM yields x-y and z image data that compare reasonably well with

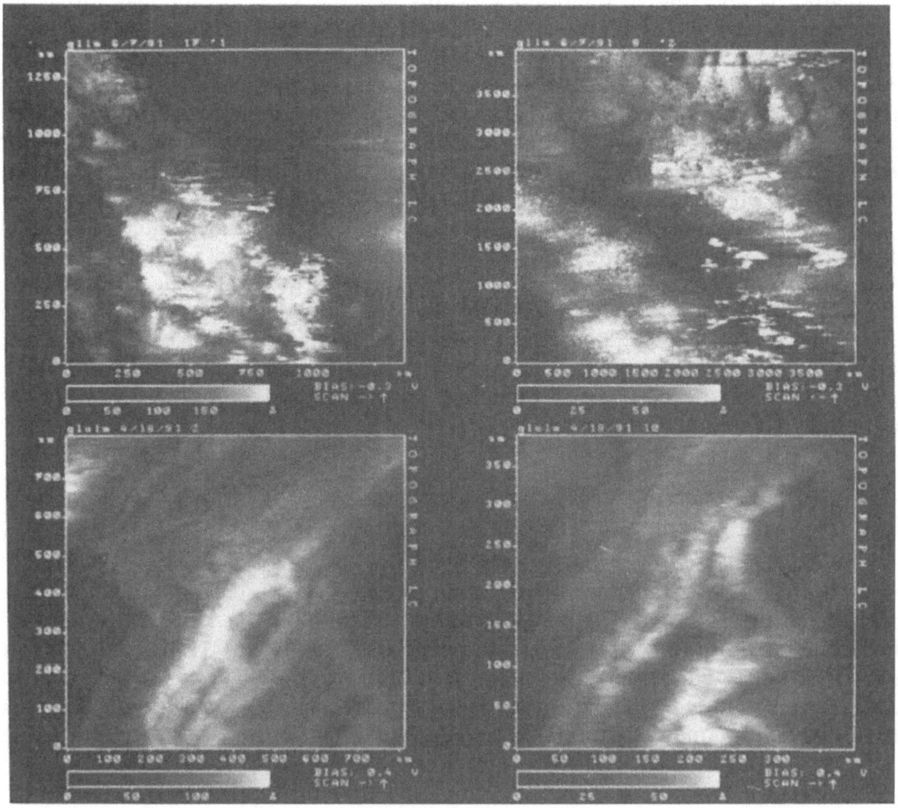

Figure 6. STM images of the surfaces of gluculase treated Adjuvax™ particles.

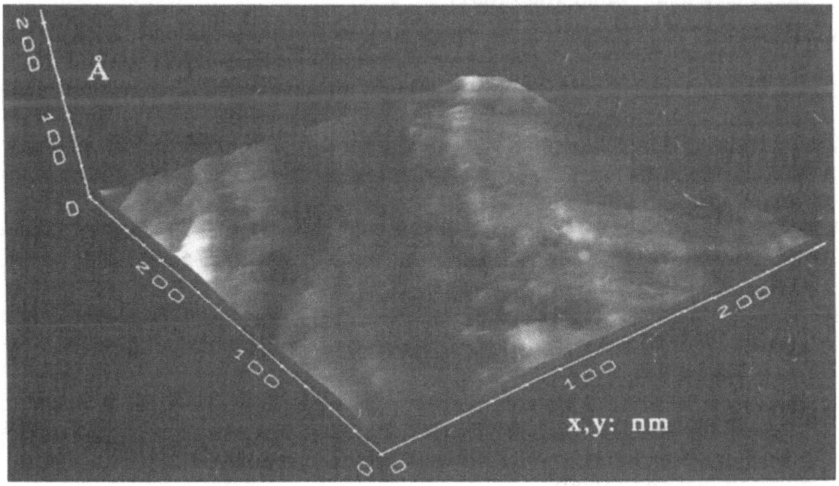

Figure 7. STM image shown as a 3-D projection of a gluculase treated Adjuvax™ particle surface.

light scattering and optical microscopy results, but the STM tip induces forces that distort the particle giving images that are too small in thickness and 2-D projected diameter. However, the AFM does not resolve the detail structure of the Adjuvax™ surface whereas the STM routinely provides images of the particle "surface". The "surface features" obtained by STM corroborate previous interpretation of yeast cell wall structure as an interwoven network of ß-glucan within a chitin microfibrillar matrix. STM images can be used to study surface morphological differences between untreated and enzymatically digested Adjuvax™ particles.

The relative ease with which STM can obtain surface electronic topography/ morphology of Adjuvax™ in an aqueous environment portends well for its use in understanding the nanoscale structure of relatively large (μm scale) biopolymers. Moreover, the STM could be a useful tool for understanding subtle electronic topographical/morphological changes produced in biomaterials processing. A critical need in developing the STM as a tool for biomaterials characterization is an understanding of the physics of tunneling with uncoated biopolymers which are traditionally viewed as "insulators".

ACKNOWLEDGMENTS

We acknowledge partial support from the Office of the Vice-President for Research at Arizona State University. One of us (A.A. Garcia) also acknowledges support from the National Science Foundation (BCS-90-09301). The AFM data was collected in the laboratory of S. M. Lindsay supported by the NSF (DIR 89-20053) and the ONR (N00014-90-J-1455). We are grateful to Lee Ann Yanez (supported by the Maricopa Community College Minority Regional Center) who compiled the statistical data on particle diameters from STM and AFM x-y image data.

REFERENCES

1. A. Engel, Biological applications of scanning probe microscopes, *Ann. Rev. Biophys. Chem.* 20:79-108 (1991).
2. S. M. Lindsay, T. Thundat, L. A. Nagahara, U. Knipping, and R. L. Rill, Images of the DNA double helix in water, *Science* 244:1063 (1989).
3. D. D. Dunlap and C. Bustamante, Images of single-stranded nucleic acids by scanning tunneling microscopy, *Nature* 342:204 (1989).
4. S. M. Lindsay, L. A. Nagahara, T. Thundat, and P. Oden, Sequence, packing and nanometer scale structure in STM images of nucleic acids under water, *J. Biomol. Struc. Dyn.* 7:289 (1989).
5. R. Emch, X. Clivaz, C. Taylor-Denes, P. Vaudauz, and P. Descouts, Scanning tunneling microscopy for studying the biomaterial-biological tissue interface, *J. Vac. Sci. Technol.* A8:655 (1990).
6. M. Amrein, R. Durr, A. Stasiak, H. Gross, and G. Travaglini, Scanning tunneling microscopy of recA-DNA complexes coated with a conducting film, *Science*, 243:1708 (1989).
7. S. Hameroff, Y. Simic-Krstic, L. Vernetti, Y. C. Lee, D. Sarid, J. Wiedmann, V. Eilings, K. Kjoller, and R. McCuskey, Scanning tunneling microscopy of cytoskeletal proteins: microtubules and intermediate filaments, *J. Vac. Sci. Technol.* A8:687 (1990).

8. A. Stemmer, A. Hefti, U. Aebi, U., and A. Engel, Scanning tunneling and transmission electron microscopy on identical areas of biological specimens, *Ultramicroscopy* 30:263 (1989).
9. A. M. Baro, R. Miranda, J. Alaman, N. Garcia, R. H. Binnig, Ch. Gerber, and J. L.Carracosa, Determination of surface topography of biological specimens at high resolution by scanning tunneling microscopy, *Nature* 315:253 (1985).
10. D. C. Dahn, M. O. Watanabe, B. L. Blackford, M. H. Jerico, and T. J. Beveridge, Scanning tunneling microscopy imaging of biological structures, *J. Vac. Sci. Technol.* A6:548 (1988).
11. R. J. Driscoll, M. G. Youngquist, and J. D. Baldeschwieler, Atomic-scale imaging of DNA using scanning tunneling microscopy, *Nature* 346:294 (1990).
12. J. A. N. Zasadzinski and P. K. Hansma, Scanning tunneling microscopy and atomic force microscopy of biological surfaces, *Ann. of N.Y. Acad. Sci.* 589:476 (1990).
13. J. N. Lin, B. Drake, A. S. Lea, P. K. Hansma, and J. D. Andrade, Direct observation of immunoglobin adsorption dynamics using the atomic force microscope, *Langmuir* 6:509 (1990).
14. P. K. Hansma, B. Northern, B. Dixon, and C. M. Peterson, Imaging molecules and cells with the atomic force microscope, in: "XIIth International Congress for Electron Microscopy", L. D. Peachey and D. B. Williams, eds., San Francisco Press, Inc. (1990).
15. S. A. C. Gould, B. Drake, C. B. Prater, A. L. Weisenhorn, S. Manne, H. G. Hansma, P. K. Hansma, J. Massie, M. Longmire, V. Eilings, B. Dixon Northern, B. Mukergee, C. M. Peterson, W. Stoeckenius, T. R. Albrecht, and C. F. Quate, From atoms to integrated circuit chips, blood cells, and bacteria with the atomic force microscope, *J. Vac. Sci. Technol.* A8:369 (1990).
16. K. H. Downing, H-J Butt, and P. K. Hansma, Imaging the membrane protein bacteriorhodopsin with the atomic force microscope, *Biophys. J.* 58:1473 (1990).
17. S. M. Lindsay, L. A. Nagahara, T. Thundat, U. Knipping, R. L. Rill, B. Drake, C. B. Prater, A. L. Weisenhorn, S. A. C. Gould, and P. K. Hansma, STM and AFM images of nucleosome DNA under water, *J. of Biom. Struc. & Dyn.* 7:279 (1989).
18. S. L. Tang, J. Bokor, and R. H. Storz, Direct force measurement in scanning tunneling microscopy, *Appl. Phys. Lett.* 52:188 (1988).
19. C. M. Mate, R. Erlandsson, G. M. McClelland, and S. Chiang, Direct measurement of forces during scanning tunneling microscope imaging of graphite, *Surf. Sci.* 208:473 (1989).
20. B. N. J. Persson, The atomic force microscope: can it be used to study biological molecules?, *Chem. Phys. Lett.* 141:366 (1987).
21. P. M. Claesson, C. E. Blom, P. C. Herder, and B. W. Ninham, Interactions between water-stable hydrophobic langmuir-blodgett monolayers on silica, *J. Coll. Inter. Sci.* 114:234 (1987).
22. H. K. Christenson and P. M. Claesson, Cavitation and the interaction between macroscopic hydrophobic surfaces, *Science* 239:390 (1988).
23. P. M. Claesson and H. K. Christenson, Very long range attractive forces between uncharged hydrocarbon and fluorocarbon surfaces in water, *J. Phys. Chem.* 92:1650 (1988).
24. J. N. Israelachvili and G. E. Adams, Measurement of forces between two mica surfaces in aqueous electrolyte solutions in the range of 0-100 nm, *J. Chem. Soc. Faraday Trans.* 174:975 (1978).

25. J. C. Eriksson, S. Ljunggren, and P. M. Claesson, A phenomoenological theory of long-range hydrophobic attraction forces based on a square-gradient variational approach, *J. Chem. Soc. Far. Trans.* 85:163 (1989).
26. Y. I. Rabinovich and B. V. Derjaguin, Interactions of hydrophobized filaments in aqueous electrolyte solutions, *Coll. Surf.* 30:243 (1988).
27. G. R. Ostroff, D. D. Easson, and S. Jamas, A new ß-glucan-based macrophage-targeted adjuvant, in: "Polymeric Drugs and Drug Delivery Systems," R. L. Dunn and R. M. Ottenbrite, eds., American Chemical Society, Washington DC (1991).
28. S. Jamas, G. R. Ostroff, and D. D. Easson Jr., Glucan Drug Delivery System And Adjuvant, Alpha-Beta Technology, Worcester, MA, U.S. Patent 5,032,401 (1991).
29. L. A. Nagahara, T. Thundat, and S. M. Lindsay, Preparation and characterization of STM tips for electrochemical studies, *Rev. Sci. Instrum.* 60:3128 (1989).
30. S. M. Lindsay, O. F. Sankey, Y. Li, and C. Herbst, Pressure and resonance effects in scanning tunneling microscopy of molecular adsorbates, *J. Phys. Chem.* 94:4655 (1990).
31. NIH Image, National Institutes of Health, Bethesda, Maryland (1991).
32. R. Garcia and N. Garcia, Electron conductance in organic chains: why are STM experiments possible on bare biological samples?, *Chem. Phys. Lett.* 173:44 (1990).
33. M. Kopecka, H. J. Phaff, and G. H. Fleet, Demonstration of a fibrillar component in the cell wall of the yeast saccharomyces cerevisiae and its chemical nature, *J. of Cell Bio.* 66 (1974).
34. A. A. Garcia, Use of colloid and interface chemistry to immobilize biopolymers for STM imaging, American Institute of Chemical Engineers Spring National Meeting, Houston, Texas (1991).

FURTHER READING (STM and AFM)

D. Sarid, "Scanning Force Microscopy: With Applications to Electric, Magnetic, and Atomic Forces," Oxford University Press (1991).

P. K. Hansma, V. B. Elings, O. Marti, and C. E. Bracker, Scanning tunneling microscopy and atomic force microscopy: application to biology and technology, *Science* 242:209-216, October 14 (1988).

BIOLOGICAL INTERFACES

INTRODUCTION

In order for microstructures that have been fabricated and characterized to be used in biology or for medical applications, they must be interfaced to biological systems. As one begins to expose these structures in biological regimes, new phenomena may result. Chemists, biologists, physicists, electronic engineers, and physicians are beginning to form teams that address this research area. This section contains articles that describe some recent work at the interface of small systems with the biology.

Schmitt et al. reports on the techniques required to functionalize surfaces for specific protein binding. The **Stayton et al.** paper on self-assembled monolayers of cytochrome describes intriguing ways in which genetic engineering can be of use in controlling the interface. The question of neural connectivity in the brain is addressed in the article by **Christie et al.** In this article a polymer based construct is proposed which has potential applicability in the implementation of interconnections in silicon-based neural networks. The last two papers in the section involve the study of lipid based interfaces. Lipid membranes are being continuously scrutinized as possible substrates for biosensors, controlled delivery systems, and osmotic membranes. The paper by **Todorov et al.** describes a study of bimolecular (black) lipid membranes as a function of pressure. **Turin** reports on the study of ion percolation through a membrane.

SPECIFIC PROTEIN BINDING TO FUNCTIONALIZED INTERFACES

F.-J. Schmitt*, R. Blankenburg+, L. Häußling+, H. Ringsdorf+, A.L. Weisenhorn§, P.K. Hansma§, D.E. Leckband‡, J.N. Israelachvili‡, and W. Knoll*

*Max-Planck Institut für Polymerforschung, Ackermannweg 10, D-6500 Mainz, F.R.G.
+Johannes-Gutenberg-Universität Mainz, Institut für Organische Chemie, J.-J. Becherweg 18-20, D-6500 Mainz, F.R.G.
§Department of Physics, University of California, Santa Barbara, CA 93106, U.S.A.
‡Department of Nuclear & Chemical Engineering, University of California, Santa Barbara, CA 93106, U.S.A.

We report on the characterization of specific binding reactions between streptavidin and biotinylated model membrane surfaces. Self-assembly techniques as well as the Langmuir-Blodgett-Kuhn method were employed to prepare reactive, functionalized surfaces on various solid supports in contact with the aqueous protein solution. Plasmon surface polaritons optical measurements as well as atomic force microscopy and studies with the surface forces apparatus give rather detailed information as to the streptavidin monolayer formation, the kinetics of this process (either binding site- or diffusion limited), the selectivity of the reaction at laterally heterogeneous membranes, and the involved intermolecular interaction potentials. The possibility to use these systems as universal binding matrices for other biotinylated species (e.g. antibodies) and the feasibility to carry out bioassays is shown.

INTRODUCTION

Surface science and supramolecular chemistry have become increasing and expanding fields of current research. Miniaturization of electronic devices (molecular electronics) and further technical progress in biosensor or diagnostic applications is strongly related to the control of molecular architecture. The properties of molecular systems depend not only on the single molecule but also on the order and structure of their supramolecular assembly. Starting with an ordered, two-dimensional system, it is then possible to build up three-dimensional structures, by adding further well defined layers.[1]

There are different approaches to obtaining these ordered layers: One possibility, used by all living species to form membranes, is the self-organization of

Synthetic Microstructures in Biological Research, Edited by
J.M. Schnur and M. Peckerar, Plenum Press, New York, 1992

lipids in water. Langmuir and Blodgett derived from this effect a method, that allowed them to organize amphiphilic molecules at the air-water interface and transfer these monomolecular films to solid supports (LB-technique).[2,3] A different method is the chemisorption of functionalized alkanes to surfaces to form self-assembled monolayers (SAM), e.g. alkanethiols or disulfides on gold or alkanesilanes on silicon oxide.[4]

We report here the use of these methods for the functionalization of interfaces and the investigations of specific recognition reactions of receptors and ligands at these interfaces.

As a model system we used the highly specific and strong interaction between biotin and streptavidin (K_{diss} = 10^{-15} mol/l). This well known system[5] is in many ways a good example for the biological "lock and key" interaction and can be chemically modified in various ways, e.g. biotin can be bound to lipids or to alkanethiols.[6,7]

Biotin was immobilized at the interface by the LB or self-assembly technique and a monolayer of streptavidin was specifically bound to it. Because of the two-fold symmetry of streptavidin, two of the four binding sites remain exposed to the surrounding aqueous medium.[8] They serve as receptors for further specific binding reactions with other biotinylated molecules. Since, in principle, most biomolecules can be biotinylated, the streptavidin monolayer acts as a universal binding film.[9,10]

In the first case we used a bifunctional biotin (bisbiotin, see Figure 1) to form intermolecular bonds between streptavidin molecules. Even more interesting is the possibility to attach a biotinylated antibody to streptavidin. This immobilization technique prevents the denaturation of the antibodies at the interface and this forms the basis of a sensitive bioassay to the corresponding antigen.

The preparation of such ultrathin films requires appropriate detection methods. We mainly used microscopic techniques, like fluorescence,[11] plasmon-surface-polaritons (PSP)[12] and atomic force-microscopies (AFM).[13] These surface sensitive techniques were used to detect small molecular changes with increasing resolution. Compared to laterally integrating methods, microscopy generally offers parallel data recording and processing of different reaction sites. As we will show, this is very important when dealing with molecular events in multicomponent systems. PSP spectroscopy was used to determine thickness changes, both in dynamic (kinetics) and static (saturation thickness) cases. With the surface forces apparatus (SFA)[14] we were able to measure directly the interaction between biotin and streptavidin. Attractive as well as repulsive forces between molecules or functionalized surfaces were obtained quantitatively.

MATERIALS AND METHODS

N-(6((biotinyl)amino)hexanoyl)dipalmitoyl-L-α-phosphatidyl-ethanolamine (lipid1) was purchased from Molecular Probes. Dipalmitoyl-phosphatidyl-ethanolamine (DPPE), Dilauroyl-phosphatidyl-ethanolamine (DLPE) and N-(7-nitrobenz-2oxa-1,3- diazol4-yl)-dipalmitoyl-L-α-phosphatidyl-ethanolamine (DPPE-NBD) were from Avanti Polar Lipids, Dimyristoyl-phosphatidyl-ethanolamine (DMPE) and biotin from Sigma and streptavidin, sulphorhodamine labelled streptavidin, the bisbiotin, the biotinylated monoclonal antibody against human chorio gonadotropin (HCG) and the antigen HCG were from Boehringer Mannheim. N-biotinyldimyristoyl-phosphatidyl-ethanolamine (lipid2), 11-mercapto-(8 biotinamido-4,7,dioxaoctyl-)-undecanoylamide (thiol1), 11-biotinyl-1-mercaptoundecane

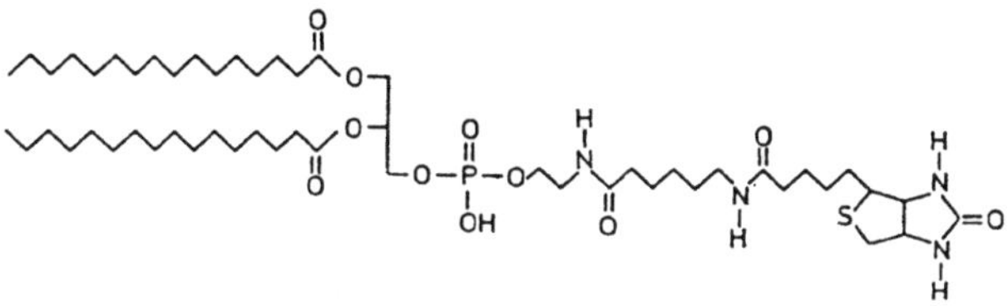

lipid1

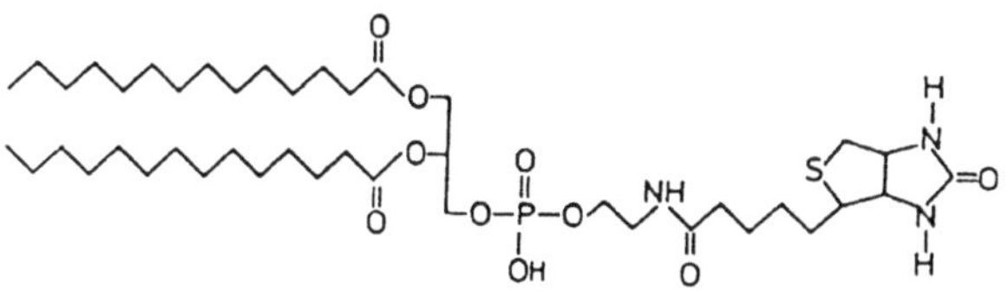

lipid2

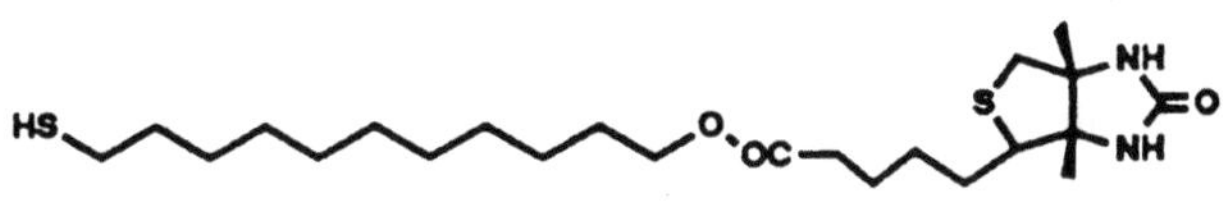

thiol2

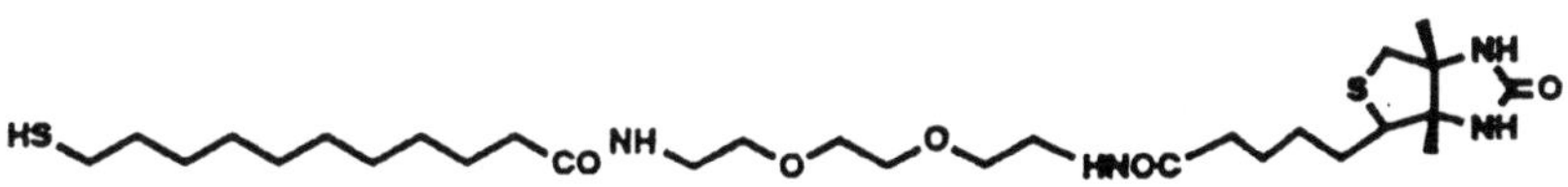

thiol1

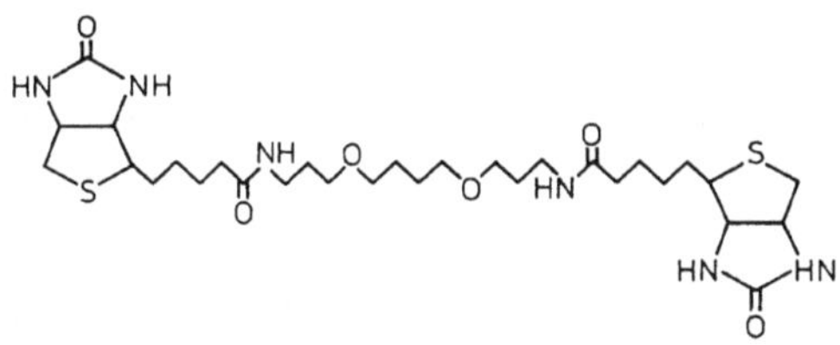

bisbiotin

Figure 1. Structural formulae of biotinlipids: lipid1, lipid2, the biotinylated alkythiols: thiol1, thiol2, and the bisbiotin.

(thiol2), ω-hydroxy-mercaptoundecane (C11OH) were synthesised as described previously.[6,7,15] For structural formulae see Figure 1.

Lipidmonolayer preparation was done either in a commercial film balance (Joyce Loebl) or in a home-made fluorescence film balance. The latter allows simultaneous measurement of π-A isotherms and microscopic observation.[11] Because most lipids are not fluorescent, a probe needs to be added, e.g. DPPE-NBD. It is known that a small amount of this probe, typically 0.5 mole%, does not alter the lipid layer properties, but allows for the visualization of the phase transition of the lipid from a liquid expanded to a liquid condensed state. The fluorescence probe is squeezed out of the growing crystalline lipid domains and enriched in the fluid lipid matrix, thus creating a contrast.[16] By changing the filter sets of the fluorescence microscope a double dye experiment can be performed. Besides the lipid fluorescence (DPPE-NBD) the sulphorhodamine fluorescence of the streptavidin can be observed. The subphase consisted of 0.5 M NaCl and the temperature was T=20°C.

Plasmon-surface-polaritons measurements were performed with a home-made apparatus, which was used for spectroscopy as well as microscopy. PSP spectroscopy measures the reflected intensity of a laser beam as a function of the angle of incidence Θ and the resulting reflexion curve is compared with Fresnel calculations.[17] The thickness of an organic layer can be obtained, with a resolution of 0.2 nm in aqueous media. In PSP microscopy a contrast is created if different spots of the sample are in resonance at different angles of incidence.[18,19] A lateral resolution of about 5μm is combined with the thickness resolution of 0.2 nm.

The monolayers were transferred from the air-water interface to a flow cuvette. The upper part was made of high index glass LASFN9 (Hellma) and precoated with 45 nm gold to carry the PSP modes. In the case of the lipid monolayers the bottom part was immersed in the subphase and the upper part was lowered nearly horizontally through the interface.[20] The closed flow cuvette was removed from the film balance and mounted into the PSP apparatus. Self-assembled monolayers of thiols were prepared by chemisorption from solution. Typically $5 \cdot 10^{-4}$ M solutions in $CHCl_3$:ethanol 1:1 were allowed to react for 6 hours with the freshly prepared gold surface. After extensive rinsing with pure solvent, the sample was dried, put together with the bottom part and mounted. Measurements in air or 0.5 M NaCl were carried out at T=20°C.

An atomic force microscope (Digital Instruments, Nanoscope II) was used to image the interfacial reaction with submicron resolution. The lipid monolayers were deposited onto molecularly flat mica, which had been hydrophobized with a crystalline layer of DPPE. A reaction chamber (as described in reference 21) allowed for online exchange of the 0.5 M NaCl subphase.

Forces measurements were made with the surfaces forces apparatus described elsewhere.[22] Several types of different surfaces were prepared:[23] symmetric biotinlipid, symmetric streptavidin and asymmetric biotinlipid streptavidin surfaces.[23,24] In this report we will focus on the asymmetric geometry. On a crystalline DPPE layer on mica, we deposited a mixture of 5 mole% lipid1 in DLPE. We chose DLPE as non-reactive lipid, because of its main phase transition of about 30°C.[23,24] By varying the temperature, it is thus possible to change the fluidity and lateral mobility of the lipids. One of the two modified surfaces was further incubated with streptavidin in a beaker and rinsed with pure solvent before mounting it into the surface forces apparatus. The experiments were carried out at T=33°C or 25°C, in phosphate buffer at pH=7.4 and in $1 \cdot 10^{-4}$ M NaCl. In all cases the solutions were saturated with DLPE to prevent desorption from the bilayer.

RESULTS AND DISCUSSION

Biotinylated lipids

Binary mixtures of the biotinylated lipids in DMPE with a composition of 0-20 mole% biotinylated lipid were spread at the air-water interface in the fluorescence film balance. If the fluorescence probe DPPE-NBD is added, the well known domain pattern is observed.[16] Subsequent recognition reactions can be carried out, by injecting sulphorhodamine-streptavidin underneath the reactive lipid monolayer. Only after adsorption to the interface the sulphorhodamine fluorescence can be seen. An example is shown in Figure 2. The same domain pattern, as in the lipid fluorescence is seen. Clearly sulphorhodamine labelled streptavidin only adsorbs to the lipid in the fluid phase. Given the specific interaction one can conclude that the biotinylated lipid is only accessible to streptavidin in the fluid lipid phase. This behaviour can be explained in a similar way to the enrichment of the fluorescence probe: the biotinylated lipid does not fit into the ordered array of DMPE molecules in the domain, is squeezed out of the growing crystals and hence enriched in the fluid phase. Thus, a phase separated system has been created with a reactive, highly specific phase and a reference phase.

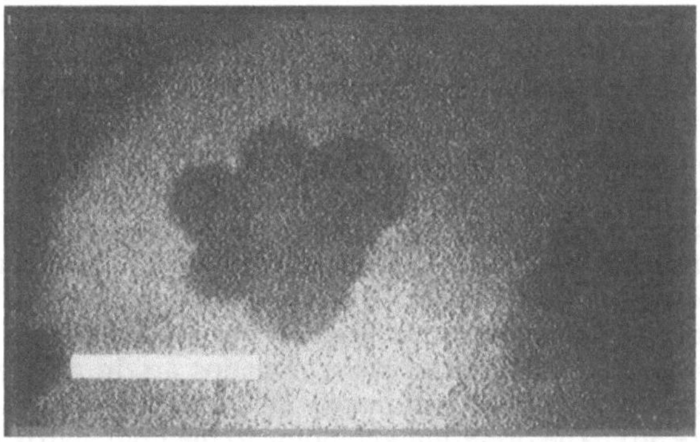

Figure 2. Micrograph of sulphorhodamin fluorescence from streptavidin adsorbed from subphase to a 5 mole% lipid1 in DMPE lipid layer in its phase coexistence region. The bar corresponds to 50 μm.

Thus, a phase separated system has been created with a reactive, highly specific phase and a reference phase.

Transfer of the lipid monolayers to a solid support offers certain advantages, e.g. the subphase can be exchanged quantitatively, the organic layer can be rinsed to remove unspecifically bound material and PSP optical experiments can be performed. Recently it has been shown that angle dependent PSP microscopy yields highly sensitive thickness information.[26] If PSP pictures are taken at different angles of incidence, the contrast changes gradually. These changes can be digitized by simple

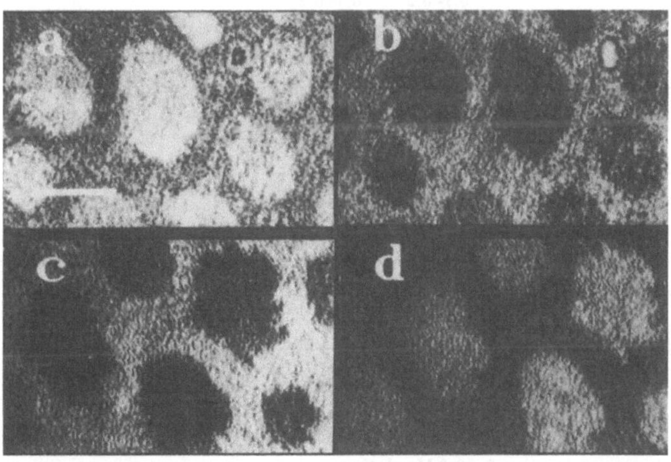

Figure 3. Plasmon-Surface-Polaritons micrograph of a transferred 5mole% lipid1 in DMPE lipid layer at an angle of incidence a) $\Theta=59.50°$ and b) $\Theta=60.10°$. Contrast inversion due to streptavidin binding in c) at $\Theta=59.70°$ and d) at $\Theta=60.50°$. The bar corresponds to 100 μm.

computer routines for each separate reaction site. The resulting plot of the obtained gray values versus the angle of incidence corresponds to a reflexioncurve. As with the PSP spectroscopy, such a curve can be compared to Fresnel calculations and thickness changes can be obtained for different spots of the same sample.

In the transferred lipid monolayer the same domain structure as at the air-water interface can be seen, but this time without any label. Figure 3a) shows the lipid matrix in resonance at $\Theta=59.50°$ and Figure 3b) the domain at $\Theta=60.10°$. After the characterization of such a reactive lipid layer is completed, the subphase is exchanged against a solution containing streptavidin. Figures 3 c) and d) show the changes after the streptavidin binding is completed: Figure 3c) was taken at $\Theta=59.70°$ and Figure 3 d) at $\Theta=60.50°$. Note that now the domains are in resonance (dark) at a smaller angle (Figure 3c) and the lipid matrix plus streptavidin is in resonance at a higher angle (Figure 3d). The preferential binding to the lipid matrix leads to contrast reversal. The obtained thickness increases for the different compositions and phases are given in Table I. The maximum thickness of the bound streptavidin layer is 3.0 nm, calculated with a refractive index of $n=1.5$, as obtained by optical interferometry with the SFA. This value is independent of the used mole fraction and streptavidin concentration and compares quite well with X ray-[25] and electron

Table I

Thickness increase on streptavidin binding, as determined by PSP microscopy with an error of ± 0.2 nm. Experiments indicated with * were not rinsed.

Lipidlayer in phase coexistence	Streptavidin thickness [nm] $5 \cdot 10^{-7}$ M solution		matrix time constants [min]
	matrix	domain	
DMPE	0.1–0.2 0.6–0.8*	0.1–0.2 0.7–0.9*	–
5mole%lipid1/DMPE	3.4*	1.8*	20, 60
10mole%lipid1/DMPE	3.0	0.6–0.7	60, 90
5mole%lipid2/DMPE	3.4*	1.8*	12
	Streptavidin thickness [nm] $6 \cdot 10^{-8}$ M solution		
5mole%lipid1/DMPE	3.0	1.4	60

diffraction data.[8] The domains also show significant thickness changes, which are further resolved in the AFM experiments (see below). As a control experiment we deposited a pure DMPE layer. The gold surface was almost completely passivated against unspecific adsorption of streptavidin by the non-reactive lipidlayer.

Observing the reflected intensity at a fixed angle of incidence, kinetic data of the binding are obtained.[27] Assuming a binding site limited reaction, one expects an exponential increase of the bound amount A with time t. Such a kinetic behaviour can be described by equation (1):

$$A(t) = A(t\infty)\cdot[1-\exp(-t/\tau)] \qquad (1)$$

A half logarithmic plot of the data should yield a straight line, from which the time constant τ is obtained. According to the model τ is given by: $(k_{ass}\cdot c_0 + k_{ass}\cdot K_{diss})^{-1}$, where K_{diss} is the dissociation constant, k_{ass} is the rate constant of the binding reaction and c_0 is the bulk streptavidin concentration.[28] As shown in Figure 4, we obtain instead a biphasic behaviour, with a fast process τ_1=20min and a slower one with τ_2=60min. Although equation (1) does fit our data only for small times t, it is a good approximation. Comparison with the literature values[29] shows, that the interfacial reaction is much slower than in solution. In the case of the 10 mole% mixture this effect is even more pronounced, with increased time constants of τ_1=60min and τ_2=90min. The corresponding experiment with the 5 mole% of lipid2 in DMPE, yields the same saturation thickness, but with a time constant τ_1=12min (see Table I). Note that the reaction of the biotinylated lipid without the additional spacer is surprisingly faster.

We conclude that the reaction rate is reduced when the interacting molecules are restricted to a 2-dimensional array. The protein can approach its receptor only from one direction and has to obtain the necessary orientation of its binding site. An increase of the receptor density from 5 to 10 mole% reduces the binding rate even more, because of increased steric hindrance of the competing receptors. It has been shown that a spacer is necessary to provide good binding between biotin and streptavidin.[6] Nevertheless the spacer should not be too long, since the number of the possible conformations increases with the spacer length: some of these conformations may be less appropriate for successful binding and thus decrease the binding rate. It can be assumed that an intermediate spacer length would provide complete binding and a fast reaction.

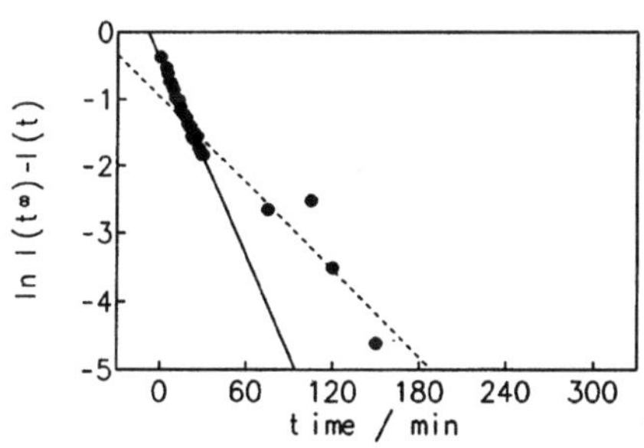

Figure 4. Streptavidin binding to a 5mole% lipid1 in DMPE layer from a 5×10^{-7} M bulk solution. The reflected intensities I(t∞)-I(t), which are directly proportional to the bound amount A(t∞)-A(t), are plotted on a half logarithmic scale. Note that only the binding to the lipid matrix is shown.

In another series of experiments a streptavidin concentration of $6 \cdot 10^{-8}$ M was applied. From Fick's 1st law the time Δt for a given mean displacement $\sqrt{(\Delta x)^2}$ and a diffusion coefficient D can be calculated:

$$\Delta t = \frac{(\Delta x)^2}{2D} \tag{2}$$

Knowing the amount of streptavidin needed to form a monolayer, the depth of bulk volume containing the necessary number of molecules can be calculated. Assuming a diffusion coefficient of $D = 5 \cdot 10^{-7} cm^2/sec$[29] one can obtain the time constant Δt for a diffusion limited binding process. For the concentration of $6 \cdot 10^{-8}$ M the time constant is $\Delta t_2 = 60$min., which is in absolute agreement to the measured value (see Table I). A solution of Fick's law is:

$$A(t) = d(t) \cdot c = c_0 \, 2/\sqrt{\pi}\sqrt{Dt} \tag{3}$$

where A(t) is the bound amount at time t and c is the concentration of streptavidin in the bound layer. Plotting the thickness increase d versus $\sqrt{t}$ should give a straight line, with the slope proportional to the diffusion coefficient D. As can be seen in Figure 5 this is true for small times t and D is obtained as $6 \cdot 10^{-7} cm^2/sec$, which is in good agreement with literature values.[30] For the higher concentration of $5 \cdot 10^{-7}$ M a time constant $\Delta t = 1$-2min is evaluated. The comparison to measured time constants as mentioned above, proves that the reactions at higher concentrations are not limited by diffusion.

Analogous experiments were carried out with the AFM.[31] A 5 mole% mixture of lipid1 in DMPE in its phase coexistence region was transferred onto crystalline DPPE to form a bilayer. The resulting surface geometry is schematically shown in Figure 6a. The corresponding AFM picture (Figure 6b) of a crystalline domain reveals, that small islands are incorporated. Their typical size of 1μm is smaller than the resolution of the PSP microscope, so their properties are averaged in the PSP experiments over the whole domain. The AFM experiments show that these fluid islands can bind streptavidin and thus decrease contrast between the matrix and the domain, see Figure 6c). A complete contrast reversal was observed, when the streptavidin binding is finished.[31] In the PSP measurements the streptavidin binding to the incorporated islands is attributed to the domain as a whole and thus interpreted as a thickness increase by binding to the crystalline domain. This leads to the relatively high values of streptavidin on the lipid domains, as mentioned in Table I.

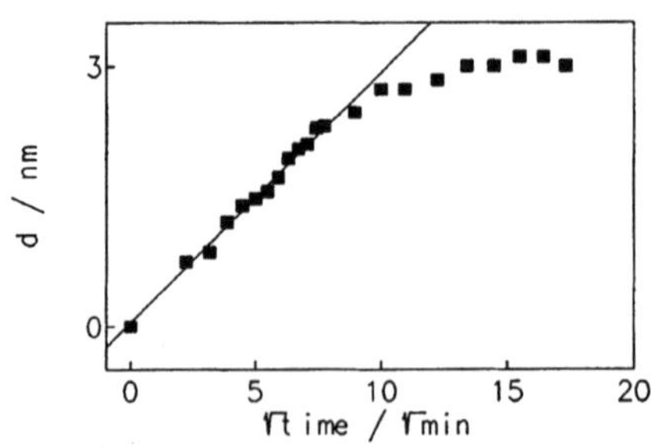

Figure 5. Streptavidin binding to a 5mole% lipid1 in DMPE layer from a 5×10^{-8} M bulk solution in a $\sqrt{t}$ plot. Note that only the binding to the lipid matrix is shown.

A

B

C

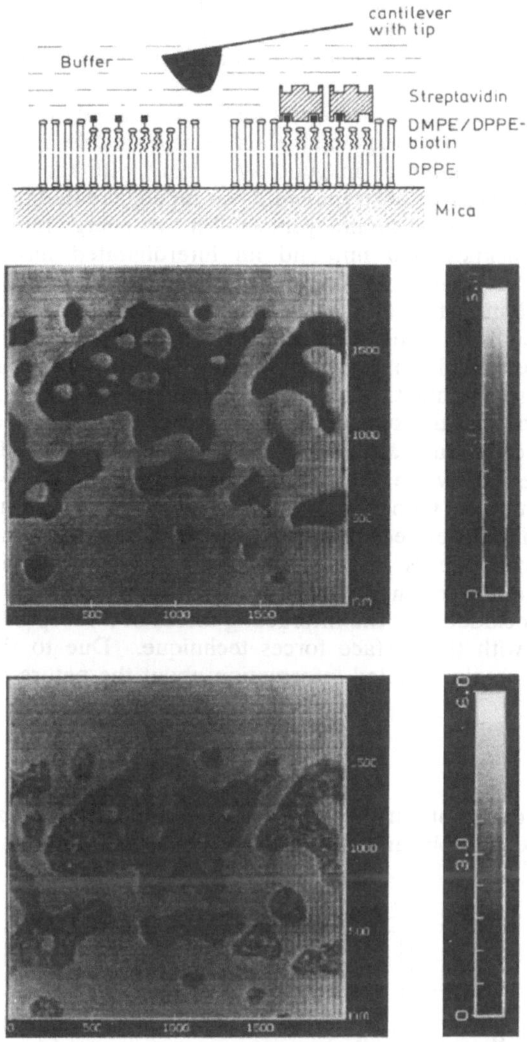

Figure 6. Schematic of the prepared sample a) and the corresponding atomic force micrographs, before b) and after c) streptavidin injection. Note the decreasing contrast in c) due to the streptavidin binding. The bar on the right relates the used grey values to the measured thicknesses with a range of 0-6 nm.

Only very few streptavidin molecules are found with the AFM on the domains. Their observed shape depends strongly on the applied AFM tip forces. With very small forces of about 10^{-11} N we resolved a single immobilized streptavidin. When the force is increased to 10^{-10} N, the molecule can no longer be detected.[31]

We conclude that the selective binding of streptavidin to the fluid lipid phase was verified with the AFM. The phase separation of the biotinylated lipid seems to be complete. Almost no (<1%) binding of the streptavidin to the crystalline domain is observed.

In the surface forces measurements we first examined the symmetric biotinlipid surfaces, which showed the expected electrostatic double layer repulsion. The characteristic Debye length was in good agreement with the theoretical value.[23,24] We showed that a monolayer of streptavidin was indeed created by adsorption. When approaching each other these monolayers exhibit two energy minima, corresponding to a bilayer of thickness 8.6 nm and an interdigitated monolayer of 4.2 nm respectively.[23,24] In Figure 7 the force-distance curve for the asymmetric biotinlipid-streptavidin surface at T=33°C and pH=7.4 is shown. For large separations the electrostatic repulsion is evident, due to the interaction of the slightly charged streptavidin and the biotinlipid surface. At a small separation the surfaces come into a very strong adhesive contact. The adhesive force of -36 mN/m is so strong, that subsequent separation of the surfaces tears the lipids out of the layer. At T=33°C, when the lipid layer is fluid and the molecules can laterally diffuse, this damage reanneals. However below the phase transition temperature at T=25°C, the surfaces are irreversibly damaged. Generally the adhesion at T=25°C is smaller, because the lipids are not mobile, fewer receptors reach their binding sites and fewer bonds are made. The binding appears to be insensitive to pH as compared with previous measurements[23] and in agreement with literature.[30]

It can be concluded that the interacting forces of a receptor ligand pair can be directly measured with the surface forces technique. Due to the defined surface geometry, one can obtain detailed information about the nature and range of these forces.

Biotinylated Thiols

In the second part of our investigations, we used functionalized alkanethiols to form self-assembled monolayers on gold.[7,32,33] The thicknesses obtained by PSP

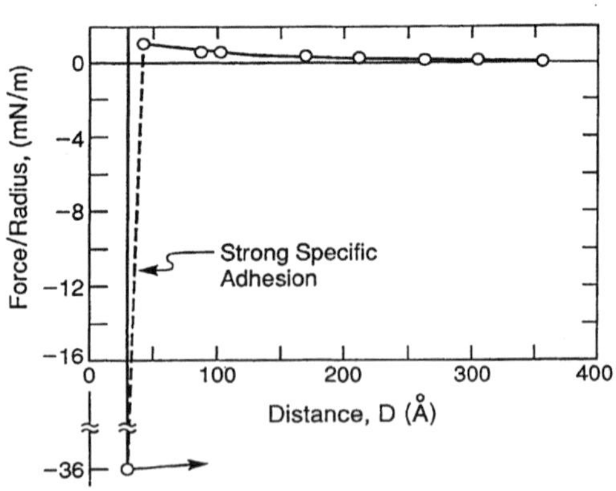

Figure 7. Force-distance curve of the asymmetric biotinlipid streptavidin surfaces at pH=7.4 and T=33°C.

Table II

Thicknesses of different self-assembled monolayers and the changes due to streptavidin binding. Determination with PSP spectroscopy using refractive indices of n(thiol) = 1.45 and n(streptavidin) = 1.50. The error is ±0.2 nm.

thiol	in air	in 0.5 M NaCl	Streptavidin thickness [nm]
C11OH	1.3	1.3	0
thiol 1	1.0	1.0	3.0
thiol 2	2.4	2.6	0.8

spectroscopy depend strongly on the molecular properties, as can be seen from Table II. Both the ω-hydroxy-(C11OH) and the biotinyl-(thiol2) thiol form complete monolayers, with thickness values in agreement with molecular models. The additional spacer in thiol1 disturbs the balance between the hydrophilic and the hydrophobic properties, so that only a relatively thin layer is obtained, see Table II. This may be explained either by an incomplete coverage or by a flat orientation of the molecules on the surface. The former interpretation can be ruled out, because biotin saturated inactive streptavidin exhibits no unspecific adsorption to the surface. Most proteins bind strongly and unspecifically to bare gold surfaces. Similarly a single monolayer of C11OH passivats the gold surface completely.

The binding capacities of the biotinylated self-assembled monolayers are inversely related to the packing density of the biotin groups.[34] See also Table II for the saturation thicknesses and Figure 8 for the kinetics. The binding to thiol1 SAM is very fast and, for a streptavidin concentration of $6 \cdot 10^{-7}$ M, diffusion limited. The diffusion coefficient is determined to be $4 \cdot 10^{-7} cm^2/sec$,[15] which compares well with the above value for the lipid case. The reaction of the thiol2 layer is less effective, much slower and can be described by the binding site limited model, see equation (1). A time constant of $\tau = 14$min is obtained by linearization.[34]

The self-assembled monolayers of alkanethiols are very stable and can be handled in air or in aqueous solution. The heterogeneity of the samples was checked by PSP microscopy, which revealed the presence of unspecifically adsorbed protein clusters at the interfaces. It was shown that rinsing with pure solvent can remove such impurities and thus decrease the error of the measurements.[15]

We have shown, that the binding of Streptavidin to biotinylated SAM is governed by the lateral packing density of the biotin groups. The same maximum thickness as in the lipid case can be obtained. Depending on the thiol used, the reaction is either diffusion or binding site limited. In the latter case, the time constant is very similar to that of the biotinlipids, but the streptavidin layer is much thinner. A possible explanation could be the internal "dilution" of one biotin per two alkyl chains in the lipid. On the other hand, the biotinylated thiols have one biotin per alkyl chain, which may sterically hinder the approach of protein molecules and lead to a thinner streptavidin layer.

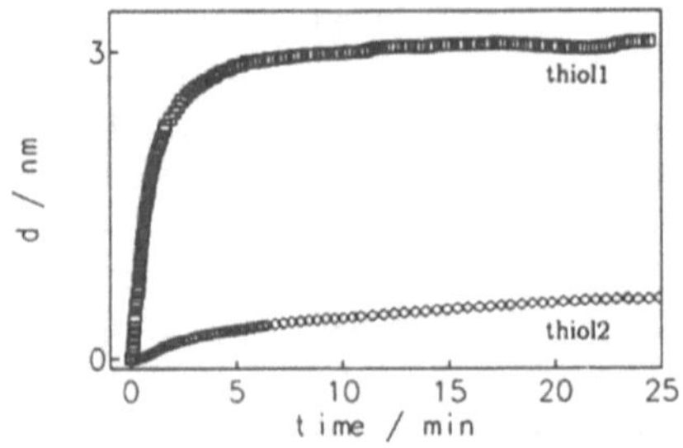

Figure 8. Kinetic of streptavidin binding to thiol1 and thiol2 monolayers, observed with PSP spectroscopy as a fixed angle of incidence.

Universal Binding Film

In the third part of our investigations, we used the immobilized streptavidin monolayer as an universal binding film for biotinylated molecules.[34] The variety of molecules which can be biotinylated seems to be unlimited. This offers a promising way to realize molecular devices. For example with a bisbiotin it should be possible to build up multilayers of streptavidin, by alternate binding of the two species. We performed such experiments with streptavidin monolayers on thiol1 and thiol2 SAMs. The bisbiotin itself provides no measurable thickness increase, due to its low surface coverage of two molecules per streptavidin. A second streptavidin binding yielded a further thickness increase of only 0.5 nm. Only a very incomplete second streptavidin layer could be built up. To examine this behaviour, we carried out surface forces measurement with symmetric streptavidin interfaces, as described previously. After a bisbiotin solution was injected between the surfaces, the curve shown in Figure 9 was observed. At a separation of D=8.5 nm a strong steric repulsion inhibits further compression. This separation corresponds to two streptavidin monolayers, similar to the symmetric streptavidin surfaces described above. But in this case these layers do not interdigitate. We conclude that the bisbiotin connected neighbouring streptavidin molecules, so that intraplanar crosslinking occurred. No measurable binding interaction to the opposing surface was observed. It seems that the bisbiotin used was too flexible and could bend over to reach the next streptavidin.

Reactions with the biotinylated antibody were carried out with streptavidin on lipid and alkanethiol monolayers. In the first case a layer, consisting of 10 mole% lipid1 in DMPE in the coexistence phase, with bound streptavidin was used as a target. The subphase was exchanged against a $7.5 \cdot 10^{-7}$ M solution of the biotinylated antibody and the reaction was observed with PSP microscopy. As can be seen in Figures 10a) and 10b) a contrast enhancement is observed, compared to Figures 3c) and d). The quantification reveals that the thickness of the streptavidin on the lipid matrix remains constant, whereas the thickness of the domain decreases. We assume, that the biotinylated antibody binds to the streptavidin and dissolves this lipid-protein complex again into the subphase. Because of this loss of lipid, the domain appears thinner. Similar to the streptavidin binding, the binding properties of the biotinylated antibody are inversely related to its receptor density. Therefore no binding occurs at the complete streptavidin monolayer on the lipid matrix. It was shown that this behaviour is also found in the reaction of the biotinylated antibody with streptavidin bound to the different thiols.[34] An additional detail of the antibody binding was revealed by PSP microscopy. Between the lipid domain and the lipid matrix exists a

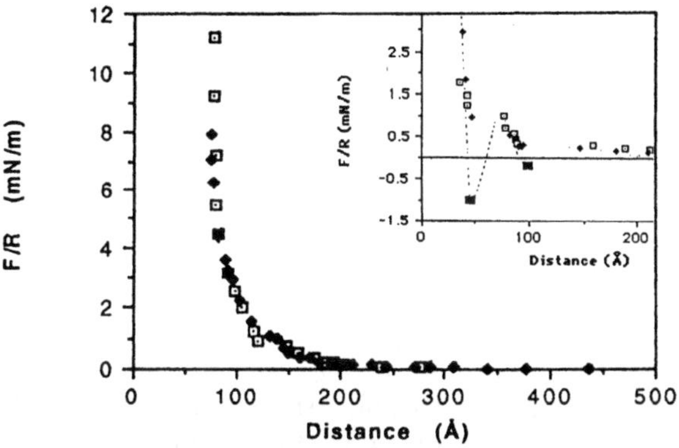

Figure 9. Force-distance curve of two symmetric streptavidin surfaces, which were treated with the bisbiotin. Measurement at T=33°C and pH=7.2. The curve without bisbiotin is shown as an inset (see text).

borderline, with an intermediate thickness, see Figure 10c. To date, we have no explanation for this observation, but again it is clear, that with a lateral integrating technique this structure would have been averaged out.

Such an immobilized antibody layer was used for a bioassay to the corresponding antigen. The surface architecture was: gold + 2.4 nm thiol2 + 0.8 nm streptavidin + 3.3 nm biotinylated antibody and a $2 \cdot 10^{-6}$ M solution of the antigen HCG was added. The resulting thickness change is depicted in Figure 11. The obtained result of 0.5 nm is less than expected for this molecule with a molecular weight of 30,000 g/mol.

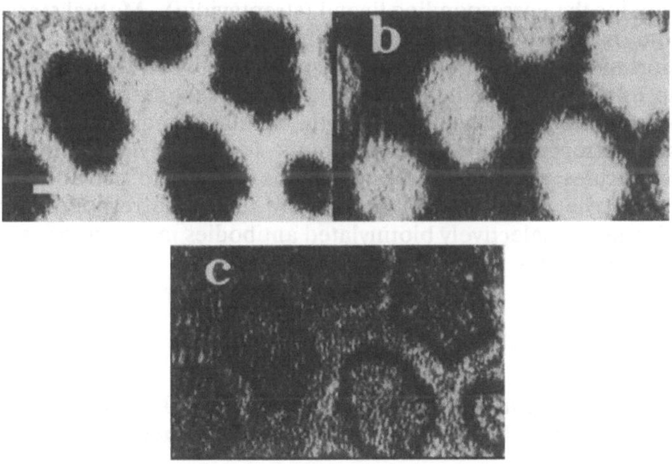

Figure 10. PSP micrographs of the same sample as Figure 3, after the injection of the biotinylated antibody. Note the contrast enhancement in a) and b) as compared to Figure 3. Figure c) was obtained at an intermediate angle of incidence and shows a borderline of the domains with an intermediate thickness. The bar corresponds to 100 μm.

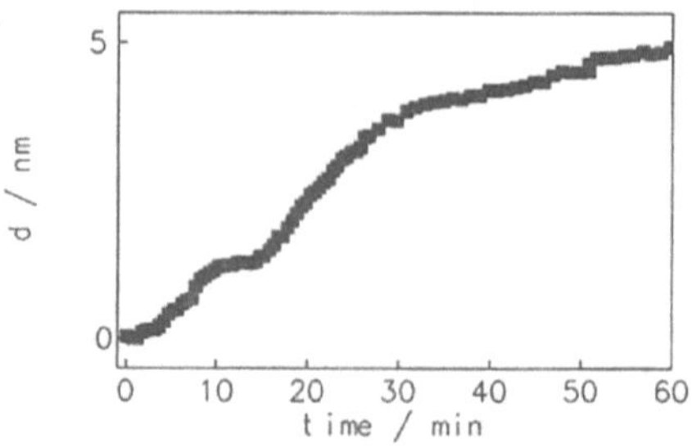

Figure 11. Thickness changes due to the antigen binding to its immobilized monoclonal antibody. This bioassay was performed without any labelling of the target molecule with PSP spectroscopy.

We assume that only a few antigen binding sites of the antibody are actually accessible to the antigens in solution. This is caused by the statistical biotinylation of the antibody, which leads to many different orientations of the antibody at the surface. Correspondingly only a few binding sites are exposed to the aqueous medium. An improvement should be obtained by the use of selectively biotinylated antibodies or antibody fragments. By this, an optimization of the antigen binding should be possible.

CONCLUSION

The deposition of monomolecular films with the LB-technique or as SAMs, allows for defined functionalization of interfaces. By selecting the endgroup of the molecule used, we obtained either a passivation or an activation (biotin) for recognition reactions. The packing density of the receptors is inversely related to their binding capacity for the corresponding ligand (streptavidin). Mutual steric interaction of the receptors increases with the number of neighbouring receptors and makes the binding reaction more difficult. The use of spacers between the attachment and the reactive group in the molecule is limited to a certain length, because the number of unsuitable conformations increases with length. Similar arguments hold, when the free binding sites of streptavidin act as receptors for biotinylated molecules. The streptavidin molecules need to be diluted at the surface to provide good and fast binding to a biotinylated antibody. A bioassay to the corresponding antigen was carried out. The use of selectively biotinylated antibodies may increase sensitivity and lead to sensor applications.

ACKNOWLEDGMENT

The financial support and the donation of several materials by Boehringer Mannheim, Werk Tutzing, F.R.G. is gratefully acknowledged.

REFERENCES

1. G. Wegner, Essay: wetting, adhesion, biocompatibility, *Adv. Mat.* 3:8 (1991).
2. G. L. Gaines, Jr., "Insoluble monolayers at liquid-gas interface", Interscience Publishers, New York (1966).

3. G. Roberts, "Langmuir-Blodgett films," Plenum Press, New York and London (1990).
4. A. Ulman, An introduction to ultrathin organic films from Langmuir-Blodgett to self-assembly, Academic Press Inc., San Diego (1991).
5. M. Wilchek and E. A. Bayer, "Avidin-biotin Technology," *Methods in Enzymology* 184, Academic Press Inc., San Diego, (1990).
6. R. Blankenburg, P. Meller, H. Ringsdorf, and C. Salesse, Interaction between biotin lipids and streptavidin in monolayers: formation of oriented two-dimensional protein domains induced by surface recognition, *Biochemistry*, 28:8214 (1989).
7. L. Häußling, B. Michel, H. Ringsdorf, and H. Rohrer, Direkte Beobachtung von spezifisch adsorbiertem Streptavidin auf Biotin funktionalisierten Monoschichten mit dem Rastertunnelmikroskop, *Angew. Chem.* 103:518 (1991).
8. S. A. Darst, M. Ahlers, P. H. Meller, E. W. Kubalek, R. Blankenburg, H. O. Ribi, H. Ringsdorf, and R. D. Kornberg, Two-dimensional crystals of streptavidin on biotinylated lipid layers and their interaction with biotinylated macromolecules, *Biophys. J.* 59:387 (1991).
9. M. Ahlers, R. Blankenburg, D. W. Grainger, P. Meller, H. Ringsdorf, and C. Salesse, Specific recognition and formation of two-dimensional streptavidin domains in monolayers: application to molecular devices, *Thin Solid Films* 160:93 (1989).
10. M. Ahlers, W. Müller, A. Reichert, H. Ringsdorf, and J. Venzmer, Spezifische Wechselwirkung von Proteinen mit funktionellen Lipidmonoschichten - Wege-Zur-Simulation von Biomembranprozessen, *Angew. Chem.* 102:1310 (1990).
11. M. Lösche, E. Sackmann, and H. Möhwald, A fluorescence microscopic study concerning the phase diagram of phospholipids, *Ber. Bunsenges. Phys. Chem.* 87:848 (1983).
12. H. Raether, "Surface Plasmons," Springer Verlag, Berlin (1988).
13. G. Binnig, C. F. Quate, and C. Gerber, Atomic Force Microscope, *Phys. Rev. Lett.* 56:930 (1986).
14. J. N. Israelachvili and G. E. Adams, Measurement of forces between two mica surfaces in aqueous electrolyte solutions in the range of 0-100 nm, *J. Chem. Soc. Faraday Trans.* 174:975 (1978).
15. L. Häußling, H. Ringsdorf, and F.-J. Schmitt, W. Knoll, Biotin functionalized self-assembled monolayers on gold: surface plasmon optical studies of specific recognition studies, *Langmuir* 7:1837 (1991).
16. M. Lösche and H. Möhwald, Impurity controlled phase transitions of phospholipid monolayers, *Eur. Biophys. J.* 11:35 (1984).
17. I. Pockrand, J. D. Swalen, J. G. Gordan II, and M. R. Philpott, Surface plasmon spectroscopy of organic monolayer assemblies, *Surface Science* 74:237 (1977).
18. B. Rothenhäusler and W. Knoll, Surface-plasmon microscopy, *Nature* 332:615 (1988).
19. W. Hickel, D. Kamp, and W. Knoll, Surface-plasmon microscopy, *Nature* 339:186 (1989).
20. F.-J. Schmitt, P. Meller, H. Ringsdorf, and W. Knoll, Specific dye adsorption at oriented monolayers, *Progr. Colloid Polym. Sci.* 83:136 (1990).
21. S. Manne, P. K. Hansma, J. Massie, V. Elings, and A. A. Gewirth, Atomic resolution electrochemistry: with the atomic force microscope: copper deposition on gold surfaces, *Science* 251:183 (1991).

22. J. N. Israelachvili, Techniques for direct measurements of forces between surfaces in liquids at the atomic scale, *Chemtracts - Analytical and Physical Chemistry* 1:1 (1989).
23. C. A. Helm, W. Knoll, and J. N. Israelachvili, Measurement of ligand-receptor interactions, *Proc. Natl. Acad. Sci. U.S.A.* 88:8169 (1991).
24. D. Leckband, J. N. Israelachvili, F.-J. Schmitt, and W. Knoll, *Science* 255:1419 (1992).
25. R. C. Ebersole, J. A. Miller, J. R. Moran, and M. D. Ward, Spontaneously formed functionally active Avidin monolayers on metal surfaces: a strategy for immobilizing biological reagents and design of piezoelectric biosensors, *J. Am. Chem. Soc.* 112:3239 (1990).
26. W. Hickel and W. Knoll, Surface-plasmon optical characterization of lipid monolayers at 5 m lateral resolution, *J. Appl. Phys.* 67:3572 (1990).
27. F.-J. Schmitt and W. Knoll, Surface-plasmon microscopic observation of site-selective recognition reactions, *Biophys. J.* 60:716 (1991).
28. M. J. Eddowes, Direct immnuochemical sensing: Basic chemical principles and fundamental limitations, *Biosensors* 3:1 (1987/88).
29. J. D. Andrade, "Biomedical Polymers," Plenum Press, New York (1985).
30. N. M. Green, Avidin, *Advances in Protein Chemistry* 29:85 (1975).
31. A. L. Weisenhorn, F.-J. Schmitt, W. Knoll, and P. K. Hansma, Streptavidin binding observed with an atomic force microscope, *Ultramicroscopy*, submitted.
32. C. Pole-Grosdemange, K. L. Prime, and G. M. Whitesides, Formation of self-assembled monolayers by chemisorption of derivatives of oligo(ethylene glycol) on gold, *J. Am. Chem. Soc.* 113:12 (1991).
33. K. L. Prime and G. M. Whitesides, Self-assembled organic monolayers: model systems for studying adsorption of proteins at surfaces, *Science* 252:1164 (1991).
34. F.-J. Schmitt, L. Häußling, H. Ringsdorf, and W. Knoll, Plasmon-surface-polaritons studies of specific recognition reactions at self-assembled monolayers on gold, *Thin Solid Films*, submitted.

CHARACTERIZATION OF SELF-ASSEMBLED MONOLAYERS OF GENETICALLY ENGINEERED CYTOCHROME b_5

P.S. Stayton, J.M. Olinger, S.T. Wollman, E.M. Thurman, P.W. Bohn, and S.G. Sligar

Beckman Institute, University of Illinois at Urbana-Champaign, 405 N. Mathews, Urbana, IL 61801

ABSTRACT

De novo gene synthesis techniques have been used to synthesize mutants of a water-soluble form of the electron transport protein cytochrome b_5 in which a unique reactive cysteine group is formed on the water-accessible surface of the protein. This cysteine is used to covalently attach the protein to a silica surface derivatized with a thiol-reactive alkylsilane. The resulting molecular assemblies are characterized by their surface coverage, antigenic activity, contact angles, and Raman spectra. These measurements indicate that organized protein assemblies are obtained with near unity coverage, but which are still relatively porous with respect to water.

INTRODUCTION

The development of a genetically engineered metalloprotein system which, through the introduction of a judiciously positioned unique reactive site, would yield oriented protein arrays at specially prepared solid substrates would have myriad applications in the generation of novel biomaterials. This goal is being pursued in our laboratory with the intent of developing an understanding of the physical, chemical, and biological principles which govern the assembly of supermolecular arrays of biological macromolecules. Heme proteins in general, and cytochrome b_5 in particular, are attractive candidates for biomaterial fabrication due to the unique optical and electronic properties associated with the heme prosthetic group.[1] With this in mind, cytochrome b_5 forms a nearly ideal model system for materials studies. It is a small, 13,000 dalton, bis-imidazole-ligated heme protein for which a high resolution crystal structure is available.[2] A bacterial over-expression system, based on an easily manipulated synthetic gene construction, has been developed which allows for production of large quantities of readily isolable recombinant protein in a water-soluble form.[3] The wild-type protein is a membrane-associated species, but *de novo* gene synthesis techniques allow for production of a completely active protein without the lipophilic tail which confers the membrane association properties. In addition, its

Synthetic Microstructures in Biological Research, Edited by
J.M. Schnur and M. Peckerar, Plenum Press, New York, 1992

globular structure should lead to efficient packing at planar surfaces, and a host of experiments have been performed to examine its participation in the formation of heterologous protein-protein complexes.[4] Finally, it is an unusually robust protein, performing favorably in response to various environmental shocks. Thus, it is a well-poised system for these materials studies.

In particular it is desirable to control the structure of organized assemblies of such biological macromolecules in both two and three dimensions. Differential orientation of the heme group, relative to an optical substrate for example, can be used to control optical, and ultimately electronic, properties of the organized assemblies in ways applicable to protein based devices. The scheme for accomplishing the ultimate orientation of a biological macromolecule involves the introduction, at the gene synthesis level, of a codon for a unique reactive site on the water-accessible surface of the protein. In particular because cytochrome b_5 lacks a native cysteine residue, site-directed mutagenesis techniques[3,5] were utilized to produce four mutants by introducing a highly specifically reactive thiol functional group on the protein surface at four independent points, replacing threonine 8 (mutant T8C) and, separately, threonine 65 (mutant T65C), threonine 73 (mutant T73C), and serine 85 (mutant S85C). In the present case, substitution of cystein residues, with their reactive thiol groups, has been chosen for anchoring to a solid substrate which has been derivatized with a complementary reactant, typically a silyl ether terminated with an alkyl halide (for the formation of a thioether surface linkage) or an alkyl mercaptan (for the formation of a disulfide bond to the surface). The reaction schemes are outlined below.

Self-assembly of cytochrome b_5 monolayers results from a nucleophilic reaction between the unique thiol of the surface cysteine and the reactive group of a derivatized substrate. Glass substrates were prepared by treatment with a silane coupling agent, either 3-iodo- or 3-bromopropyltrimethoxy silane as illustrated in Scheme 1. The resulting covalent bond is a thioether linkage between mutant protein and the substrate. Surface coverage was estimated to be 90% by quantitating the heme content with a pyridine hemochrome assay,[6] demonstrating that nearly complete monolayer coverage has been achieved. Covalent attachment of cytochrome b_5 to the surface through specific thioether linkages was verified by extensive buffer washings to remove physisorbed protein and comparisons to control experiments with wild-type b_5, which showed no detectable absorption in surface absorption experiments. After fabrication, the organized protein assemblies were characterized by a series of experiments including surface titrations (to generate surface coverages for both linker and protein), antigen binding, contact angle measurements, and surface enhanced Raman spectroscopy (SERS).

SURFACE COVERAGE MEASUREMENTS

Initial measurements of surface coverages of the derivatization agent were carried out using the thiopropyl analogue of the bromo- and iodopropylsilanes eventually used for the majority of the protein immobilization experiments. This molecule was chosen for the ease with which surface coverages could be determined using a titration with dithionitrobenzoate (DTNB, or Ellman's reagent). Titrations performed in specially designed low-volume cells generated surface coverage values for the linker molecule of ca. 64 Å^2/molecule. Molecular models show that the size of an individual alkylsilane should be dominated by the cross-sectional area of the alkyl chain which, for example, in an all-*trans* configuration is ca. 25 Å^2. Using this value the surface coverage of the linker moiety is $\Gamma_{linker} = 0.39$. Of course, it is

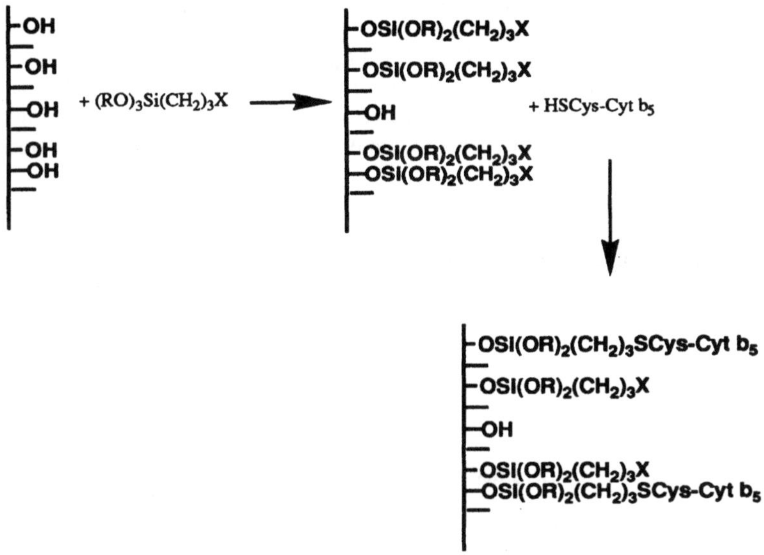

Scheme 1

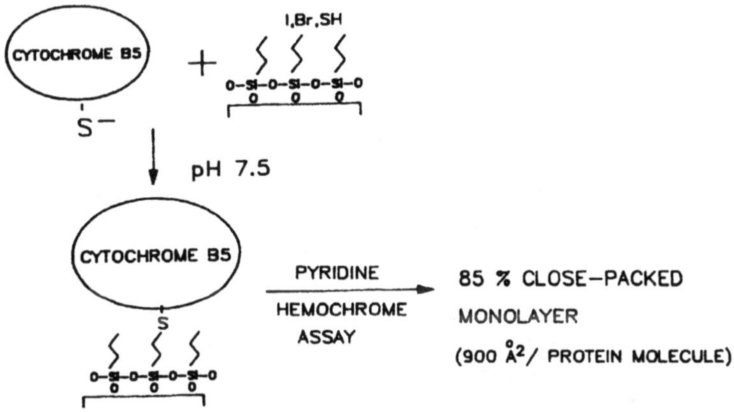

Protein Derivatization and Surface Coverage

Figure 1. Schematic illustration of the steps associated with the pyridine hemochrome assay for surface derivatized cytochrome b_5. Surface coverages vary from sample-to-sample in the range 750 $Å^2$/molecule $\leq \Gamma_{cyt} \leq$ 900 $Å^2$/molecule, i.e. $0.85 \leq \Gamma_{cyt} \leq 1.00$.

assumed that the surface coverage is independent of the distal reactive group on the propyl chain, an assumption for which there is no simple check.

In order to assess the coverage of cytochrome b_5 it is necessary to carry out the surface titration in a different fashion. This is accomplished by a surface release experiment followed by a pyridine-heme chromagen assay as indicated in Figure 1.

Surface coverages were found to vary from sample-to-sample in the range 750 $\mathring{A}^2$/molecule $\leq \Gamma_{cyt} \leq$ 900 $\mathring{A}^2$/molecule. Since the projected area of a single cytochrome b_5 molecule obtained from the high resolution crystal structure is ca. 25 Å x 30 Å, this corresponds to a surface coverage in the range $0.85 \leq \Gamma_{cyt} \leq 1.00$. It is especially important to note that surface coverages of protein approaching 1.00 can be obtained from surfaces on which the linker molecules are present at significantly lower coverages.

Table I. Contact angle measurements for various derivatized surfaces. Each trial consisted of a completely new sample preparation with fresh derivatizing agent and fresh protein.

Surface	Contact Angle	Number of Trials
clean Ti:Zn glass	3.5° ± 0.7°	2
3-bromopropylsilane/Ti:Zn glass	51.5° ± 2.6°	8
n-propylsilane/Ti:Zn glass	54.5° ± 1.5°	3
wild-type cyt b_5/3-bromopropylsilane/ Ti:Zn glass	48.0°	1
T65C cyt b_5/3-bromopropylsilane/ Ti:Zn glass	65.6° ± 5.1°	8

CONTACT ANGLE MEASUREMENTS

These results are also reflected in the surface contact angle measurements. Using a home-built contact angle goniometer, contact angles were measured for a variety of surfaces, prepared under varying conditions for water as solvent, using the sessile drop method.[7] The results are given in Table I. Clearly the surface structures obtained from the bromopropylsilane are very open as can be judged from the low contact angles. For comparison, the contact angles expected for a close-packed surface of a reasonably hydrophobic species, such as iodine atom, would lie in the range 110° - 115°. In fact, the contact angles for the n-propylsilane derivatized surface are statistically indistinguishable from those for the bromopropylsilane surfaces. These results establish the relatively open nature of the derivatized surfaces without protein, since access to the hydrophilic surface through monolayer defects is the most reasonable explanation of the contact angles intermediate between hydrophilic and completely hydrophobic surfaces.

Each of the protein-derivatized surfaces give contact angles in the same general range as the linker-only surfaces. There are two possible explanations. First, there is a significant degree of variability in the protein coverage values obtained from the pyridine-hemochrome assays described above. The results could indicate that a relatively low surface coverage was obtained and again access to the surface by water molecules was obtained. Alternatively, molecular graphics show that the water-accessible (i.e. hydrophilic) portion of the cytochrome b_5 protein surface is a majority fraction of the solvent-accessible amino acids. Therefore, the contact angles for the

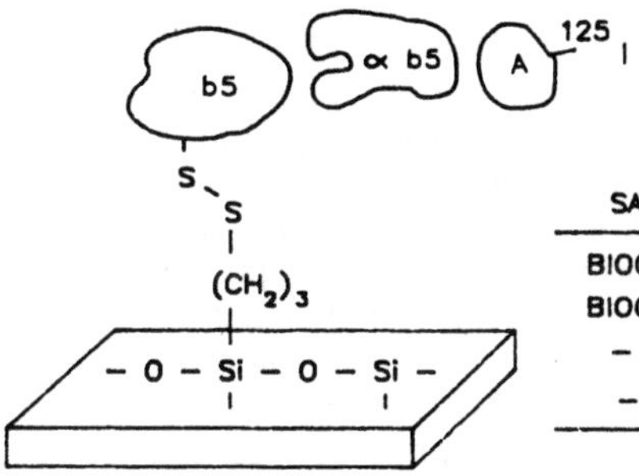

SAMPLE	BOUND LABEL
BIOCHIP-1	14,350
BIOCHIP-2	11,550
- α b5	126
- b5	118

Cytochrome b_5 (T8C) Derivatized Silicon Surface is Antigenically Active

Figure 2. Schematic illustration of the surface antigenic activity experiment involving polyclonal antibodies to the T8C mutant of cytochrome b_5 and radiolabelled protein A. The table in the insert shows the results from two replicate trials and the appropriate control experiments.

protein covered surfaces may simply reflect the intrinsic water wettability of these surfaces.

ANTIGENIC ACTIVITY

Measurements of antigenic activity of cytochrome b_5 derivatized surfaces were carried out as indicated in Figure 2. A thiopropyl derivatized surface was reacted with the T8C genetically engineered mutant. Following protein derivatization the surface was exposed to polyclonal antibodies raised against the mutant protein. Surface binding of the antibody was followed by extensive washing with buffer solution to remove physisorbed antibody species; and finally the entire surface complex was derivatized and tagged with ^{125}I -labelled protein A. As the table in the inset of the figure shows, the mutant protein-covered surfaces gave radiocount levels ca. two orders of magnitude higher than the two control samples, consisting of mutant cytochrome b_5 on which no antigenic binding reaction had been carried out, i.e. cytochrome b_5 and ^{125}I -labelled protein A, and a surface for which no cytochrome b_5 was present, i.e. T8C antibody and ^{125}I -labelled protein A. These experiments confirm the results of the surface titration experiments (*vide supra*), in that mutant protein is present on the surface which is chemisorbed (not removed by buffer washes) and which is sufficiently intact to retain its antigenic activity. Although this latter observation is not as strong an indication of structure retention as might be desirable, it does indicate that gross disruption of the primary structure has not occurred.

SURFACE RAMAN SPECTROSCOPY

We have examined two separate aspects of surface Raman scattering from molecules at dielectric surfaces in some detail: (1) the development of structures which permit the study of surface chemistry characteristic of dielectric surfaces while retaining the signal enhancement advantages of surface enhanced Raman spectroscopy, and (2) the acquisition of polarized scattering from such structures and the use of this information to extract the molecular orientation of the adsorbate in a semi-quantitative manner.[8,9] To summarize, a structure consisting of 50 Å Ag or Au islands/50 - 100 Å SiO_2 spacer/adsorbate is prepared on the surface of a masked SiO_2 substrate, and excitation radiation is then coupled into the structure just above the critical angle for total internal reflection (calculated for a metal-free interface), as shown in Figure 3. A series of exhaustive experiments including measurements of chemical resistance, Auger spatial mapping and depth profiling, coverage dependences of Raman signals, and studies of the signal intensity vs. spacer layer thickness were undertaken which showed that the molecules giving rise to the majority of the Raman signal were located at the SiO_2 surface and not in defect channels to the metal surface. However, there remains a contribution to the signal which is probably associated with defects.

The defect considerations are not as important for the molecular assemblies under consideration in the present work. For the protein molecular assemblies the mere size of the individual globin molecules make all but the largest defects unimportant insofar as considerations of physical location are concerned. In addition, the questions to be addressed by these experiments center around the use of the peak positions in nonresonant surface scattering to characterize the state of the heme prosthetic group. Thus, the questions of where the enhanced molecules are situated is of less importance. In particular the heme oxidation state is indicated by bands in the region 1340 - 1370 cm^{-1}, the spin state by bands in the region 1480 - 1510 cm^{-1}, and the core size, indicating the possible effects of doming, or non-planarity, in the region ca. 1550 - 1600 cm^{-1}. These markers are summarized in the frequency correlation diagram shown in Figure 4, and the spectra for the T8C and T65C mutants are shown in Figure 5.

Clearly the marker bands for the two mutant proteins are very similar. If we assume that the positions of the marker bands are unaffected by binding to the substrate, then we can make the following observations about the state of the heme proteins after derivatization. The oxidation state marker band appearing just below 1370 cm^{-1} in both cases is characteristic of an oxidized heme species. Although this

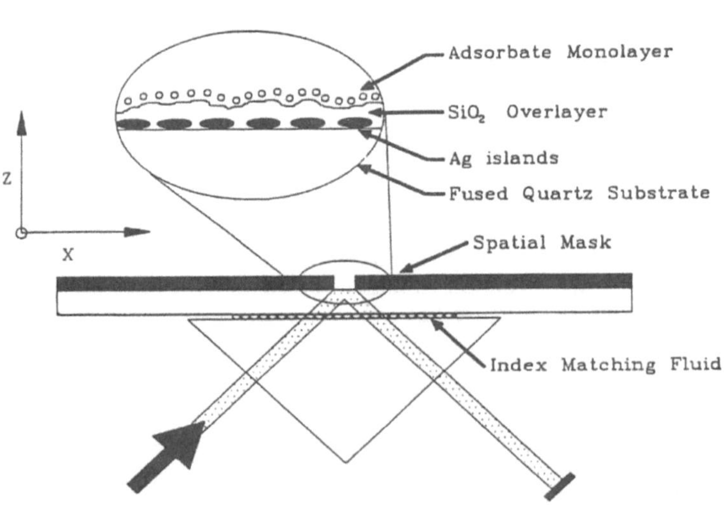

Figure 3. Excitation geometry and composition of the structures used for total internal reflection Raman scattering from organized assemblies of the mutant proteins.

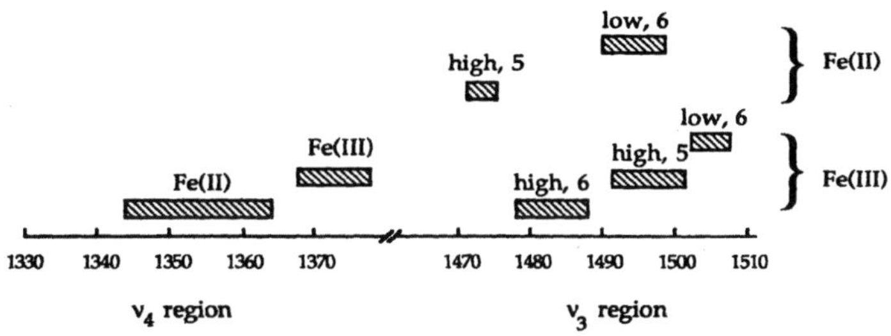

Figure 4. Plots of the observed positions of marker bands associated with particular variants of the heme state for proteins in homogeneous solution in a frequency correlation diagram.

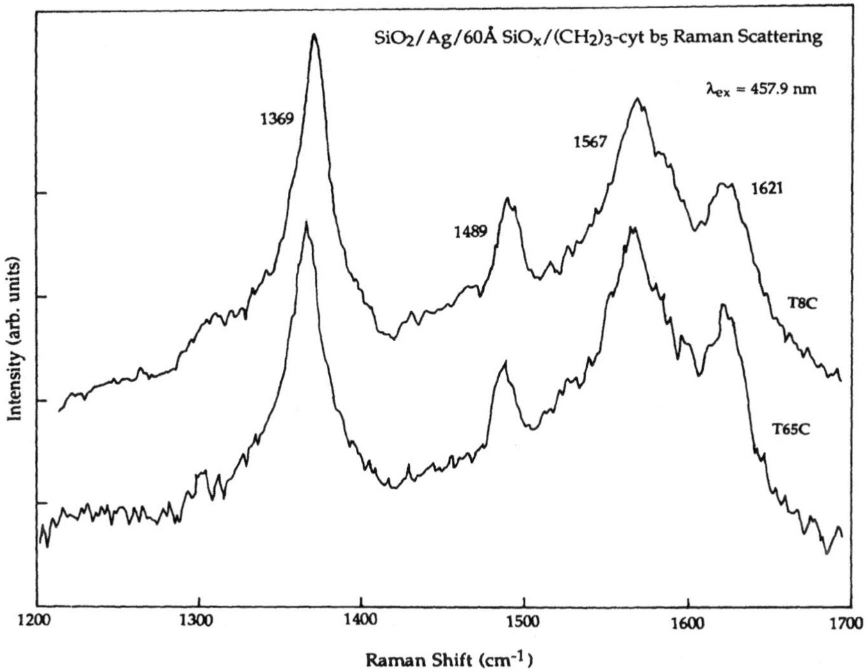

Figure 5. Raman spectra of T8C and T65C mutants of cytochrome b_5 in the fingerprint region. Laser excitation is at 457.9 nm in the surface enhanced scattering configuration shown in Figure 3.

is the form which is prepared and is expected, it has been observed that heme can be photoreduced under sufficiently strong laser fluences. The spin state marker bands are consistent with an interpretation in which the heme is high spin, but six-coordinate. There are two possible physical interpretations to this rather unlikely observation. First, there could be a slight alteration in the bonding in the heme pocket for all molecules, leading to an average band position characteristic of the high-spin six-coordinate species. Alternatively, there could be two populations on the surface, high-spin five-coordinate and low-spin six-coordinate, with the resulting band position the weighted average of the two component bands. The width of the observed bands, though, argues against this latter interpretation.

CONCLUSIONS

A series of single and double mutants in the cytochrome b_5 family have been prepared by *de novo* gene synthesis and expression in *E. coli*. In the mutants various surface amino acids have been replaced with specifically reactive cystein residues, which are then used to anchor the proteins to a prederivatized surface. The surface derivatization reactions can proceed either via formation of a disulfide linkage or a thioether bond. The resulting structures have been characterized by antigenic activity and surface titration experiments to determine coverage, contact angle measurements to determine water wettability, and Raman spectroscopy to learn about the average environment of the heme prosthetic group. Surface coverages for protein approach full monolayer, even though the linker is present $\leq$ 0.5 monolayer. This observation is attributed to the large size discrepancy between protein and linker. Raman experiments are most consistently interpreted in terms of a weakening of an axial ligand binding heme prosthetic group to the cytochrome cleft.

REFERENCES

1. M. Gouterman, Optical spectra and electronic structure of porphyrins and related rings, <u>in</u>: "The Porphyrins," D. Dolphin, ed., Academic Press, New York (1978).
2. F. S. Mathews, M. Levine, and P. J. Argos, Three dimensional fourier synthesis of calf liver cytochrome b_5 at 2.8Å resolution, *J. Mol. Biol.* 64:449 (1972).
3. S. B. von Bodman, M. A. Schuler, D. R. Jollie, and S. Sligar, Synthesis, bacterial expression, and mutagenesis of the gene coding for mammalian cytochrome b_5, *Proc. Natl. Acad. Sci. USA* 83:9443 (1986).
4. P. S. Stayton, M. T. Fisher, and S. G. Sligar, Determination of cytochrome b_5 association reactions, *J. Biol. Chem.* 263:13544 (1988).
5. B. A. Springer and S. G. Sligar, High level expression of sperm whale myoglobin in *escherichia coli*, *Proc. Natl. Acad. Sci.* 84:8961 (1987).
6. K. G. Paul, H. Theorell, and A. Akeson, The molar light absorption of pyridine ferroprotoporphyrin (pyridine haemochromogen), *Acta Chem. Scand.* 7:1284 (1953).
7. P. C. Heimenz, "Principle of Colloid and Surface Chemistry," Marcel Dekker, New York, pp. 317-322 (1986).
8. D. J. Walls and P. W. Bohn, Surface enhanced raman spectroscopy of molecules on dielectric surfaces, *J. Phys. Chem.* 93:2976 (1989).
9. D. J. Walls and P. W. Bohn, Enhanced raman scattering at dielectric surfaces 2. molecular orientations from polarized raman scattering, *J. Phys. Chem.* 94:2039 (1990).

LAPLACIAN GROWTH MODELS OF NEURAL CONNECTIVITY

P. Christie, D. J. Ennis, and V. R. DiLosa

Electrical Engineering Department, University of Delaware, Newark, DE 19716

ABSTRACT

A Laplacian growth model of neural connectivity in the rudimentary brain is investigated by computer simulation and polymer-based growth experiments. The model assumes that neurons grow by following the concentration gradient of tracer chemicals emitted by other neurons. Computer simulations of the resulting hierarchical branching structures are compared to the interconnection arrangements observed in fully optimized electronic computer systems. It is shown that the interconnection complexity of both systems may be characterized by a fractal dimension near two and that this is a direct consequence of minimizing the spatial resources devoted to wiring the system. We conclude by proposing a polymer-based analog of the Laplacian interconnection process which may find applicability in the implementation of interconnections in silicon-based neural networks.

1. THE CONSTRUCTION PROBLEM

The brain may well be one of the most complex objects ever constructed. Amazingly, this feat is repeatedly achieved though a largely unsupervised self-assembly process. We argue that, while genetic information must play a role in the determination of the large scale features of the brain, the complexity of the brain is far too great for the spatial location of each synapse to be encoded within the structure of DNA. Neurophysiological studies have indicated that the brain is structured so that there are many regions devoted to specific tasks, such as speech, cognition, emotion, etc. Any plausible interconnection model has to reconcile this localized structure with the current distributed processing paradigm of neural computation.

It is possible to envisage a characteristic length below which processing is achieved by distributed computation and above which higher level functions are implemented using more localized computational techniques. However, there is no experimental evidence to support this viewpoint and it is much more likely that the

Synthetic Microstructures in Biological Research, Edited by
J.M. Schnur and M. Peckerar, Plenum Press, New York, 1992

brain's connectivity is organized hierarchically. This structure allows for a high degree of distributed computation between neighboring neurons while simultaneously allowing high-level localized structures to exist through the hierarchical aggregation of highly interconnected neuron clusters.

We argue that this hierarchial self-assembly process must be driven by a thermodynamic mechanism due to the chemical reactions intrinsic to the growth process. The implementation of a rich hierarchical interconnection structure is then a consequence of maximizing the entropy of the system in a manner consistent with the hierarchical boundary conditions. Direct evidence for this construction hypothesis is difficult to obtain, given the problems encountered in observing the growth of such small filaments *in vivo*. We therefore begin our discussion of growth mechanisms in Section 2 with an analysis of the readily available evidence collected from modern computer systems. Our intent is to show that the problems encountered in constructing today's complex computer systems have forced designers to adopt many of the design compromises that nature itself has used in the construction of the brain. Specifically, we argue that the total wire length of the interconnections within a hierarchical structure can be minimized by ensuring that the interconnections are characterized by a fractal interconnection dimension near two.

Section 3 introduces a physical mechanism to implement this strategy at the fetal stage of development. Following the mechanism suggested by Kalil[1] we propose that growing axons make their way to specific parts of the rudimentary brain by following a chemical trail whose production is probably determined genetically. In this manner axons "sniff out" the presence of other neurons. The dynamic evolution of the interconnection structure is therefore modeled by a gradient descent mechanism. For the purpose of this paper it is convenient to formulate this dynamical process in terms of the chemical potential μ associated with the tracer chemical. (The chemical potential may be interpreted as the potential energy associated with the concentration of the tracer chemical at constant temperature.[2])

The concentration gradient associated with the chemical trail emitted by other neurons is then proportional to $\nabla\mu$ and the maximum gradient may be determined by solving

$$\nabla \cdot \nabla \mu = 0 \, .$$

This expression for gradient descent is more formally written as

$$\nabla^2 \mu = 0$$

and in this form it is usually called the Laplace equation. However, while plausible at the molecular level, it is not at all clear whether this simple Laplacian growth mechanism is sufficient to organize the structure of the interconnections within the brain in a manner which ensures large-scale constructibility. Therefore, we present computer simulations of the Laplacian growth that indicate that the intricate hierarchical branching structure of interconnections within the brain is compatible with this process. Moreover, we demonstrate that interconnections can use gradient descent to precisely locate the sources of the tracer chemical and thus could initiate the formation of synaptic junctions.

These self-assembly principles are then used to justify an electrostatic analog of the neuron growth process, which is introduced in Section 4. It is shown that the polymerization rate of the conducting polymer polypyrrole is dependent on the electric field at the growth interface. The growth of the polymer is therefore governed by a gradient descent mechanism where the chemical potential is replaced by an electrostatic potential V. The results obtained from the computer simulations are shown to be reproduced by the electropolymerization of polypyrrole and it is suggested that this process may find application in the construction of silicon-based neural networks.

2. PROPERTIES OF HIERARCHICALLY ORGANIZED SYSTEMS

The efficient utilization of the system's spatial resources dictates that modern computer systems employ hierarchical fabrication techniques. This results in the familiar rack-board-chip partitions of the system and even at the gate level this approach is maintained due to the emphasis Computer Aided Design (CAD) packages place on cell-based design techniques. We argue that modern computer systems are constructed under strong interconnect-limited conditions[3] and the readily available data on their interconnection structure will help us in understanding the problems encountered in the construction of other complex systems such as the brain.

The number of interconnections that can be implemented for communication between the partitions is strictly limited by their ability to support the required number of terminals. Measurements of the connectivity of a wide variety of commercial processing systems[4,5,6] have established an empirical relationship, known as Rent's rule, between the number of terminals, T, a partition of K processing elements requires to communicate with the rest of the system:

$$T=AK^{p},$$

where p is the Rent exponent and A is the average number of terminals associated with a single processing element. In general the number of interconnections will be somewhat smaller than the number of terminals. We may take this into account by defining the number of interconnections as $I=\alpha T$, where α is assumed to be approximately independent of p and K, and less than unity. The range of p is normally bounded by $0<p<1$. The upper bound is readily understood since the total number of terminals leaving a block of K processing elements cannot increase faster than K if A is a constant and independent of system size. Conversely, the number of terminals required for K elements cannot decrease with increasing K if system-wide communication is to be maintained. Table 1 lists a range of Rent exponents characteristic of different technologies.

The geometrical structure of the interconnections within the system will be very complex. Different hierarchies will contain different numbers of interconnections, each with their own characteristic range of lengths. This makes the characterization of the system very difficult using standard geometrical techniques. However, the descriptive complexity of the problem can be greatly reduced if we consider the interconnections to be distributed on a support whose dimension is not necessarily equal to that of the two-dimensional surface of the partitioned substrate. It may be shown that the dimension which characterizes the distribution of interconnections among hierarchies is given by[7]

$$D_i=3-2p.$$

This dimension is, in general, non-integer and varies from $D_i=1$ when $p=1$ to $D_i=3$ when $p=0$. The interconnection dimensions associated with modern computer systems are listed along with the Rent exponents in Table 1.

Table 1. Summary of Rent exponents and interconnection dimensions for a variety of system technologies. Adapted from Reference 5, pg. 418.

System type	Rent exponent p	Dimension D_i
Static memory	0.12	2.76
Microprocessor	0.45	2.10
Gate array	0.50	2.00
High-speed computer	0.63	1.74

Apart from its geometrical significance, which we shall return to in Sections 3 and 4, the interconnection dimension plays an important role in characterizing the distribution of interconnection lengths within the computer, since it may be shown that the number of interconnections of length k to $k+dk$ is given by[7,8]

$$n(k)dk = Bk^{-D_i}dk$$

where B is a constant. It is therefore convenient to re-express the concept of fractal dimension in terms of an effective system temperature.[9,10,11] Systems characterized by a large D_i will have a greater percentage of the total number of interconnections allocated to lower level partitions of the system, and this may be interpreted as a low temperature, since the average wire length (energy) per interconnection will be small.

Conversely, systems characterized by a small D_i will have a large percentage of the total number of interconnections allocated to higher level partitions. This will correspond to a high temperature, since the average wire length (energy) per interconnection is large. Following the analysis of Kohmoto[12] we may therefore define an effective system temperature by

$$\tau = 1/D_i$$

The concept of an effective system temperature is useful since it is then possible to relate the characteristics of the interconnection structure to a widely used algorithm for wire length optimization called Simulated Annealing (SA).[13,14] The SA algorithm attempts to optimize interconnection placements by allowing the conditional acceptance or rejection of pairwise swaps of processing element positions. If a swap results in a decrease in total system wire length the swap is retained. If the swap results in an increase in total system wire length the swap is retained according to a conditional probability which is defined by a system annealing temperature. Initially, the annealing temperature is set to a high value and then slowly reduced according to one of a variety of annealing schedules. As the annealing temperature is reduced to zero the system is assumed to settle in an arrangement which is close to its globally optimum arrangement. This approach successfully avoids local wire minima and achieves close to optimum results in a relatively short amount of processing time.

The characteristics of systems of varying fractal dimension have been investigated experimentally by attempting to optimize their interconnection placements with an SA algorithm.[7,15] Experimental fractal systems were constructed by allocating interconnections to a quadratically partitioned array of processing elements. Valid circuit configurations were ensured by permitting only one interconnection between the same pair of processing elements. A constant number of interconnections was allocated so that the number at each level scaled in a manner consistent with a constant interconnection dimension, D_i. At each hierarchical level the interconnections were allocated randomly and no attempt was made to optimize the layout within a given hierarchical level. The arrays were then rearranged by an SA algorithm in an attempt to obtain a placement which possesses a lower average wire length. Since the SA algorithm picks the elements to be swapped randomly from a list of all elements within the system the hierarchical structure is lost after annealing.

Figure 1 plots the resulting percentage decrease in total wire length as a function of the interconnection dimension, for a constant number of allocated interconnections. The number of processing elements in the system was 1024 and 1550 interconnections were allocated to each of the systems. We observe that as the dimension of the system increases (or, alternatively, as the effective system temperature decreases) the difference between the original total wire length and the total wire length after annealing decreases. For an interconnection dimension $D_i \approx 2$ we observe no reduction in total wire length, indicating that the fractal structure is close to the minimum wire length arrangement, even though the interconnections were allocated randomly within each level.

3. THE LAPLACIAN GROWTH MODEL

The results of the last section indicate that there may be significant advantage in using fractal structures for wiring the brain. However, it may be argued that the brain, unlike a modern computer system, is constructed in three dimensions and so the structures examined in Section 2 are not appropriate. However, there are many structures within the brain which share the quasi two-dimensional structure of modern computer systems. For example, the cortex, if stretched out, is about 1 m on a side and about 1 mm thick.[8] This region of the brain therefore possesses a length to depth ratio comparable to that of a chip, which is about 1 cm on a side and has circuitry occupying a depth of about 10 μm.

The convoluted dendritic structure of cortical interconnections are now interpreted as an attempt to implement minimum wire length arrangements in a manner consistent with a hierarchical organization strategy. We propose that these structures may be generated by a self-assembly process whose dynamics are governed by the Laplace equation. In order to determine if this model is capable of generating the complex fractal structures necessary for neural connectivity we have developed a computer simulation of the growth process.

The first task in developing the simulation is to re-express the Laplace equation in a discrete form more suited to computer manipulation. Using standard techniques,[16] it may be shown that the two-dimensional Laplace equation,

$$\frac{\partial^2 \mu}{\partial x^2}+\frac{\partial^2 \mu}{\partial y^2}=0,$$

may be approximated by a difference equation acting over a discrete array of points

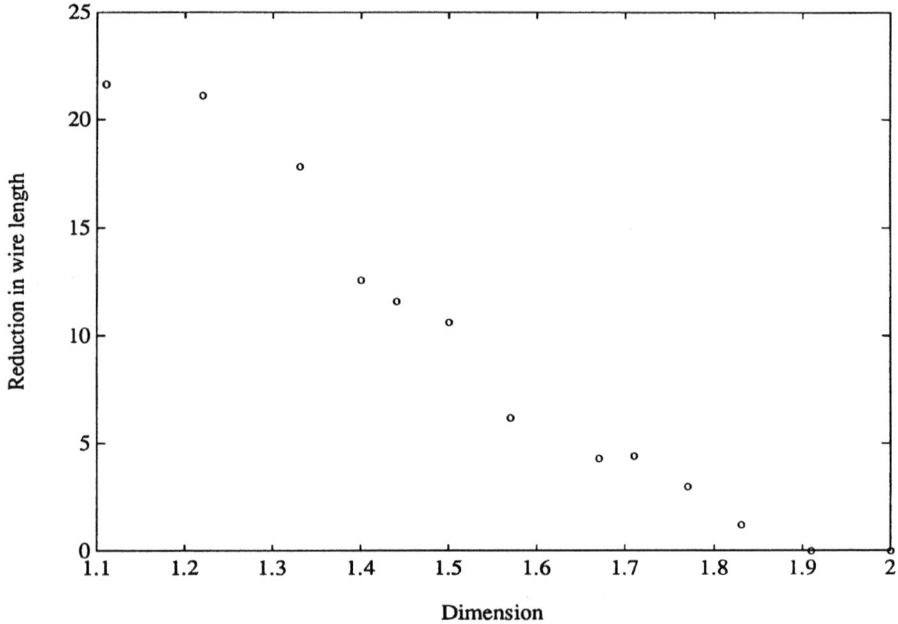

Figure 1. Percentage reduction in total wire length as a function of the system's interconnection dimension.

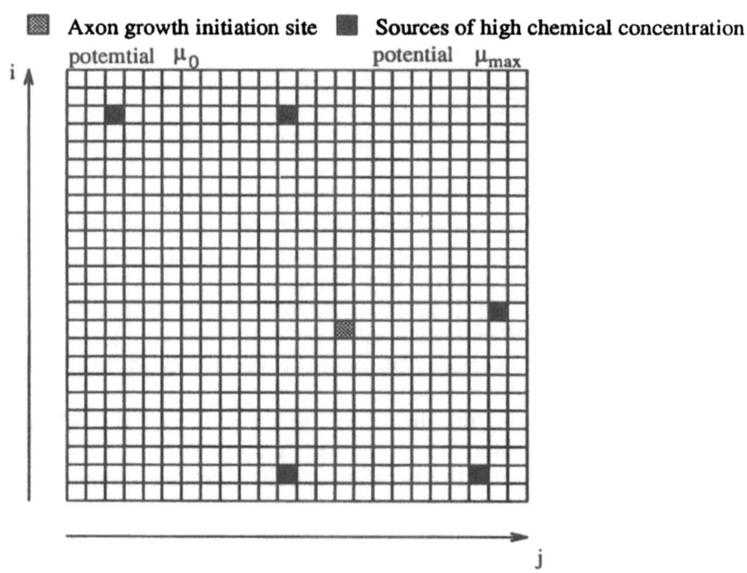

Figure 2. Notation used for pixel array in Laplacian growth simulations.

$$\mu_{i,j}=\frac{1}{4}(\mu_{i+1,j}+\mu_{i-1,j}+\mu_{i,j+1}+\mu_{i,j-1})$$

where $\mu_{i,j}$ is the chemical potential at the discrete location *i,j* expressed in terms of the potentials of its four nearest neighbors in the north, south, east, and west directions, see Figure 2. This difference equation states that we may determine the solution of the Laplace equation, for a given set of initial conditions, by ensuring that each point in the plane is characterized by a chemical potential which is the mean of its four nearest neighbors.

Once the Laplace equation has been solved for the initial conditions we may begin interconnection growth by associating a growth probability with the chemical potential gradient associated with each pixel around the perimeter of an initiation site. Since each pixel has a constant width in each direction the potential gradient will be proportional to the potential at each growth site. The growth probability associated with each growth site is then defined as:

$$P_{k,l}=\frac{\mu_{k,l}}{\sum_{GS}\mu_{k,l}}$$

where the $\mu_{k,l}$ are potentials associated with the locations of the active growth sites at the perimeter of the initiation site and the summation is over all growth site potentials.

The interconnection growth process proceeds according to the following algorithm. Steps 1 through 4 are then repeated until the growth is terminated.

1. A growth site with an associated growth probability $P_{k,l}$ is selected at random from the active growth site list.

2. A random number R is chosen within the range $0<R<1$.

3. If $R<P_{k,l}$ this site is removed from the active list and is converted to a growth site by appropriate shading of the pixel. The chemical potential associated with the growth site is changed from the calculated value to that associated with the initiation site. If $R>P_{k,l}$ the growth site is kept on the active list and another growth site is chosen at random until growth occurs.

4. Because of the change in chemical potential the solution of the Laplace equation is re-calculated for the new boundary conditions associated with the new interconnection growth and the active growth list is updated.

The choice of the initial conditions that correspond to the chemical potential of neural cells before growth is, of course, unknown. We have speculated that at the initial stages of growth the chemical gradients observed by a single neuron will appear isotropic due to the approximately uniform distribution of neural cells. We have therefore simulated the early stages of growth by a radially symmetric initial conditions. The chemical potential, measured in arbitrary units, was set to +10 at the circular boundary and to zero at the central growth initiation site. The resulting growth is shown at the top of Figure 3. We observe that, although the initial conditions are radially symmetric, the growth itself is not and that the Laplacian growth model appears to reproduce the characteristic fractal branching patterns observed in real neural cells. The fractal dimension of the structure at the top of Figure 3 has been measured using a box counting technique and is given by D_f=1.7. We note the dimension associated with the computer simulation corresponds to the fractal dimension associated with high-performance computer systems, see Table 1.

After the initial growth stage within a radially symmetric potential field it is assumed that the dendrite will eventually make its way to an isolated source of the tracer chemical. At this point the chemical potential gradient will become highly anisotropic and the dendritic tip will grow directly towards the neuron. The bottom of Figure 3 shows growth simulations produced by the close proximity of initiation sites to an array of neuron emitter sites. The growth initiation sites and the neural emitter are organized in four layers, the first and last layers consist of two growth-initiating sites and the two middle layers consist of four emitter sites. (The actual emitter and growth sites are difficult to distinguish from the polymer growth but are located in the middle of the circles superimposed on the diagram.) We observe that branching growth still occurs, although to a lesser degree, and that the dendritic tips grow directly towards the sources. This final stage of the growth would ultimately result in the formation of synaptic junctions in the actual fetus.

Although there are many details of the neural growth process which are not incorporated into this model, some of which may be very important, the simplicity of this experiment should not detract from the fundamental consequences of the result: complex branching structures similar to those observed in complex neuronal interconnections may be generated by simple growth mechanism based on gradient descent. This fine scale structure includes both the rich interconnection structure initiated by the initial fractal growth and also the final stages where individual tips of the growth home in on the neuron cells to form the synaptic junctions.

4. SYNTHETIC MICROSTRUCTURES FOR INTERCONNECTION MODELING

As well as being simple, the Laplace equation is observed to characterize a wide range of physical phenomena, from heat flow,[17] to electrostatics.[18] This universal behavior suggested a series of experiments which employed an electrostatic model of the growth process, where an electrostatic potential energy V replaces the chemical potential energy μ. These experiments focussed on the growth of a conducting polymer by a process of electropolymerization. The polymer polypyrrole was employed in these experiments since the polymer reaction, in an appropriate solution, is dependent on the local electric field at the growth interface. We would therefore expect the polymer growth to be along the directions of largest electrostatic potential gradient.[19]

Figure 4 shows the results of polymerizing a 0.1 M pyrrole/ 0.5 nM sodium para-toluenesulphonate (p-Ts)/acetonitrile solution in a radially symmetric electric field. A Pt anode and a Ni cathode (approximately 7 cm in diameter) were mounted concentrically in a petri dish containing approximately 5 mm depth of the monomer solution. We observe the same characteristic branching structures obtained from the computer simulations. The measured fractal dimension of this polymer growth was approximately 1.36 and does not compare well to the value of 1.74 quoted by Kaufman et al.,[20] but this may be attributable to the different electric fields used in the two experiments.

It is our intention to use this synthetic polymer as a tool for modeling the complexity of synapse formation in the brain. However, the basic utility of the polymer

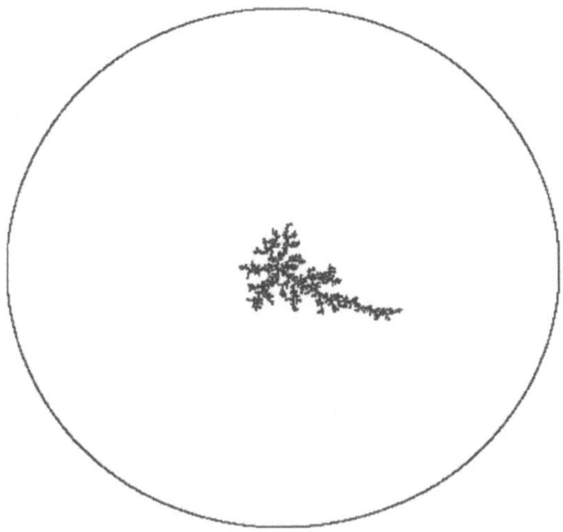

Figure 3. Computer simulation of the Laplacian growth process with radially symmetric initial conditions.

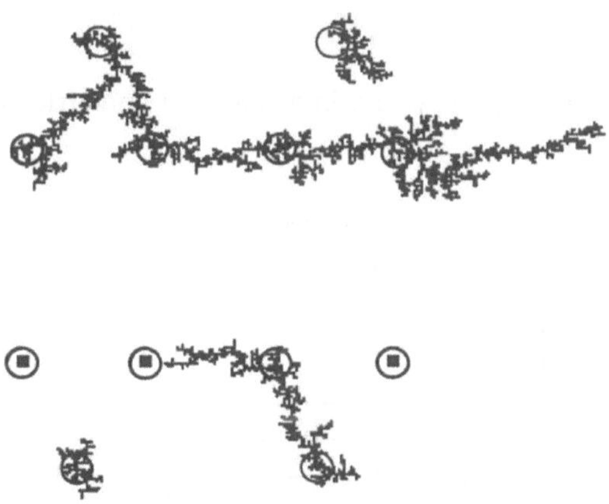

Figure 4. Computer simulation of the growth process using discrete sources of tracer chemical.

growth mechanism suggests that it may have applications far beyond our initial intentions. Preliminary experiments have indicated that the polymer forms good electrical contacts with silicon. This suggests that this growth process may be used to produce self-organizing electrode arrays for artificial neural networks. We propose that artificial synaptic junctions may be fabricated using conventional VLSI techniques in silicon. The interconnections between the junctions could then be implemented by simply covering the surface of the silicon array with a solution of the monomer. Upon applying a suitable voltage to the array the neurons would spontaneously self-interconnect, see Figure 6.

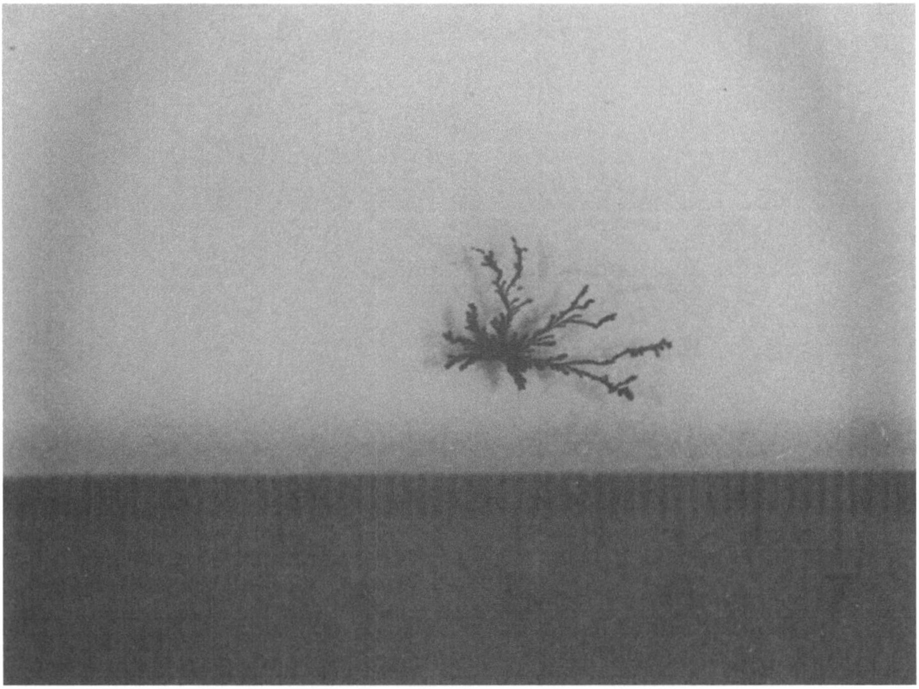

Figure 5. Growth of polypyrrole in a radially symmetric electric field.

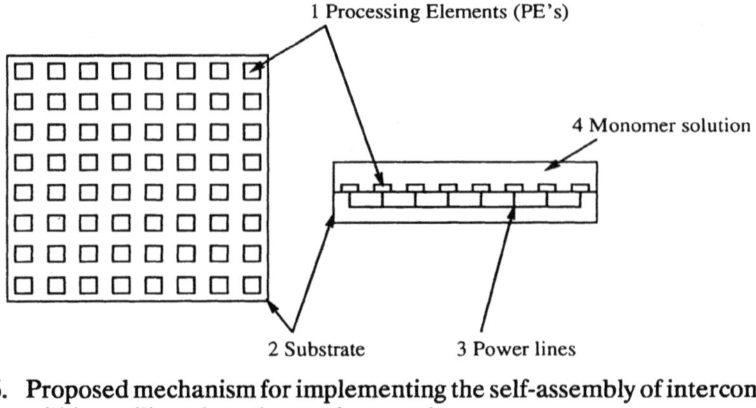

Figure 6. Proposed mechanism for implementing the self-assembly of interconnections within a silicon-based neural network.

Experiments have also shown the polymer to be a strongly anisotropic one-dimensional conductor. Thus, its lateral conductivity is many times less than its longitudinal conductivity along the carbon backbone. This property is crucial for this application since individual polymer growths from neighboring electrodes may cross and intertwine with minimal cross-talk. The unsupervised self-interconnection of this simple test structure suggests that it may be possible for interconnects based on conducting polymers to massively self-interconnect much larger structures based in silicon.

CONCLUSIONS

A statistical analysis of silicon-based processing systems suggests that the structure of the brain's dendritic arbor is an attempt to implement an optimal interconnection placement in a manner consistent with a hierarchical processing strategy. We propose that a Laplacian growth model is sufficient to explain how such an optimal placement could be achieved without prior planning and computer aided design packages.

In order to test this hypothesis computer simulations and a polymer-based analog of the dendritic growth process have been developed. Although the growth process is clearly idealized, the same broad features observed in neuronal interconnects are also observed in the polymer growths. This suggests that the Laplacian growth model may be a useful tool for the analysis of interconnection issues within the brain.

The fact that polypyrrole is a well understood conducting polymer which is known to form good conducting contacts with silicon and other metals suggests that this polymer may also be used to implement the connectivity in silicon-based neural networks. Clearly, much work needs to be done in this area but initial growth simulations suggest that large arrays of silicon processing elements could be massively interconnected using one-dimensional conducting polymers.

The implications are potentially far-reaching. Over the last two decades the trend of silicon technology has been towards ever smaller and more power-efficient devices. The projection of this trend into the next century indicate that molecular scale devices will be within our reach within the next twenty to thirty years. Yet we are always left with the nagging question of how to interconnect such devices at the molecular level. Perhaps, the future role of organic technology is not to compete with silicon in device implementation but rather to compliment it by using its inherent capacity for implementing interconnection complexity.

ACKNOWLEDGEMENTS

D. J. E. acknowledges the financial support of the University of Delaware Honors Program.

REFERENCES

1. R. E. Kalil, Synapse formation in the developing brain, *Sci. Am.* 76:December (1989).
2. N. H. Sabah, The concept of electrochemical potential and its application to semiconductor systems, *Int. J. Elect. Eng. Educ.* 13:176 (1976). [The concept of the chemical potential is explained in most textbooks on thermodynamics but a simple introduction to the basic concepts relevant from an engineering perspective may be found in this reference.]
3. R. W. Keyes, The wire limited logic chip, *IEEE J. Solid State Circuit* SC-17:1232 (1982).

4. B. S. Landman and R. L. Russo, On a pin versus block relationship for logic graphs, *IEEE Trans. Computers* C-20:1469 (1971).
5. H. B. Bakolgu, Section 9.4, Machine Organization and Rent's Rule, in: "Circuits, Interconnections, and Packaging for VLSI," Addison-Wesley, Reading MA (1990).
6. D. K. Ferry, L. A. Akers, and E. W. Greeneich, Section 5.2, in: "Ultra-Large Scale Integrated Microelectronics," Prentice-Hall, Englewood Cliffs NJ (1988).
7. P. Christie, A fractal analysis of interconnection complexity, submitted to *Proc. IEEE*. Special Issue on Applications of Fractals in Electrical Engineering (Late 1992). Currently available as a University of Delaware Technical Report, Udel-EE 92-3-2.
8. C. Mead, Neuromorphic electronic systems, *Proc. IEEE.* 78:1629 (1990). [An excellent overview of interconnection scaling issues is presented]
9. T. Tel, Fractals, multifractals, and thermodynamics, *Z. Naturforsch.* 43:1154 (1988).
10. T. Bohr and T. Tel, The thermodynamics of fractals, in: "Directions in Chaos," H. Bai-Lin, ed., World Scientific (1987).
11. H. E. Stanley and P. Meakin, Multifractal phenomena in physics and chemistry, *Nature* 335:405 (1988).
12. M. Kohmoto, Entropy function for multifractals, *Phys. Rev. A* 37:1345 (1988).
13. P. J. M. van Laarhoven and E. H. L. Aarts, Section 7.2 in: "Simulated Annealing: Theory and Applications," D. Reidel Publishing Company, Dordrecht Holland (1988).
14. D. F. Wong, H. W. Leong, and C. L. Liu, "Simulated Annealing for VLSI Design," Kluwer Academic Publishers, Norwell MA (1988).
15. P. Christie, Modeling the complexity of interconnect-limited systems, submitted to *IEEE Trans. Circuits and Systems* special issue on Simulation, Modeling, and Electrical Design of High-Speed and High-Density Interconnects (late 1992). Currently available as a University of Delaware Technical Report, Udel-EE 92-4-1.
16. G. de Vahl Davis, "Numerical Methods in Engineering and Science," Allen and Unwin, London (1986).
17. H. S. Carslaw and J. C. Jeager, Ch. 1, in: "Conduction of Heat in Solids," Clarendon Press, Oxford (1959).
18. W. H. Hayt, Jr., Ch. 7, in: "Engineering Electromagnetics," McGraw-Hill, New York (1989).
19. M. Jujii and K. Yoshino, Observation of electrochemical polymerization of conducting polymer and its interpretation by fractal dimension, *Jap. J. App. Phys.* 27:L457 (1988).
20. J. H. Kaufman, C. K. Baker, A. I. Nazzal, M. Flickner, O. R. Melroy, and A. Kaputnik, Statics and dynamics of the diffusion-limited polymerization of the conducting polymer polypyrrole, *Phys. Rev. Lett.* 56:1932 (1986).

FURTHER READING

A. L. Goldberger, D. R. Rigney, and B. J. West, Chaos and fractals in human physiology, *Sci. Am.* February:43 (1990).

P. Christie, Ch. 5 Clouds, Computers, and Complexity, in: "Frontiers of Computer Systems Research, Vol. 2," S. K. Tewksbury ed., Plenum Publishing Corporation, New York (1991).

INTERFEROMETRIC INVESTIGATIONS OF BILAYER LIPID MEMBRANE DEFORMATION AND FLEXOELECTRICITY

Angelio T. Todorov, Alexander G. Petrov, Michael O. Brandt, Gilles Picard, Normand Denicourt, and Janos H. Fendler

Département de Chimie, Université de Montréal, C.P. 6128, succursale A, Montréal, Québec, Canada H3C 3J7

Bimolecular thick, bilayer or black, lipid membranes (BLMs) provide the thinnest man-made semipermeable barrier which separates two compartments containing aqueous solutions.[1,2] Advantage has been taken of BLMs to model the functioning of the biological membrane by incorporating synthetic and natural ion carriers.[3,4] Sensitive electrical measurements, voltage clamping, and single-channel recording have formed the bases of our current understanding of biological transport mechanisms.[4,5] More recently, BLMs have been utilized as matrices for supporting size-quantized semiconductor and magnetic particles which mimic bulk photoelectrical and magneto-optical devices.[6,7] Continued utilization of BLMs for biophysical and solid-state modeling requires an understanding of their physical-mechanical properties.

We have recently launched investigations into the consequences of steady-state and oscillating-pressure-induced BLM deformation. Advantage was taken of optical interferometry to obtain continuous information on pressure- and applied-potential-induced deformation of glyceryl monooleate (GMO; Sigma) and phosphatidylserine (PS) BLMs. This allowed the first time visualization of *both* translational (lateral) movement and curvature changes of the BLM. Capacitance and voltage-dependent capacitance, resistivity, and voltage-dependent resistivity changes were determined simultaneously with the optical interferometric measurements.

EXPERIMENTAL SECTION

Glyceryl monooleate, bovine brain phosphatidylserine (PS; Sigma), and decane (Phillips Petroleum Company) were used as received. Water was distilled, passed through an ion-exchange column (Barnstead), and finally filtered using a Millipore system (water resistivity = 18 MΩ.cm).

BLMs were formed across a 1.10 ± 0.01 mm diameter conical hole drilled in a 0.52 ± 0.01 mm thick black Tefzel film at room temperature (*ca.* 20°C). The cone angle with respect to the film surface was determined to be 37°. The Tefzel film was inserted diagonally in a 1.00 cm optical path fluorescence cuvette and sealed with silicone rubber glue. The cuvette was filled with 0.10 M KCl solution. BLMs were made by smearing the BLM-forming solution (50 mg GMO or 25 mg PS per mL of decane) across the Tefzel aperture with a steel needle. The needle was rinsed with methanol after each dipping. Surfactant solutions were maintained in a nitrogen atmosphere between usage.

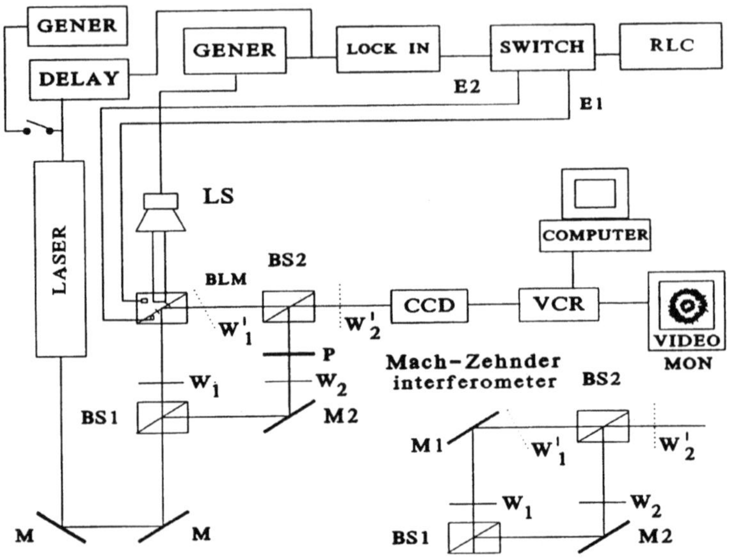

Figure 1. Schematics of the experimental set used for simultaneous electrical and real-time stroboscopic interferometric measurements of BLMs. A Quantronix 4116 Q-switched, mode-locked and pulse-picked Nd-YAG laser provided second-harmonic 532 nm, 120 ps, 40 μJ pulses at 50-5000 Hz repetition. Delay was provided by a 111 AR Avionics Inc. delay generator. A Hewlett Packard 8116 A function generator was used. The lock in was a Dynatrac 391 A Ithaco lock-in amplifier. The switch box was home built. RLC = 1689 M GENRAD digibridge, LS = a commercial 8 W loudspeaker, BLM = the cell compartment for the bilayer lipid membrane, BS1 and BS2 = beam splitters, M and M_2 = mirrors, CCD = an NEC *color* camera, VCR = a commercial VHS video recorder, VIDEO MON = NEC *color* monitor, and COMPUTER = a Hewlett Packard IBM-compatible PC. For comparison, the scheme of a simple Mach-Zehnder interferometer is included.

Thinning of the membrane to a BLM was monitored by both capacitance measurements and reflectivity. With thinning, the intensity of reflected light decreased and interference patterns appeared upon the formation of a BLM. Special care was taken to obtain as flat a BLM as possible. Specifically, the size and the geometry of the pinhole in the BLM-supporting Tefzel film were carefully optimized.

Schematics of the experimental set-up used for steady-state and real-time stroboscopic interferometric measurements of BLMs are shown in Figure 1.

RESULTS AND DISCUSSION

BLM Formation and Characterization

Stringent requirements were placed on BLM preparations. Measurements were performed only on BLMs which had relatively small Plateau-Gibbs borders (PG border < 10% of total BLM surface), survived for four hours or longer, and were relatively flat. Specific capacitances, C_S ($C_S = C/S_o$ where C is the measured capacitance and S_o is the surface area of the planar BLM, determined by careful measurement of the calibrated video images of the BLM), were determined to be 0.387 ± 0.010 $\mu F/cm^2$ and 0.391 ± 0.010 $\mu F/cm^2$ for the GMO and PS BLMs. For a given BLM, the C_S value remained within 10% of its initially determined value during flexoelectric measurements and subsequent to them. This necessitated the monitoring of capacitances for four to five hours. It is important to recognize that slight (<10%) variations in the BLM radius, in the torus-to-bilayer ratio, or in the specific capacitances did not affect hydrostatic-pressure-induced changes or the value of the flexoelectric coefficient, determined at a given frequency.

Steady-State Hydrostatic-Pressure-Induced BLM Deformation

Optical interferometry has been found to be particularly valuable for monitoring BLM deformation.[8] Interference of two plane waves occurs when the light beam follows more than one path from its source to the point of detection, such as in the Mach-Zehnder interferometer or in the system used in the present work (Figure 1).[9]

Introduction of 100 μL of liquid into one side of the BLM resulted in an interferogram, shown in Figures 2a and 3a, with rapidly appearing fringes (on the time scale of seconds), which were counted at a defined position on the BLM and related to the hydrostatic-pressure-induced lateral movement of the BLM.[8] This BLM was assessed to be laterally displaced by 5.55 μm in the x and 7.83 μm in the y direction (Figure 2b). The altered lateral position is clearly seen in the three-dimensional image of the displaced BLM (Figure 3b).

A good relationship was obtained between the extent of BLM curvature changes and the applied hydrostatic pressure responsible for them. A delicate balance between opposing forces is responsible for maintaining all surfactant assemblies, including BLMs.[10] The repulsive force between headgroups are counteracted, in part, by the interfacial pressure of the aqueous solution on the surfactants. Since this attractive force (by unit length) is acting in the plane of the monolayer (*i.e.*, one-half layer of the BLM), it is called interfacial surface tension and is designated by γ_m for monolayers and by γ for BLMs ($\gamma = 2\gamma_m$). Thermodynamically, γ is defined as the force (by unit length) required to increase the volume of the spherical cup by an infinitesimally small value, dV_c, by stretching it by an infinitesimally small area, dA_c at a hydrostatic pressure, P_{hp}, and it is obtained from:

$$\gamma = \frac{R_m^2}{4}\left(\frac{P_{hp}}{F_c}\right)$$

where P_{hp}/F_c is the inverse of the slope of the lines obtained on plotting the extent of BLM curvature changes against the applied hydrostatic pressure and R_m is the radius of the BLM (excluding the Plateau-Gibbs border) and F_c is the distance of the curvature displacement of the BLM. Determination of R_m for each BLM preparation

a

b

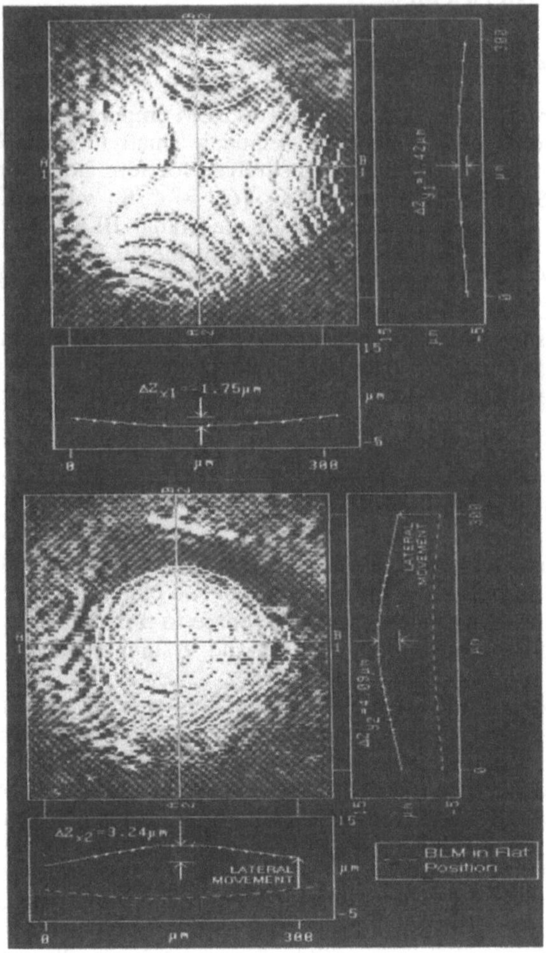

Figure 2. Computer-enhanced interferograms of a GMO BLM (radius of the BLM = 147 μL, bathed in 0.10 M KCl) prior (a) and subsequent (b) to the application of static 18 mNm^{-2} hydrostatic pressure.

led to mean interfacial surface tensions (γ) of 0.20 ± 0.02 mN/m and 0.29 ± 0.03 mN/m for GMO and PS BLMs, respectively. The value of γ for GMO BLMs agrees well with that obtained previously by two-exposure interferometric holography (γ = 0.2 mN/m) or by capacitance measurements as a function of applied hydrostatic pressure (γ = 0.35 mN/m).[11]

Oscillating Hydrostatic-Pressure-Induced BLM Deformation

Flexoelectricity manifests in the development of microvolt-range transmembrane potential upon subjecting one side of the BLM to an oscillating hydrostatic pressure at a given frequency. The phenomenological definition of the flexoelectric coefficient, f, is the ratio between the bending-induced transmembrane potential, U_f, and the change of curvature (c) which accompanies the bending of the membrane:[12]

$$f = \frac{U_f \varepsilon \varepsilon_o}{2c}$$

where ε_o is the absolute dielectric permittivity of vacuum and ε is the permittivity of the hydrocarbon region of the BLM. f was evaluated by real-time stroboscopic interferometric measurements.

Selected interferograms capturing a complete cycle of 300 Hz oscillating-pressure-induced real-time movement of a GMO BLM are illustrated in Figure 4. The sixteen interferograms, representing different positions during the movement of the

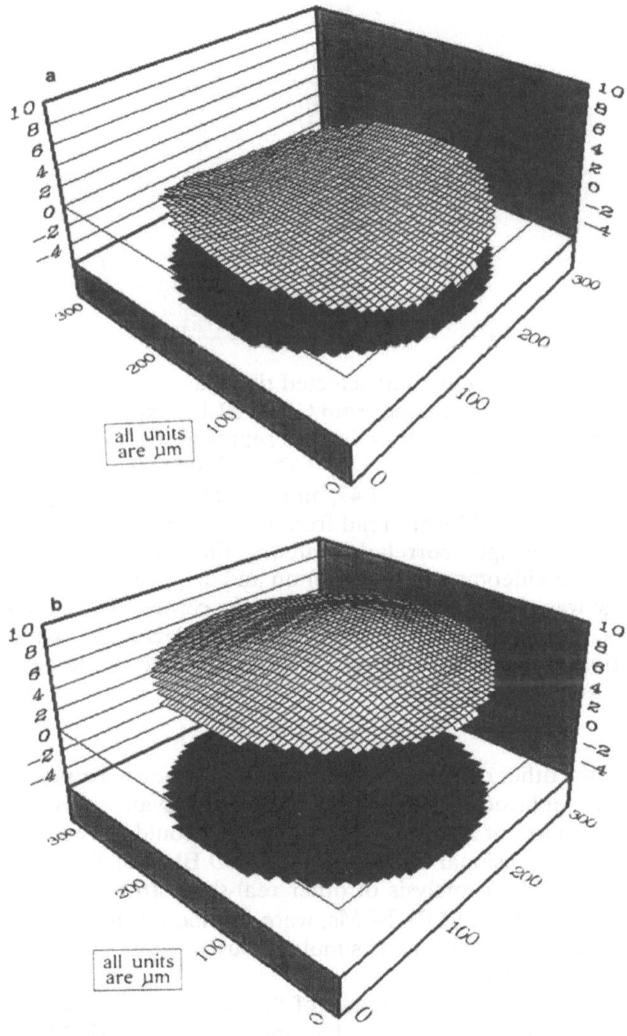

Figure 3. Three-dimensional presentations of the GMO BLM shown in Figure 2 prior (a) and subsequent (b) to static bending. Notice the difference between the scales (0 to 300 µm and -4 to 10 µm)!

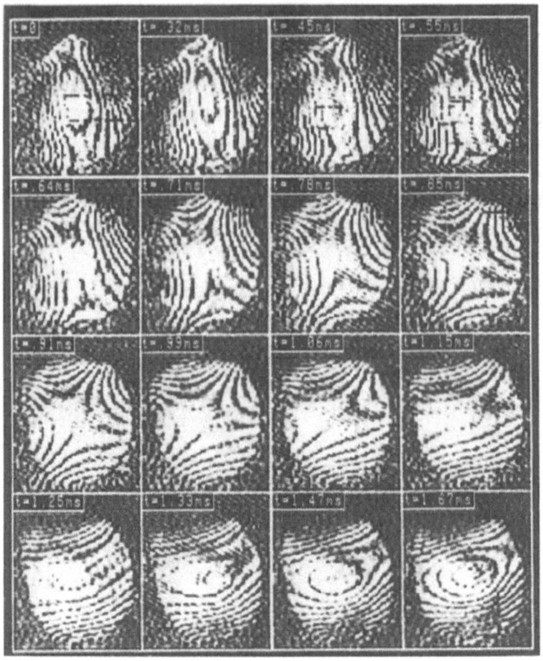

Figure 4. Interferograms obtained at selected times during the real-time movements of a GMO-BLM (prepared from GMO BLM) exposed to 300 Hz oscillating pressure. The sixteen static images representing a complete movement of the BLM, were taken at t = 0, 0.32, 0.45, 0.55, 0.64, 0.71, 0.78, 0.85, 0.91, 0.99, 1.06, 1.15, 1.25, 1.33, 1.47, and 1.67 ms. The time series of images of the oscillating BLM were read from the videotape and stored as a series of disk files. Height correlation among the images was accomplished by playing the videotape in slow motion and acquiring a new image each time a fringe was created at, or moved past, the center. Thus, the absolute order of the most central fringe increased (or decreased, depending on the direction of movement) by one from each image to the next, indicating its absolute height.

BLM from one to the other extreme, were obtained by capturing *distinct* videorecorded images from those replayed in slow motion. Taking advantage of computer image analysis, the entire course of a given BLM's movement could be reconstructed in three dimensions. Three-dimensional images of the GMO BLM at the three positions are shown in Figure 5. Similar analysis of other real-time stroboscopic interferometric images, obtained for GMO and PS BLMs, were carried out at different frequencies. U_f was determined routinely and simultaneously with real-time stroboscopic interferometric measurements.

The two limiting frequency regimes of the flexoelectric coefficients have been experimentally realized for both the PS and the GMO BLMs in the optical measurements. Mean f values of 24.1 x 10^{-19} Coulombs and 16.5 x 10^{-19} Coulombs were obtained for PS BLMs. For GMO BLMs, the obtained mean f values were 0.87 x 10^{-19} Coulombs and 0.30 x 10^{-19} respectively.

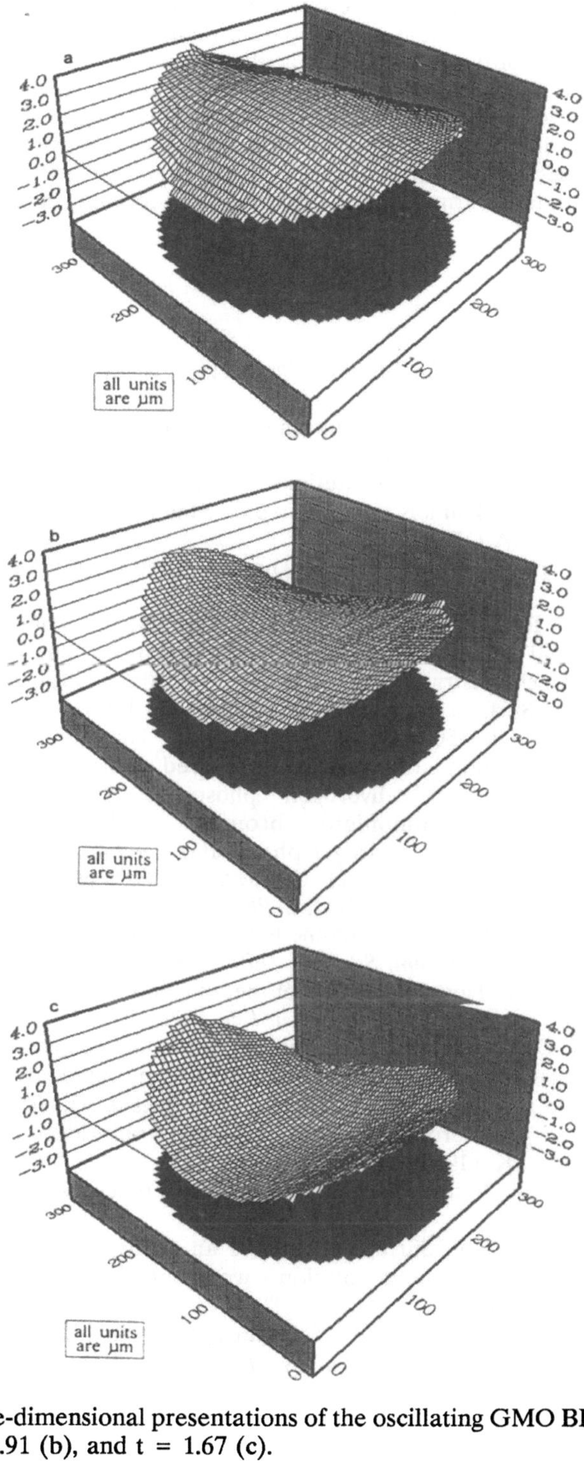

Figure 5. Three-dimensional presentations of the oscillating GMO BLM at t = 0 (a), t = 0.91 (b), and t = 1.67 (c).

Methodologies developed here open the door to systematic investigations of mechanosensitive ion transport and impulse transmission in membrane biology, as well as to the exploitation of flexoelectricity in artificial systems. Flexoelectricity in BLMs is a unique mechanoelectric phenomenon simultaneously involving two degrees of freedom of the molecular system–electrical and mechanical. Utilization of this phenomenon should allow, at least conceptually, the construction of molecular elements with mechanical input-electrical output and/or with electrical input-mechanical output, thus permitting the fabrication of supramolecular sensors and/or actuators.

REFERENCES

1. J. H. Fendler, "Membrane Mimetic Chemistry," Wiley-Interscience, New York (1982).
2. H. T. Tien, "Bilayer Lipid Membranes (BLM) Theory & Practice," Marcel Dekker, New York (1974).
3. C. Miller, "Ion Channel Reconstitution," Plenum, New York (1986).
4. B. Hille, "Ionic Channels of Excitable Membranes," Sinaver Associates, Sunderland, MA (1984).
5. B. Sakmann and E. Neher, "Single-Channel Recording," Plenum, New York (1983).
6. X. K. Zhao, S. Baral, R. Rolandi, and J. H. Fendler, Semiconductor particles in bilayer lipid membranes (BLMs). Formation, characterization, and photoelectrochemistry, *J. Am. Chem. Soc.* 110:1012 (1988); X. K. Zhao and J. H. Fendler, Surface enhanced spectroscopy on bilayer lipid membranes, *J. Phys. Chem.* 92:3350 (1988); Y. Yuan, P. Tundo, and J. H. Fendler, Photopolymerization of vesicles prepared from n-Hexadecyl 11-(4-vinylbenzamide)undecyl hydrogen phosphate and from mixtures of dioctadecyldimethylammonium bromide and n-Hexadecyl 11-(4-vinylbenzamide)undecyl hydrogen phosphate, *Macromolecules* 22:29 (1989); X. K. Zhao, P. J. Hervé, and J. H. Fendler, Magnetic particulate thin films on bilayer lipid membranes (BLMs), *J. Phys. Chem.* 93:908 (1989); S. Baral and J. H. Fendler, Cadmium-sulfide-mediated photoelectric effects in bilayer lipid membranes, *J. Am. Chem. Soc.* 111:1604 (1989).
7. J. Kutnik and H. T. Tien, Deposition of metallic and semiconducting layers onto bilayer lipid membranes, *Photochem. Photobiol.* 46:413 (1987).
8. G. Picard, N. Denicourt, and J. H. Fendler, Simultaneous electrical and optical interferometric measurements of pressure- and applied-potential-induced bilayer lipid membrane deformation, *J. Phys. Chem.* 95:3705 (1991).
9. W. H. Steel, "Interferometry," 2nd Edition, Cambridge University Press, Cambridge, U.K. (1903).
10. J. N. Israelachvili, "Intermolecular and Surface Forces," Academic Press, New York (1982).
11. A. Moran and A. Ilani, Surface tension of an artificial bileaflet membrane in comparison to parent lipid solution - water interfacial tension, *Chem. Phys. Lipids* 4:169 (1970).
12. A. G. Petrov and V. S. Sokolov, Curvature effect in black lipid membranes: dynamic characteristics, *Eur. Biophys. J.* 13:139 (1986).

ION PERCOLATION THROUGH A CUBIC LIPID/WATER PHASE

Luca Turin

Laboratory of Cellular and Molecular Pharmacology, NIEHS, Research Triangle Park, NC 27709 and Unité Associée 671 du CNRS, Station Marine, F-06230 Villefranche/Mer, France

INTRODUCTION

X-ray diffraction studies have shown that mixtures of certain lipids with water yield exotic regular phases in which water is arranged in interpenetrating cubic networks of channels.[1,2] Though electrical properties of amorphous lipid-water systems have been extensively studied (see reference 3 for review), regular phases have so far received little attention. A cubic phase formed in a saline solution will conduct electricity if the network of water channels containing dissolved salts spans the sample. By contrast, the unhydrated lipid will have a very low conductance. Starting with pure lipid, the time evolution of conductance as the proportion of water within the lipid increases could provide insight into the kinetics of formation of porous phases otherwise difficult to obtain from structural measurements. Here I report continuous measurements of the ionic conductivity of monoolein samples as water concentration increases from near zero to its equilibrium value in presence of free water. A definite conductance threshold is observed at an intermediate water concentration, after which conductance follows a power law. The results are discussed in the framework of percolation theory.

METHODS

Monoolein (1-monooleoyl-*rac*-glycerol, Sigma) is a convenient lipid to use, because it forms a cubic phase (Q^{224}) stable in the presence of excess water (Mariani et al., loc. cit.). To measure the conductance of water-MO samples, a glass capillary of 1.1 mm internal diameter was filled with 3 mM NaCl and connected to the headstage of a patch-clamp amplifier (Axopatch 1-C, Axon Instruments, Burlingame, CA) via an Ag/AgCl half-cell consisting of a wire running down the length of the capillary down to about 10mm from the open end. The principle of the measurements was as follows: 1-2 mm of the pipette medium were expelled from the tip, and replaced with liquid monoolein from a dry sample held at 55°C. The end of the capillary was then dipped in a water bath containing the same NaCl solution used to

Synthetic Microstructures in Biological Research, Edited by
J.M. Schnur and M. Peckerar, Plenum Press, New York, 1992

fill the pipette, and the bath connected to earth via a second Ag/AgCl half-cell. A small sinusoidal voltage (2 mV amplitude, f=80 Hz) was then added to the command voltage of the patch clamp amplifier. The resulting current signal was demodulated by a two-phase lock-in amplifier referenced to the command sinewave. The lock-in amplifier then produces two DC signals proportional respectively to pipette conductance and capacitance. The data was sampled at 16-bit resolution with a Macintosh, using Labview software (National Instruments, Austin, TX) and analysed with Igor software (WaveMetrics, Lake Oswego, OR).

Early experiments revealed that it was essential to avoid mixing of the lipid and water during the pipette filling phase in order to obtain a stable baseline value for the nonconducting lipid. The filling procedure was modified as follows. Medium was first expelled from the pipette, and 1mm or so of air was then sucked into the free end. The pipette was then dipped into the lipid, and a further 2mm of lipid was aspirated. The air bubble acted as a spacer preventing contact with water until desired. The capillary was tapped to dislodge the air bubble which then rises past the chlorided silver wire. The wire shunts the air bubble and ensures electrical continuity. With this arrangement, measurements could be obtained from approximately ten seconds after the lipid was put in contact with water on both its faces. At the end of each run, the setup was calibrated with a resistor, and the series resistance of the medium between the tip of the silver wire was measured by breaking off the part of the capillary filled with lipid and measuring the conductance of the shortened tube. Conductance records were compensated for series resistance. No compensation was made for the swelling of the lipid plug during hydration, but calculations show that this introduces no more than a few percent error. All experiments were performed at temperatures of 21-23°C.

RESULTS

In 45 out of a total of 53 experiments, conductance began to rise within 2-3 minutes of the lipid plug being in contact with water, went through an inflection point and smoothly rose to a maximum. At much longer times, it sometimes relaxed slowly to a level at 80% or so of maximum. Figure 1 illustrates a typical record.

Measurements were usually stopped when conductance had reached a steady value, i.e. at the peak of the record in figure 1. In every case, the lipid plug inspected at the end of the experiment was found to have undergone an approximately twofold increase in volume, and appeared optically clear and homogeneous, with no visible birefringence when viewed between crossed polarizers. A striking observation was that the hydrated plug was very stiff, whereas the unhydrated lipid flowed easily out of the pipette under pressure. These properties are consistent with the notion that most of the lipid is in a viscous, conducting and isotropic phase containing 50% or so water in volume. These are also the properties of the bicontinuous cubic phase Q^{224} of monoolein, the only one known to be in stable equilibrium with excess water. At very long times of the order of several hours, the ends of the lipid plug began to scatter light as if the phase was slowly disaggregating.

The evolution of conductance with time exhibited an interesting discontinuity occurring within the first few minutes. Figure 2 plots the square root of conductance vs time in one experiment, starting just after lipid-water contact was established. A slow upwards creep in conductance from the very start is evident. After 2.8 minutes, however, there is an abrupt change of slope and for the next 3 minutes, $g(t)^{1/2}$ follows a straight line, indicating that g(t) is parabolic. This feature, though often less clearly delineated than in Figure 2, was present in all the records. After the threshold point, $g(t)^{1/2}$ appeared to follow first-order kinetics, usually with two time constants.

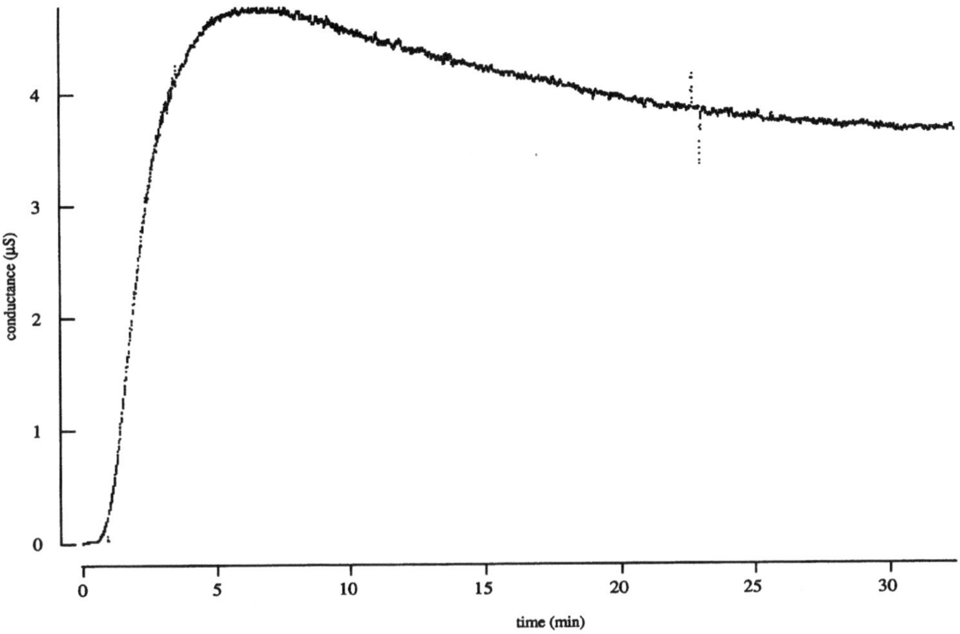

Figure 1. Conductance of a cylindrical monoolein sample as a function of time. The origin of the time axis corresponds to the moment of first contact with water. Three phases in the evolution of conductance are visible: a short latency of less than 1 minute, followed by a smooth sigmoidal rise to a peak, and a slow decay to a lower, stable conductance level. Expt 8.

This suggested that the evolution of conductance with time might be approximated by a simple relaxation process raised to a power close to 2. The conductance threshold was therefore taken as the origin of time and conductance coordinates, and records were fitted with the following equation

$$g(t)=a[(1-\exp(-t/\tau_1))+(1-\exp(-t/\tau_2))]^m$$

Table 1 lists the parameters τ_1, τ_2 and m for the 10 digitized traces. The mean exponent was 2.08, and its range was 1.64-3.06. Seven out of ten values fell between 1.9 and 2.1. Two fits from this set of 7 are shown in figure 3, showing a good agreement between data and percolation theory (see discussion).

In the remaining 8 experiments, a very different behaviour was observed. Conductance rose very slowly and linearly for up to 4 hours. Whenever this happened, a large section in the middle of the plug, amounting to between 1/3 and 3/4 of the plug was found to be strongly birefringent. Birefringence and low but nonzero conductance strongly suggest that the middle plug may be in one of the hexagonal phases. Conversion to the cubic phase occurs very slowly at room temperature. In some experiments, however, the birefringent plug was repeatedly heated for a few seconds to 50°C. It became thinner at each heating until it disappeared completely. After this, conductance values typical of nonbirefringent plugs were obtained.

Table 1

expt. number	τ_1	τ_2	m
2	11.33	54.57	1.64
5	3.18	139.7	2.47
6	2.32	3.8	2.03
8	.70	1.14	3.06
9	10.04	52.29	1.92
12	5.5	31.05	1.9
13	4.86	55.29	1.95
14	5.18	32.06	1.77
15	7.53	183.12	2.05
16	8.08	32.43	1.99

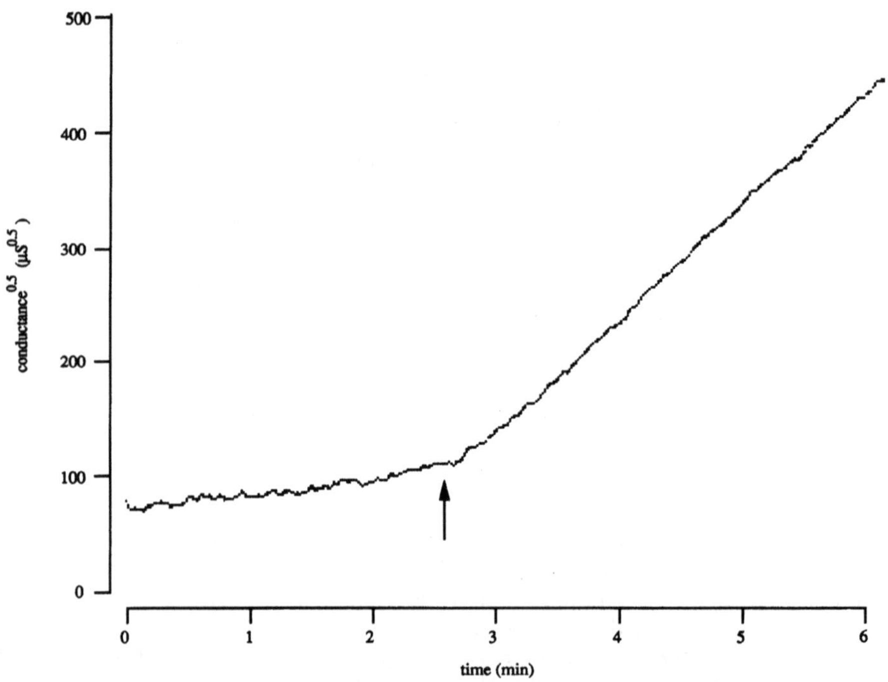

Figure 2. Early portion of a conductance record shown at higher resolution. The square root of conductance is plotted against time. The origin of the time axis corresponds to the moment of first contact of the monoolein sample with water. A slow upward creep in conductance is evident. After 2.8 minutes, a sharp change in slope is seen (arrow), after which $g^{0.5}$ follows a straight line, indicating a parabolic rise in conductance. Expt 12.

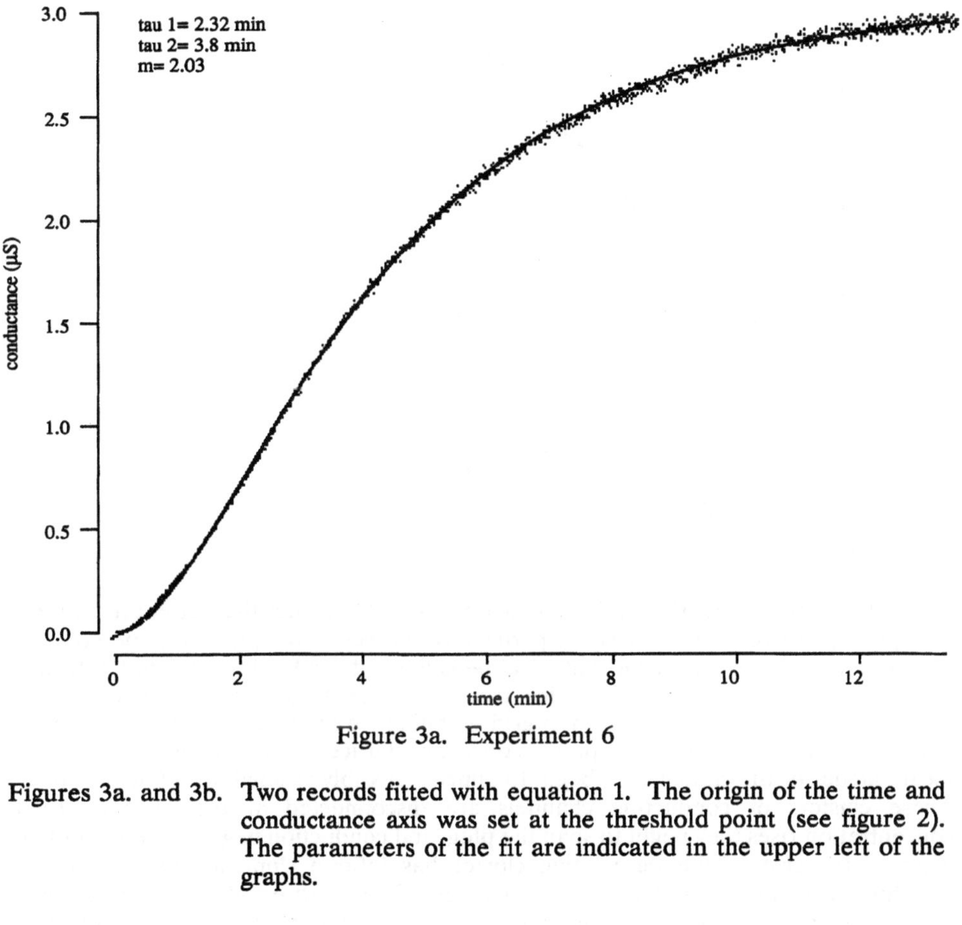

Figure 3a. Experiment 6

Figures 3a. and 3b. Two records fitted with equation 1. The origin of the time and conductance axis was set at the threshold point (see figure 2). The parameters of the fit are indicated in the upper left of the graphs.

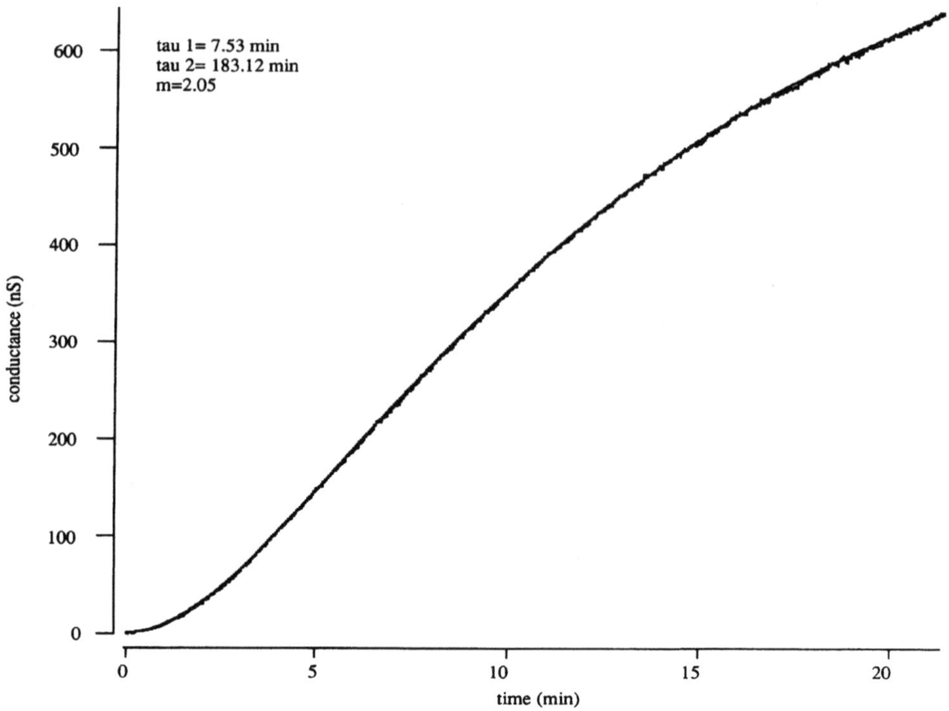

Figure 3b. Experiment 15

DISCUSSION

Both the definite threshold and the power-law behaviour thereafter are features strongly suggestive of an underlying percolation process. We recall that percolation theory seeks to account for the statistical behaviour of randomly populated networks of links connecting sites on a lattice (see reference 4 for review). Clearly, the establishment of the conducting water grid spanning the lipid plug shares features with a three-dimensional percolation process on a cubic lattice. As water activity increases, local patches of the cubic phase begin to appear, possibly as small as a few unit cells. These clusters of conducting channels are disconnected at first, but as their concentration rises they begin to span the plug, and conduction begins. We recall that whereas the size of the percolating cluster has infinite slope at the percolation threshold, conductance is expected to have zero slope at threshold, as we observe.

An unusual feature of the monoolein-water system is that the conducting network will possess a highly regular, predictable geometry, as witnessed by the fact that crystal X-ray diffraction patterns can be obtained from samples of dimensions comparable to ours. This makes it possible to use the exact results obtained for the 3-D cubic lattice by Kirkpatrick.[5] He has shown analytically that conductance of the cubic lattice is proportional to the square of the site occupation probability. Considering the simplicity of the experimental setup, the agreement between experiment and theory is encouraging.

In summary, the monoolein-water phase appears to be an unusually simple and regular system exhibiting percolation properties. There remain a number of features, however, which require further exploration. The physical meaning of the multiple time-constants and of the conductance droop at long times is still unclear, as is the nature of the initial conductance "creep" before the threshold. No doubt edge effects, such as the existence of a conducting layer close to the wetted glass, will have to be kept into account in more refined modelling. Further experiments would also benefit from direct measurements of other relevant variables: for example, the percolation threshold could be studied more precisely by measuring water concentration inside the lipid either from volume changes or by infrared absorption.

ACKNOWLEDGEMENTS

David Armstrong and James Putney are thanked for their hospitality and advice. The experiments described above were suggested to me by Vittorio Luzzati, whose help is gratefully acknowledged. This work was supported by the Centre National de la Recherche Scientifique.

REFERENCES

1. W. Longley and T. J. McIntosh, A bicontinuous tetrahedral structure in a liquid-crystalline lipid, *Nature* (London) 303:612-614 (1983).
2. P. Mariani, V. Luzzati, and H. Delacroix, Cubic phases of lipid-containing systems, *J. Mol. Biol.* 204:165-198 (1988).
3. R. Larter, Oscillations and spatial nonuniformities in membranes, *Chem. Rev.* 90:355-381 (1990).
4. S. Kirkpatrick, Percolation and conduction, *Rev. Mod. Phys.* 45:574-588 (1973).
5. D. Stauffer, "Introduction to percolation theory," Taylor & Francis, London (1985).

APPLICATIONS

INTRODUCTION

Applications of synthetic microstructure technology to biological research extends well beyond the fabrication of probes of electrical or chemical state. In this section, three papers appear dealing with novel applications of microstructure technology. The first of these papers (**Breckenridge et al.**) deals with the use of microstructures in the study of cell colony morphogenesis. As cells organize into colonies, they look for topological cues to provide assembly directions. These cues are of near- or sub-micron dimensions. As such, they can be approximated using the fabrication techniques discussed in this volume. Breckenridge et al. use such artificial microstructures to gain an understanding of the importance of topography in mammalian cell organization.

Similarly, cells look for a variety of chemical cues and "adhesives" to bind to various sites. An understanding of this binding process and an ability to manipulate it is of prime importance in interfacing cells to more conventional sensing systems. The incorporation of these interface materials in bio-probes and sensors is an essential element in the application of these devices. Furthermore, the installation of the requisite layers is clearly in the domain of microstructure science and technology. The paper by **Pizziconi and Page** outlines the fundamentals of this technology.

As a further extension of this theme of microstructure-induced ordering in biological assemblies, we have the paper by **de Medinaceli and Wyatt**. Here, we learn how fabrication of microstructural guiding sheets could be fabricated to aid in nerve re-alignment in neurosurgery.

ARTIFICIALLY INDUCED NERVE CELL PATTERNING OR REAL NEURAL NETWORKS

Lorna Breckenridge, Peter Clark, Patricia Connolly, Adam S.G. Curtis, Julian A.T. Dow, Richard Wilson, Robert Lind, and C.D.W. Wilkinson

Department of Electronics and Electrical Engineering and Department of Cell Biology, The University, Glasgow, Scotland, United Kingdom

ORGANISATION OF CELLS INTO PATTERNS

Biological cells respond to many cues in their environment.[2] The topography of the surrounding space and the presence of chemicals which cause adhesion both affect their behaviour. The topographic cues have been recognised for many years. Harrison[3] in 1911 described the behaviour of cells grown on spider web fibres, noting that the cells tended to follow grooves or spaces between more solid organs while Dunn[4] studied the reaction of cells to cylindrical fibres and found that they were sensitive to the curvature of the surface and would align on fibres of less than 200μm diameter.

The advent of lithographic techniques capable of patterning substrates to give a vast range of shapes has opened up the range of topographies which can be studied. In a series of experiments, Clark et al.[5-7] have studied the reaction of cells to steps, micron-sized grooves and very fine submicron grooves. The protocol adopted for the study of the reaction of cells to steps was to observe, using time lapse photography, the number of cells crossing a step of given height. The control for no step was the number of cells crossing an arbitrary line on the recorded micrograph. The steps were made by dry etching Perspex (poly(methylmethacrylate)) in oxygen, using an aluminium mask defined by photolithography. To ensure an uniform chemistry of the surface, some samples were etched overall after the step had been formed and the aluminium removed.[5]

The results for a cell line, BHK baby hamster kidney cells are shown in Figure 1. It can be seen that while no height of step completely inhibits crossing of the step, the probability of a successful crossing reduces as the step height increases. Cells which do not cross may align with the step and again the probability of such alignment increases with the step height. the response to the step varies with cell type. While the probability of crossing a 5μm step is much the same for baby kidney hamster cells, chick heart fibroblasts and chick cerebral hemisphere cells (neural cells), rabbit peritoneal exudate neutrophils were much less inhibited by the step, in line with the role of leucocytes in the body.

Thus the response to steps can be expressed in terms of the probability of the cell making a protrusion in a given direction. This ability to make protrusions is a

Synthetic Microstructures in Biological Research, Edited by
J.M. Schnur and M. Peckerar, Plenum Press, New York, 1992

function of the organisation of the cyto-skeleton -- the framework of micro-filaments, microtubules and intermediate filaments which control the motion or extension of a cell.

The response of cells to grooves[6] also shows the effects of the cyto-skeleton. The method of study of the reaction of cells to grooves was to plate the cells onto the surface and observe, after a given time, the degree of alignment of the cell to the axis of the groove. Typical data for the degree of alignment of BHK cells is shown in Figure 2. It can be seen that in grooves which are 2μm wide and have 2μm spacing, the alignment is almost complete for 2μm deep grooves. Even for 0.2μm deep grooves a significant degree of alignment occurs. In general, it is found that the groove depth has the strongest effect on alignment, the effect of groove width being weaker, but tending to increase for narrower grooves.

Cultured chick embryo neurons (from the dorsal root ganglia) plated onto 8μm grooves with 20μm wide ridges showed little alignment when the grooves were 1μm deep. However on 2μm deep grooves, the neurite outgrowth was markedly aligned to the groove direction.

Given these results it is natural to reflect on the behaviour of cells when presented with grooves which are less than their 'width'. To answer this question we made 260nm period square wave gratings in fused silica;[7] such gratings present a series of grooves to the observed cells. The gratings were defined in photoresist using crossed beam of blue light and the resulting surface corrugation shadowed by evaporation of metal at a shallow angle.[8] This metal acted as a mask during reactive oxygen ion etching of the photoresist to give a deep high aspect ratio resist pattern which itself could be used as a mask for etching the fused silica in CHF_3.

Isolated cells from both cell lines aligned on such gratings; but islands of the epithelial MDCK cells did not. Moreover, the morphology of the MDCK cells was markedly affected by the presence of the gratings. However, such gratings had little effect on chick embryo nerve cells.

EXTRACELLULAR ELECTRODES FOR OBSERVATION OF ELECTRICAL ACTIVITY

To make a neural network of specified connectivity,[1] both a means of guiding the nerve cells processes and a mechanism for recording the electrical activity of the cells is required. Conventional intracellular electrodes give an excellent signal to noise

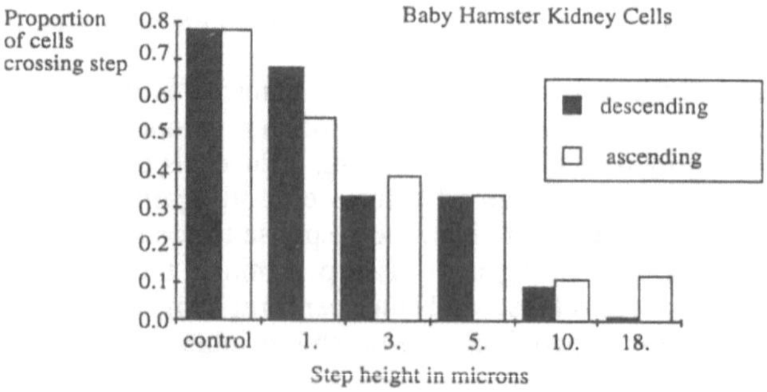

Figure 1. Probability of BHK cells crossing step of given height.

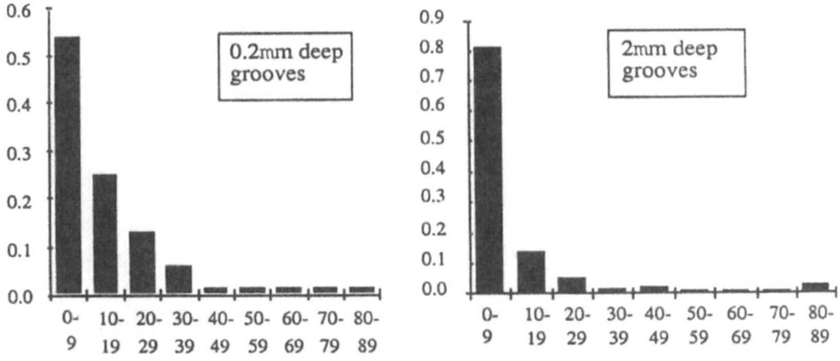

Figure 2. Proportion of cells with long axis in direction of grooves. BHK cells 0.2μm grooves and spaces.

ratio, but they are fragile and it is difficult to manipulate a large number of them simultaneously. The insertion of the electrode is inevitably traumatic to the neuron, and so observation over extended periods of time, say days, is not possible. For these reasons, one would like to use an essentially passive method, either extracellular electrodes or voltage sensitive dyes.

Extracellular electrodes have been used for observation of the nervous signals *in vivo*[9] and *in vitro*.[10-11] The signal strength is some 100 times smaller than is observed with an intracellular electrode. In our work[12] the electrode is a 60nm thick layer of gold and is insulated by a 1μm layer of Silicon Nitride deposited by plasma chemical vapour deposition. As shown in Figure 3, the Silicon Nitride is opened to reveal a 8μm by 8μm area of electrode by reactive ion etching in a C_2F_6 plasma. For experimentation with live cells, a cloning ring is cemented over the patterned area so that saline, medium and cells can be introduced above the electrodes. The electrodes are then platinised by passing a 1μA current through each electrode individually for 10 seconds. The platinisation reduces the impedance of the electrode, increasing the level of the recorded signal.

When an action potential is passing through a neuronal process, there is ionic transport through the membrane. This potential gives rise to ionic currents in the neighourhood of the membrane. Thus, if an electrode is placed a small distance from the cell, a voltage will be detected in an amplifier attached to the electrode. The waveform of this extracellular disturbance due to the passage of the action potential depends on the extracellular ionic currents and on the filtering characteristics of the electrode and electrode-electrolyte interface (which both can be represented as resistor-capacitor networks). It is by no means a replica of the voltage waveform of the action potential observed with an intracellular electrode.

It is important to maximize the signal to noise ratio from an extracellular electrode. The voltage picked up by the electrode can be simulated by calculating the current flow from an idealised cell. Taking the cell as a sphere emitting current and using finite element analysis, Lind et al.[13] showed that the topography of the space surrounding the cell was important in increasing the current and that a cell on an electrode in a deep well would give the most voltage. The calculation showed that levels as high as 2mV were possible.

The alternative passive method for observing electrical activity is to use dyes which change their absorption as a function of the local voltage. This technique is in

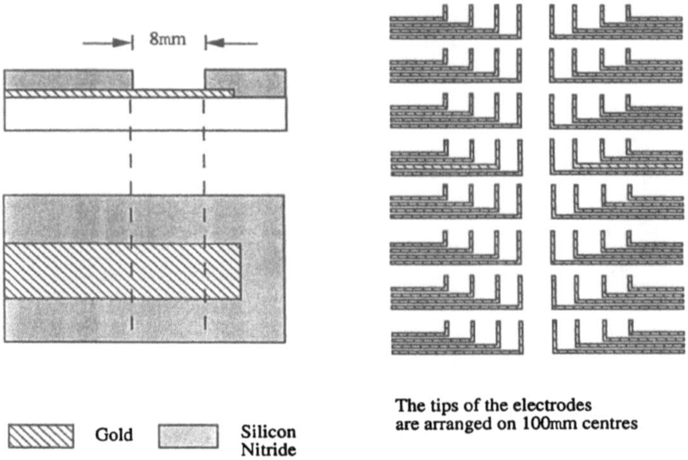

Figure 3. Diagram of one extracellular electrode and layout of electrode array.

principle very attractive as potential can be sensed all over a sample. Unfortunately, there are difficulties in three areas: the toxicity of the dye, bleaching of the dye under optical illumination, and the magnitude of the change in absorption. In recent elegant observations, Parsons et al.[14] observed 50 by 50μm areas (with a photodiode measuring each area) and obtained a signal to noise ratio of 10 or more. Observation was maintained for periods of 3 hours or more. This should be compared with extracellular electrodes which can, in principle, be placed as close as 10μm apart, have signal to noise ratios of 100 or more and can be used for days. Better voltage sensitive dyes could change this comparison.

PROGRESS TOWARDS NEURAL NETWORKS

The central nervous system of the pond snail *Lymnaea stagnalis* has been the subject of much study using classical neurophysiological techniques in recent years. The identity, function, and characteristics of many individual neurons and their involvement in circuits controlling simple rhythmatic behaviour patterns such as feeding and locomotion is known.[15] Identified dopaminergic neurons from the central nervous system of this animal were cultured and formed active electrotonic connection, reproducing *in vitro*, the behaviour seen *in vivo*. This was observed using extracellular electrodes and the behaviour persisted for a week.

Again the leech provides a useful experimental vehicle as neurons can be removed with long processes still attached. Figure 4 shows the stimulation of such a neuron by the injection of a current pulse using an extracellular electrode. The corresponding signals picked up from the process as the action potential propagates down the process past two more extracellular electrodes. The neuron was aligned in a groove which crossed a series of electrode tracks thus defining a series of microelectrodes. An intracellular electrode is inserted into the cell body. This experiment shows that both stimulation and recording can be done using extracellular electrodes and, thus, that the basic concept that using extracellular electrodes long-term recording of the behaviour of neuron networks of a chosen configuration can be achieved.

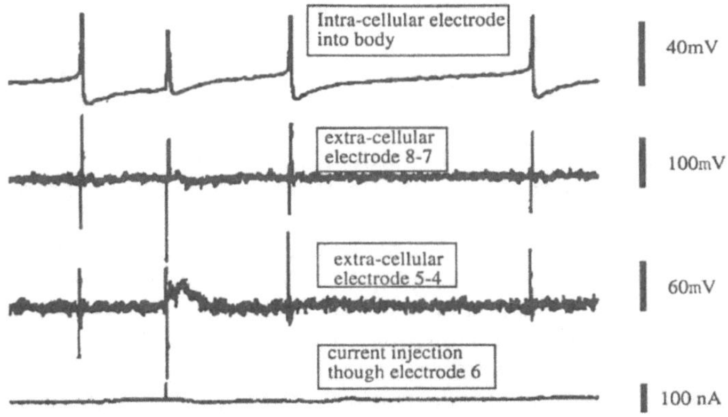

Figure 4. Stimulation of a process from a single identified leech Retzius neuron.

CONCLUSIONS

Considerable progress has been made towards constructing neural networks in culture. The use of optimised extracellular electrodes in structures with grooves to guide the neurons has been shown to allow both the construction of two neuron circuits and to stimulate single neurons. Signals have been observed from the processes of leech neurons as well as from the cell body. It seems very likely that extracellular electrodes in guiding structures will play a role in the elucidation of the functioning of the nervous system.

ACKNOWLEDGEMENTS

The expert technical assistance of Mr. Bill Monogan, Mrs. Lois Hobbs, and Mr. Robert Harkness in the photolithographic definition and manufacture of the structures and of Ms. Mary Robertson and Mr. Mike McGrath in the culture and growth of cells was extremely important in the progress of this work. The British Petroleum Venture Research Unit gave the major financial support to this programme from 1984 to 1990 and their constant help and assistance must be warmly acknowledged. The support of the Science and Engineering Research Council and the University of Glasgow has been vital for this work. The present support of one of us (L. B.) by the Welcome Trust is extremely helpful.

REFERENCES

1. J. G. White, E. Southgate, E. N. Thomson, and S. Brenner, The structure of the nervous system of the nematode *Caenorhabditis Elegans*, *Phil. Trans B* 314:1-340 (1986).
2. A. S. G. Curtis and P. Clark, The effects of topographic and mechanical properties of materials on cell behaviour, *Critical Reviews in Biocompatibility* 5:343-362 (1990).
3. R. G. Harrison, On the stereotropism of embryonic cells, *Science* 34:279 (cited in reference 1) (1911).

4. G. A. Dunn and A. F. Brown, A new hypothesis of contact guidance in tissue cells, *Exp. Cell. Research* 101:1-14 (1976).
5. P. Clark, P. Connolly, A. S. G. Curtis, J. A. T. Dow, and C. D. W. Wilkinson, Topographical control of cell behaviour. 1 Simple step cues, *Development* 99: 439-448 (1987).
6. P. Clark, P. Connolly, A. S. G. Curtis, J. A. T. Dow, and C. D. W. Wilkinson, Topographical control of cell behaviour: II. multiple grooved substrata, *Development* 108, Biologists Ltd. (1990).
7. P. Clark, P. Connolly, A. S. G. Curtis, J. A. T. Dow, and C. D. W. Wilkinson, Cell guidance by ultrafine topography, *J. of Cell Science* 99:73-77 (1991).
8. K. M. Thomas, S. P. Beaumont, C. D. W. Wilkinson, Soft x-ray contact printing of periodic structures for distributed feedback (DFB) semiconductor, lasers and integrated optics, *SPIE Integrated Circuit Engineering VI* 993:58-65 (1988).
9. D. J. Edell, A peripheral information transducer for amputees: long term multichannel recordings from rabbit peripheral nerves, *IEEE Trans. Biomed. Eng.* 33:203-214 (1986).
10. P. Connolly, P. Clark, A. S. G. Curtis, J. A. T. Dow, and C. D. W. Wilkinson, An extracellular microelectrode array for monitoring electrogenic cells in culture, *Biosensors & Bioelectronics* 5:223-234 (1989).
11. C. A. Thomas, P. A. Springer, G. E. Loeb, Y. Berwald-Netter, and L. M. Okun, A miniature microelectrode array to monitor the bioelectric activity of cultured cells, *Expl. Cell Research* 74:61-66 (1972).
12. P. Connolly, P. Clark, A. S. G. Curtis, J. A. T. Dow, R. Lind, and C. D. W. Wilkinson, Extra-cellular electrodes for monitoring cell cultures, in: "Electric Field Phenomena in Biological Systems", I.O.P. Short Meetings Series 21, (1989).
13. R. Lind, P. Connolly, C. D. W. Wilkinson, and R. D. Thomson, Finite element analysis applied to extra-cellular micro-electrodes, *Sensors and Actuators* 3:23-30 (1991).
14. T. D. Parsons, B. M. Salzberg, A. L. Obaid, F. Raccuia-Behling, and D. Kleinfeld, Long-term optical recording of patterns of electrical activity in ensembles of cultured *Aplysia* neurons, *J. of Neurophysiology* 66:316-333 (1991).
15. N. I. Syed, A. G. M. Bulloch, and K. Lukowiak, In vitro reconstruction of the respiratory central pattern generator of the mollusk *Lymnaea*, *Science* 250:282-4 (1990).

IMAGING AND CHARACTERIZATION OF MACROMOLECULAR INTERFACE STRUCTURES FOR WHOLE CELL BIOSENSORS

Vincent B. Pizziconi and Darren L. Page

Department of Chemical, Bio & Materials Engineering, Arizona State University, Tempe, Arizona 85287-6006

INTRODUCTION

Biosensors

Biosensor development continues to receive wide attention from researchers in the biomedical and biotechnology communities. The sustained level of interest in biosensors is due to the steady and marked evolution in biosensor technology which incorporates the latest advances in electrochemical, optoelectrical, piezoelectric, field effect transistors, and thermal transducer design. Even more promising approaches are now being explored resulting from advances made in the new frontiers of nanofabrication and molecular biology. One example of the latter is the exploitation of biological structures as the sensing elements of biosensors. This 'hybrid' approach represents a significant departure from 'traditional' chemical sensors which utilize strictly synthetic materials to detect biochemical events. The integration of advanced transducer design with complex, biologically active sensing elements holds great promise due to the superior detection capabilities offered by biological materials. Moreover, the newly emerging fields of molecular bioengineering and nanoengineering may ultimately provide the basis for the rational design of hybrid biosensors and other molecular-based devices as demonstrated by this feasibility study and other companion papers in this proceeding.

Hybrid biosensor prototypes have been developed which incorporate a variety of active biological sensing elements including proteins, subcellular organelles and even whole living cells, with the latter utilized as either populations of free cells or as intact tissue. One of the primary goals in the design of whole cell-based biosensors is to exploit natural cell sensing processes directly (e.g., molecular recognition) or by transducing related cellular responses. The likelihood in achieving optimal transduction of cellular responses is enhanced by immobilizing cells directly onto transducer surfaces. Thus, a key factor in hybrid biosensor design is the choice of immobilization technique employed and is the central focus of this paper.

The approach taken herein is to first discuss the various ways biological components can be 'attached' to synthetic transducer substrate surfaces. Secondly, an

Synthetic Microstructures in Biological Research, Edited by
J.M. Schnur and M. Peckerar, Plenum Press, New York, 1992

application of a more recent cell immobilization technique currently being evaluated by these investigators for the design of a hybrid immunobiosensor will be presented. Specifically, the technique exploits molecular biological adhesion processes to interface living whole cells to a thin-film thermopile transducer. For completeness, other whole cell biosensor systems currently under study are briefly reviewed.

BACKGROUND

Whole Cell Biosensors

Sensing is an aptitude which is highly developed in biological cells. Since they continuously monitor and respond to their constantly changing extracellular environment. For this reason, intact biological cells and tissues may eventually be proven to be the ultimate biological sensing elements for hybrid biosensor design.

Whole cell biosensor candidates are virtually unlimited. Thus far, cells from such diverse classes as microorganisms, plant cells, aquatic cells, and mammalian cells have been exploited. Reported applications of whole cell biosensors include on-line control of biochemical processes such as fermentation,[1] toxicology testing,[2] environmental monitoring[3] and clinical diagnostic assay.[4,5] The cellular responses that have been transduced include the assimilation of organic compounds, changes in respiration activity, the production of electrochemically active metabolites, bioelectric responses[6] and metabolically-related pH[7] or thermal responses.[8,9]

Specific cell types studied to date include microorganisms ranging from yeast cells[1] to photosynthetic bacteria[3] which have been used in the design of microbial-based biosensors. The higher eucaryotic, plant and animal cell lines have been typically used in an intact tissue slice form rather than as single cell populations. For example, plant tissues utilized for prototype biosensors thus far include banana, beet, corn kernel, squash, jack bean, potato, cucumber and eggplant tissue,[4,10] as well as, in vitro cultured tobacco callus tissue.[11] In the first animal tissue based biosensors, kidney, mucosal and lung were exclusively utilized.[4,12] More recently, an adenosine biosensor using rabbit thymus has been reported.[13] Additionally, antennules from the aquatic species, such as the crab,[6] crayfish,[14] and other crustaceans[15] have been exploited in the design of prototype chemoreceptor-based biosensors. In concert with these cellular approaches, we are currently exploring the utility of mammalian immune cells as a design basis for the development of an immunobiosensor.[6]

Immobilization of Biological Structures to Biosensor Surfaces

Independent of the specific biological element of interest, it is typically necessary to properly 'interface' the active sensing component with the synthetic transducer component by methods which are able to retain their full biological activity and provide stable, optimal signal transduction.[16] Many approaches are potentially able to satisfy the requirements of the interface region as depicted conceptually in Figure 1. However, there are distinct advantages and disadvantages to many of the conventional immobilization methods and are discussed in more detail below.

There are many techniques that have been developed for immobilizing biological components to synthetic supports (substrates). They have been developed for a variety of applications including bioprocessing (i.e., biocatalysis and bioseparations), disease therapies (e.g., extracoporeal detoxification), medical diagnostics (e.g., biosensors) and other related technologies. Biological components

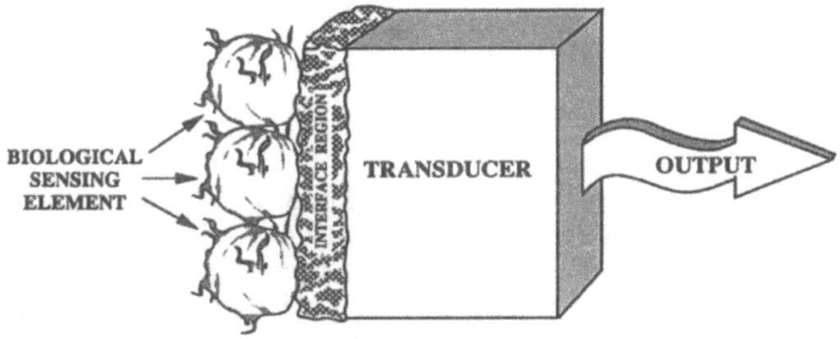

Figure 1. Schematic of a hybrid biosensor.

immobilized thus far consist of proteins (e.g. enzymes, receptors, antibodies), whole organelles (e.g., mitochondria, chloroplasts, microsomes), as well as, intact structures including liposomes, microorganisms, tissue cells and tissue slices. When using whole cells, they can either be used in their metabolically active, viable state or in a permeabilized,[14] nonviable state.

A brief review of the various biological immobilization methods is given below. More complete descriptions on specific details of the diverse methods that have been used for the immobilization of proteins and cells for various applications are reviewed elsewhere.[17-25]

Methods of Immobilization

Methods of immobilization can be separated into three major classes, i.e., covalent binding, entrapment, and physical adsorption, although in many applications combinations of these groups are commonly used. Each of these immobilization methods has advantages and disadvantages that should be weighed carefully for a given application. Independent of the specific immobilization method used, the basic design goals are that (a) the biological sensing component must maintain a stable, minimally acceptable biological activity after immobilization (b) the immobilization method should withstand application-specific environmental conditions (e.g., mechanical forces, pH, temperature) and (c) the interface should allow optimal transduction of the biological property of intent.

Covalent Binding

Covalent binding is considered one of the most stable forms of immobilization but it can also lead to the greatest loss in biological activity.[26] Covalent immobilization requires reactive functional groups on both the substrate and the biological component. Organic substrates (e.g., polymers) typically have groups that can be activated by chemical reagents for subsequent covalent attachment to specific protein functional groups (i.e., amino acid sequences). Protein functional groups include lysine t-amino groups and N-terminal amino groups, L-cysteine thiol groups, L-tyrosine phenolic groups, L-arginine guanidine groups, L-histidine imidazole groups, L-cysteine disulfide groups, L-tryptophan indole groups, L-methionine thioester groups, and L-serine and L-threonine hydroxyl groups.[21] Inorganic substrates, such as metals and glass usually have relatively inert oxide or silanol surface groups, respectively. Thus for these materials, it is usually necessary to add organic functional groups by silanation in order for covalent binding to occur.

Although direct covalent attachment is harmful to living cells,[26] viable cells can be immobilized by the covalent pre-immobilization of extracellular matrix (ECM) proteins, their molecular recognition peptide fragments,[5] or other bioaffinity ligands to stimulate subsequent cell adsorption.

Entrapment

Entrapment is a method of immobilization in which biological components are physically restricted within barriers that prevent their migration. Encapsulation is a form of entrapment in which the biological components are enclosed inside liposomes or semipermeable membranes or fibers. Due to their relatively small size, proteins (e.g., ligands or enzymes) must be trapped in polymer matrices with very small lattice dimensions to prevent leakage. Synthetic polymer matrices used for protein entrapment include polyacrylamide, silastic, and polyvinyl alcohol. Natural protein matrices include collagen, albumin, calcium alginate, and gelatin, among others. A similar approach is to entrap proteins within porous microfibers made of cellulose triacetate, cellulose, or poly-gamma-methyl-L-glutamate, etc. Cross-linking of trapped proteins using glutaraldehyde or other difunctional groups can be used to further hinder protein leakage.

The larger biological components such as liposomes, organelles, or whole cells can be trapped within similar gels or fibers having the same or much larger lattice dimensions. Polymers useful for cell entrapment include agar/agarose, calcium alginate, polyacrylamide, kappa-carrageenan, cross-linked proteins, and photo-linkable resins, among others. The choice of polymer depends on whether viability is desired and the preferred lattice dimensions. These choices, in turn, are dependent on the biological sensing component (e.g., organelles versus eucaryote cells), the ligand, as well as reactant or reaction product size. A simple approach described in a companion paper in this proceeding utilizes surface corrugations to trap cells under the simple influence of gravity (i.e., by sedimentation).[27]

The major disadvantage of entrapment with respect to biosensors is that it may not provide the intimate association between biological components and the substrate that is oftentimes necessary to attain optimal signal transduction. For this reason, it may be desirable to entrap biological components in membranes with special properties (e.g., conductive polymers) in order to optimize transduction of some specific activity of the biological component.

Physical Adsorption

Physical adsorption is due to a variety of weak (i.e., relative to covalent binding) atomic or molecular interactions between the biological component and the substrate. These interactive forces include electrostatic (e.g., ionic) and electrodynamic (e.g., van der Waals) forces, hydrogen, and 'hydrophobic' forces. The electrostatic and electrodynamic forces may contribute to both long-range and short-range interactions while hydrogen and hydrophobic forces are short-range in nature. The long-range forces have been associated with non-specific cell adsorption processes[28,29] whereas the short-range forces are thought to govern 'biologically-directed' cell adhesion allowing for cellular attachment to evolve into a developed cellular assembly relationship. Thus, the molecular structure dominates at the short range and the forces involved are correspondingly large and specific.[30,31] For any given system, the relative occurrences of each type of interaction differ and therefore provide differing contributions to the overall strength of the immobilization. The strength of adsorption is determined by the composition and total number of the available binding sites, which are somewhat dependent on the 'closeness' of the steady state interaction (i.e., the existence of long-range versus short-range interactions). The physical nature of adsorptive forces makes them very sensitive to environmental conditions such as temperature, pH, ionic strength, and dissolved organics.

In general, the physicochemical properties of interfaces govern the nature of adsorption forces. These include interfacial free energy, wettability, and surface charge. In practice, information on interfacial free energy is obtained from tests of wettability, i.e., contact angle measurements. Another useful parameter related to the contact angle is the critical surface tension. These parameters have been used to correlate various levels of interfacial bioactivity including protein adsorption,[32] cell spreading,[33] and cell adhesion.[34] For example, the substrate surface-free energy, degree of wettability and the critical surface tension of various plastic substrates have been shown to play an important role in both protein adsorption and cellular adhesion processes to these polymeric materials.[35,36] Conversely, protein adsorption markedly alters surface properties. Thus, the presence or absence of proteins, either as an intrinsic component of the substrate surface or adsorbed to the substrate surface from solution, dictates whether a biological cell undergoes nonspecific or specific adsorption to the substrate and whether cell spreading occurs. As an example of a specific adsorption system, it has been shown that cell spreading correlated well with the amount of fibronectin, a key cell adhesion protein, that was preadsorbed to the substrate.[37] Thus, the degree of cell anchoring to synthetic surfaces relies heavily on the presence of an adsorbed protein provided from media containing sera, from cell microexudate, from other complex biological fluids (e.g., blood) or from preadsorption of selected proteins, such as ECM proteins.

Adsorption is substrate specific and depends on the type and density of binding sites and on the total available surface area. Several types of substrates have been found that exhibit high protein adsorption including charcoal, clay, ceramics, cellulose, kaolinite, collodion, silica gel (i.e., hydrogel), porous glass, hydroxyapatite, collagen, alumina, titanium, and the ion exchange resins (e.g., DEAE Sephadex and DEAE cellulose). Additionally, adsorbed proteins can be covalently cross-linked thus creating a more stable gel-like structure. The strength of adsorption becomes increasingly important with increasing size of adsorbed materials. For example, organelles and whole cells are more sensitive to mechanical fluid stresses partly because a lower percentage of their total surface area is in intimate contact with the substrate (assuming the native shape is maintained) and subsequently more affected by shear forces.

Thus, tissue cells can be very difficult to effectively immobilize via adsorption unless they belong to a special class of cells i.e., naturally adherent cells. Naturally adherent cells utilize metabolically driven processes to both adsorb to the surface and spread. Cell spreading increases contact with the substrate while minimizing the effect of shear forces. Adherent cells can also increase the total number of available binding sites by both moving membrane proteins to and secreting proteins at the cell/substrate interface. In this way, the cell may create a strengthened gel-like structure of cross-linked glycoproteins to stabilize its immobilization.

Unlike covalent and entrapment immobilization techniques, physical adsorption requires little to no reagents or matrix substrates, thus making it simple and fast and causes less distortion of natural protein shape.[26] The major disadvantage of adsorption is that it is highly sensitive to the environmental conditions as mentioned previously.

Biological Immobilization Processes

The special interface properties required in biosensor design impose limitations on the choices of substrates. Typically, biosensor transducer substrates are chosen for their unique physical properties, e.g. electrical (graphite, silicon, metal), optical (glass, quartz, silicon, metal films) or thermal (metal, diamond) properties rather than any biological property related to immobilization. Thus, the substrates require the addition of binding sites either through silanation or other surface modification.

This is the case with our thermopile biosensor which utilizes a Mylar® substrate. In this case, we opted to augment cell adhesion with a novel surface modification technique using natural cell-substrate linker structures. Our work focused on the immobilization of living immune cells to polymer substrate surfaces using selected cellular-adhesive, ECM proteins. In this case, immobilization involves both pre-adsorption of the appropriate ECM protein to the thermopile substrate followed by cell binding onto specific ECM protein binding regions. Advantages of this approach are attachment specificity, maintenance of biological activity and stability (static and dynamic), among others. These are important factors when attempting to immobilize viable cells to synthetic polymer substrate surfaces. A brief review of biological immobilization processes will make this method of choice more apparent.

Cellular Adhesion

The attachment of cells to their surroundings is prerequisite to their maintaining proper cell function and shape, as well as, tissue integrity.[38] The most noticeable change associated with attachment is the dramatic shape transformation from spherical to a flattened cell morphology. Cell adhesion is now known to occur in a series of steps which has been described as a multistep paradigm of adsorption, contact, attachment, and spreading[39] that is depicted schematically in Figure 2.

Non-adherent cells under normal conditions will not securely adhere to most surfaces since they do not utilize the metabolically driven processes needed to provide for stabilized adsorption. However, they can be stimulated by specific ligands to undergo stable adsorption. This is due to the expression of specific cell membrane protein integrins that bind to specific ligands (e.g., proteins) and trigger cellular adhesion. In multicellular organisms, integrin ligands tend to be the ECM proteins, such as laminin, collagen, fibronectin, etc. A more comprehensive review of cell adhesion processes can be found elsewhere.[39,40]

Extracellular Matrix Proteins

The biological substrata to which cells adhere to in most natural cellular environments is the ECM. The ECM is a biological composite structure comprised of an organized meshwork of structural and adhesive fibrous proteins embedded in a polysaccharide gel. The construction of the ECM is accomplished by molecular self-assembly and cellular deposition of materials. The cellular adhesive proteins are primarily responsible for cell attachment, migration and growth on the ECM. Cellular adhesion occurs via receptor-ligand interactions between cell receptor proteins

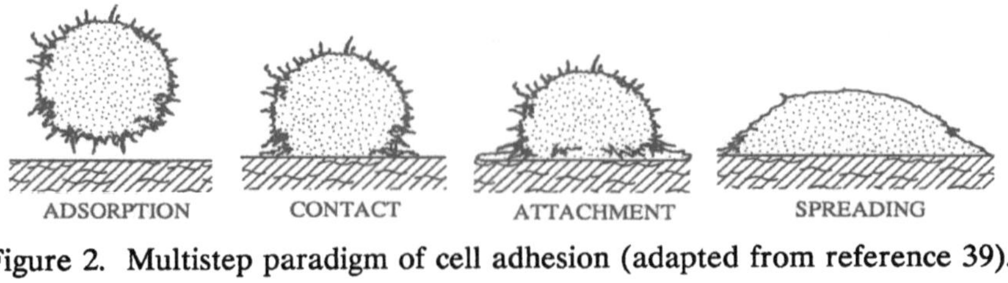

Figure 2. Multistep paradigm of cell adhesion (adapted from reference 39).

(integrins) and ECM adhesive proteins. A model of cell adhesion shown schematically in Figure 3 depicts the molecular arrangement of ECM proteins to cell receptor proteins.

Universal Molecular Recognition Cell Binding Sequences

Cell attachment to extracellular matrices centers on the molecular recognition between specific amino acid sequences that reside along regions of extracellular matrix proteins and complementary sequences that reside along regions of cell membrane receptor proteins or integrins. The ECM adhesive proteins are encoded with, what are considered to be, universal cell binding molecular recognition sequences, i.e., minimal, generic amino acid sequences for cellular attachment and also protein self-assemblies. Reported binding sequences for several, diverse cellular systems are summarized in Table 1.

The ECM proteins and/or cell-binding sequences (natural or synthetic) may be of great utility in providing novel materials for molecular device design. In particular, they provide highly specific mechanical linkages between synthetic microstructures and biological systems. This is an extremely attractive approach to the development of whole cell biosensors which presently require relatively large (high cell densities) cellular assemblies to be kept in direct contact with a transducer. Two ECM proteins relevant to the development of a whole cell immunobiosensor presently undergoing feasibility studies in our laboratories are laminin and fibronectin because they specifically bind to mast cells as shown in Table 1. Further, the newly emerging field of tissue engineering[59] is exploiting similar features of these and other biological structures to create novel, smart materials for the design of sophisticated devices for medicine,[36,60] process biotechnology[61] and even for whole organ regeneration.[62] A more detailed review of the molecular biology of extracellular matrix proteins is given elsewhere.[63]

ECM Adhesive Protein Nanostructure

In vivo, laminin promotes the attachment of epithelial and other cells to the basal lamina. Laminin is a large glycoprotein of about 900-kDa and is primarily associated with the basement membrane. Based upon transmission electron microscopy (TEM) studies employing rotary shadowing techniques, laminin is composed of three chains arranged in a cruciform shape as shown in Figure 4a. The dimensions determined by TEM micrographs shown in Figure 4b are approximately

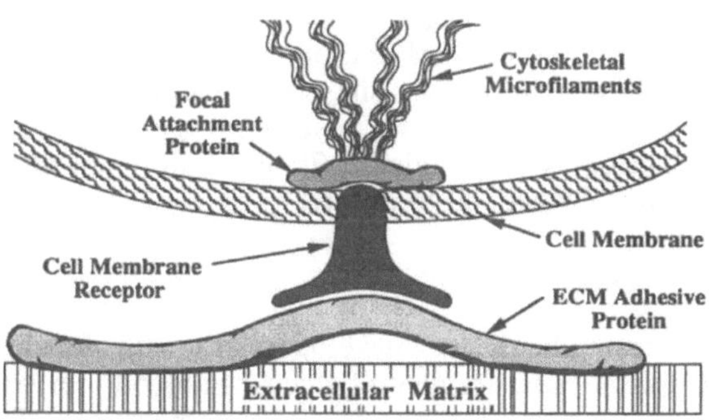

Figure 3. Schematic showing macromolecular structures responsible for cellular assembly (adapted from reference 38).

Table 1
Comparison of ECM cell binding sequences for various cell assembly systems.

Cellular Assembly	Cellular Adhesive Protein	Cell Recognition Sequence
Platelet Adhesion	Fibronectin[41]	ARG-GLY-ASP [RGD]
Platelet Adhesion	Vitronectin[41]	ARG-GLY-ASP [RGD]
Platelet Adhesion	Fibrinogen[41]	ARG-GLY-ASP [RGD]
Platelet Adhesion	Von Willebrand Factor[41]	ARG-GLY-ASP [RGD]
Endothelial Cells	Thrombin[42]	ARG-GLY-ASP [RGD]
	Collagen IV[43]	[TAGSCRKFSTM]
Fibroblasts	Fibronectin[41]	ARG-GLY-ASP [RGD]
	Serum Spreading Factor[44]	ARG-GLY-ASP [RGD]?
	Elastin[45]	VAL-GLY-VAL-ALA-PRO-GLY [VGVAPG]
Bacteria	Fibronectin/Fibrinogen[46]	ARG-GLY-ASP [RGD]
Mast Cells	Laminin/Merosin[47,48]	TYR-ILE-GLY-SER-ARG [YIGSR]
		ILE-LYS-VAL-ALA-VAL [IKVAV]
	Fibronectin[49]	ARG-GLY-ASP [RGD]
Epithelial Cells	Laminin/Merosin[50]	TYR-ILE-GLY-SER-ARG [YIGSR]
	Epinectin[51]/Epibolin[52]	ILE-LYS-VAL-ALA-VAL [IKVAV]
	Epiligrin[53]	TYR-ILE-GLY-SER-ARG [YIGSR]?
Neuronal Cells	Tenascin[54]	not yet determined
	Laminin/Merosin[55]	ILE-LYS-VAL-ALA-VAL-[IKVAV]?
Bone Cells	Osteopontin[56]	ARG-GLY-ASP [RGD]
	Bone Sialoprotein[57]	ARG-GLY-ASP [RGD]
Leukocytes	Complement C3bi[58]	ARG-GLY-ASP [RGD]

35 nm for the three short arms and 75 nm for the long arm. Several specific sites on the laminin molecule, representing minimal amino acid binding sequences, have been determined. These sequences serve as functional interaction points for both proteins and cells. Figure 5a depicts the molecular organization and orientation of a laminin-cell adhesion system. Figure 5b illustrates the self-assembly characteristics of laminin.

In vivo, fibronectin attaches fibroblast and other cells to connective tissues. Fibronectin is also a large glycoprotein of approximately 500-kDa which exist in two distinct forms; i.e., either in its solubilized form as a dimer in blood where it plays a major role in the formation of cryoprecipitates, or in its insoluble multimer form as it exists in the ECM. Fibronectin is composed of two chains approximately 60 nm in length linked end-to-end at a 60° angle, forming a 'cotter pin' shaped molecule as depicted in Figure 6a based upon TEM and other studies. A TEM of fibronectin is shown in Figure 6b. Fibronectin contains sites for cell attachment, as well as, sites for fibrin and collagen. Figure 7a depicts the molecular organization and orientation of a fibronectin-cell adhesion system. Figure 7b illustrates how the molecular self-assembly of fibronectin is envisioned to occur.

Whole Cell Thermopile-Based Biosensor

A thin-film thermopile device, developed originally in our bioengineering laboratories at ASU for the measurement of glucose in blood,[8] served as the synthetic microstructure that was targeted for whole cell immobilization. The thermopile is constructed by evaporating antimony and bismuth thin-films, forming regular patterns

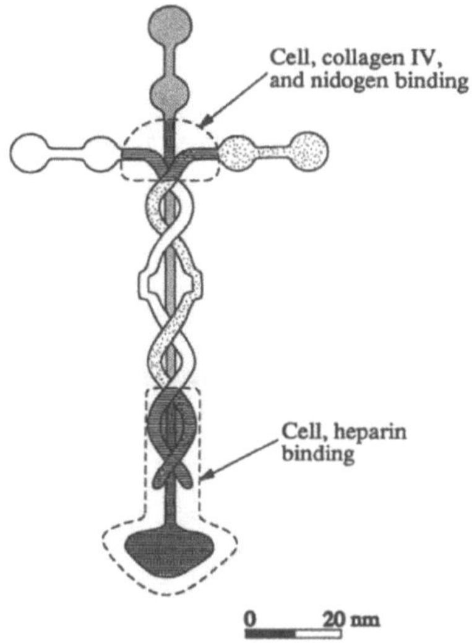

Figure 4a. Schematic representation of laminin based upon electron microscopy and related studies (adapted from reference 64).

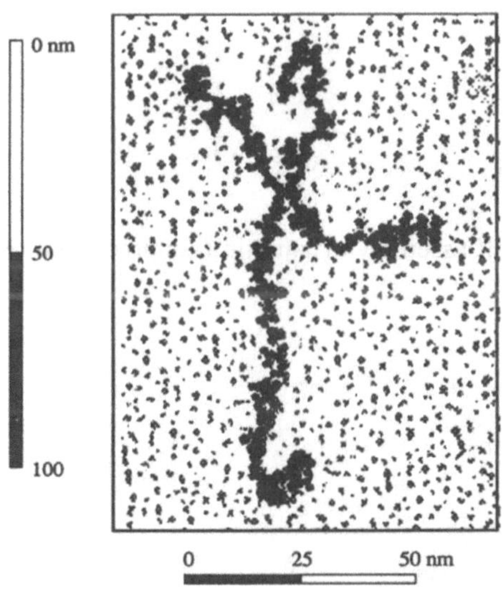

Figure 4b. Transmission electron micrograph of laminin (adapted from reference 64).

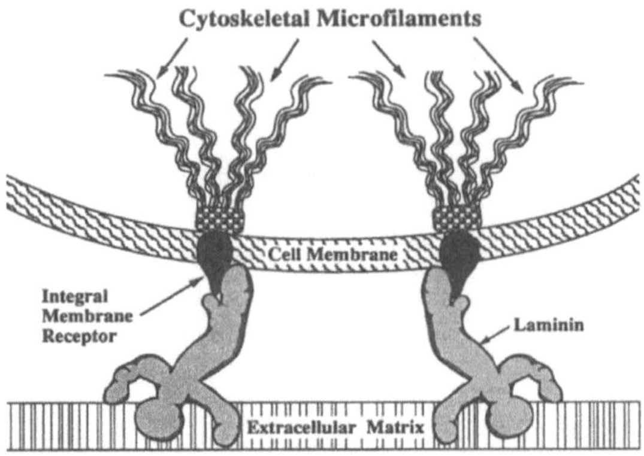

Figure 5a. Depiction of laminin in situ (adapted from reference 65).

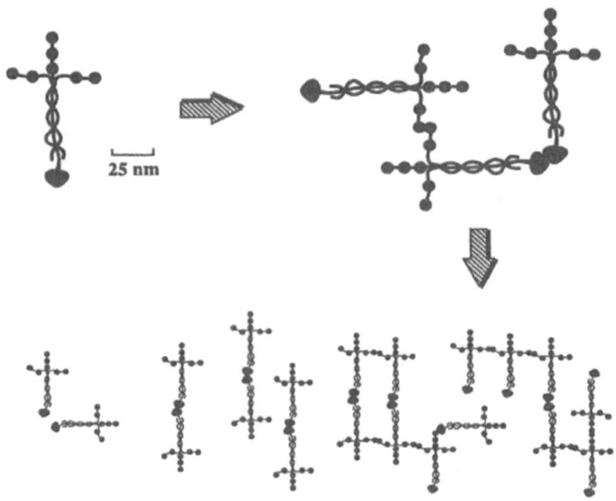

Figure 5b. Self-assembly of laminin (adapted from reference 66).

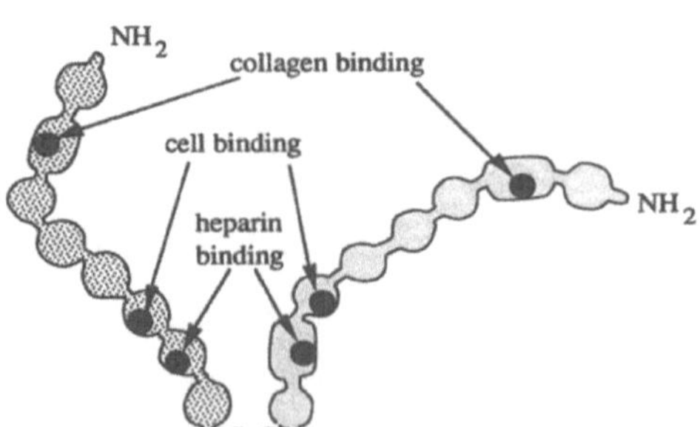

Figure 6a. Schematic representation of fibronectin (adapted from reference 64).

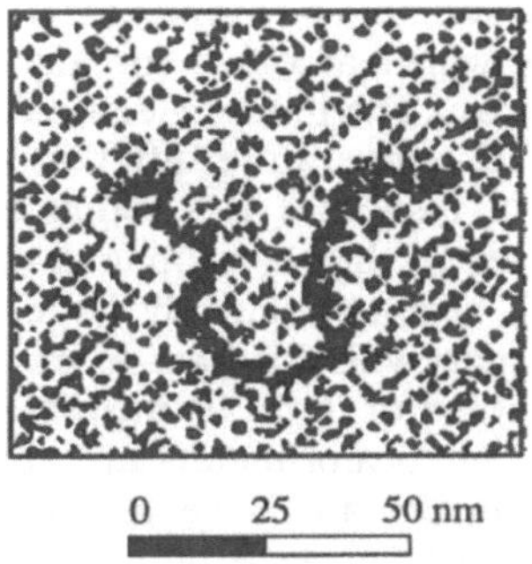

Figure 6b. TEM of fibronectin (adapted from reference 64).

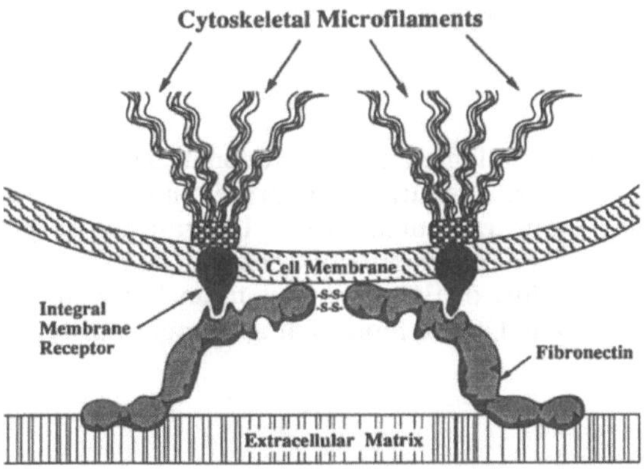

Figure 7a. Fibronectin in situ (adapted from reference 67).

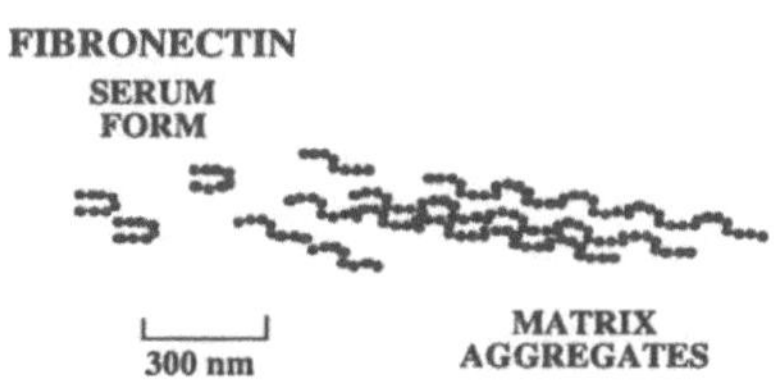

Figure 7b. Fibronectin self-assembly (adapted from reference 66).

of thermocouple junctions, on a thin Mylar® support. A schematic depicting immobilized cells on the active junctions of the thermopile is shown in Figure 8.

As can be seen in Figure 8, two sets of thermopile junctions (active and reference junctions) are utilized in order to measure relative changes in temperature between the thermopile surface and fluid environment. Thus, the thermopile measures differential thermal responses which accounts for any ambient thermal shifts not associated with cellular responses. The thermoelectric EMF for this thin film device is approximately 100 μV per °C temperature differential per couple. Thus, a fifty junction device gives a 5 mV/°C response. It is estimated that temperature changes as small as 10^{-5}°C can be detected by this thermopile design. However, even with this sensitivity, the thermopile will require relatively large (high density) immobilized cellular assemblies in direct contact with this transducer. This is based upon reported cell thermal outputs on the order of 1 to 10 picowatts per cell, as measured by microcalorimetry.[68] Although microcalorimeters are able to measure the heat flux of biological systems, these devices are large and expensive and are not easily amenable nor practical for diagnostic assay or on-line or continuous biosensing. This is particularly true when multiple or serial assaying requiring relatively short assay times is desired.

MATERIALS AND METHODS

Experimental Design

Two different but complementary experimental paths were undertaken in this study to image and characterize natural ECM protein interface structures. The ECM proteins are currently being evaluated in our laboratories as potential interfacial immobilization components of an immunobiosensor. The more macroscopic phase of this work addressed the ability of ECM proteins to anchor whole cells onto synthetic polymer surfaces. The second, more molecular approach, utilized scanning tunneling

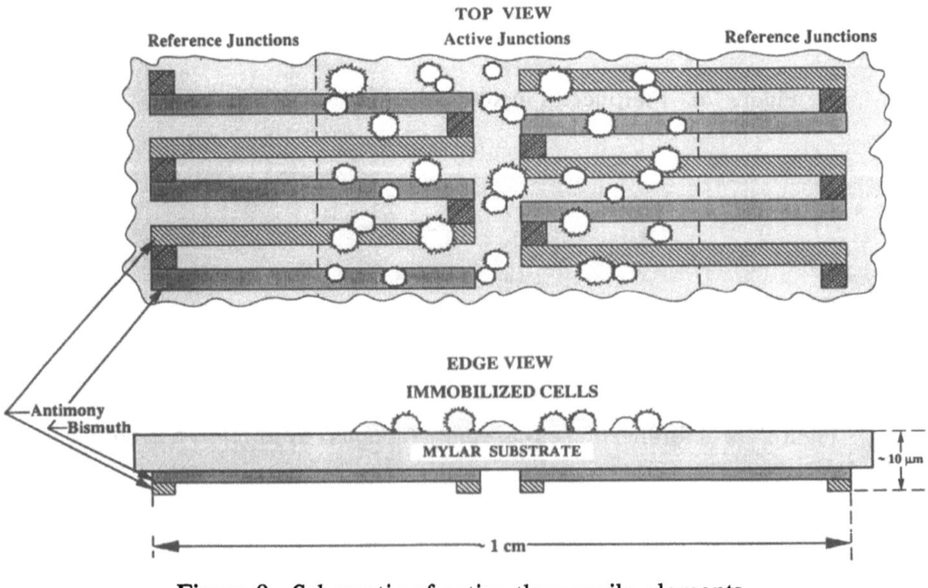

Figure 8. Schematic of active thermopile elements

microscopy (STM) to study and characterize adhesive protein structures that were deposited at interfaces in an aqueous environment. Although these protein structures have been studied extensively with TEM, scanning probe microscopy, i.e., STM and atomic force microscopy (AFM), offers distinct advantages of imaging biological structures. Specifically, STM and AFM provide the unique ability of imaging uncoated biological structures at potentially extremely high resolution (1.0 Å lateral resolution and 0.01 Å vertical resolution) at ambient pressure in their fully hydrated state, i.e., with the sample completely immersed in water. A more complete discussion on the physics of STM[69-72] and AFM[73] is given elsewhere. Also, a very comprehensive review of the widely diverse applications of STM, AFM, and related techniques has been recently published.[81] In addition, a brief review of the reported applications and limitations of STM for imaging biological structures is available.[9]

Thin-Film Thermopile Biosensor

The basic test apparatus shown schematically in Figure 9 is comprised of a small (~1cm²) thermopile deposited on an electrically-insulative plastic film (Mylar®). The face of the thermopile is electrically and thermally insulated with styrofoam and placed downward. The plastic insulating material (i.e., Mylar®) also provides a convenient substrate surface for interfacing cells to the device. Cells can be either cultured directly onto the thermopile or immobilized with the appropriate cell adhesive protein system depending on the cell type, i.e., naturally adherent or suspension cells.

Preliminary Cell Adhesion Studies

MC/9 and WEHI-3 (ATCC, Rockville, MD) and RBL-2H3 (kindly supplied by Dr. Juan Rivera, NIAMS/NIH) cell lines, cultured in Dulbecco's modified Eagles media (DMEM), were placed on polystyrene culture plates precoated with either laminin or fibronectin (Sigma Chemical, St. Louis, MO) or placed on control surfaces (polystyrene culture plates with no ECM proteins present). Some MC/9 cells were

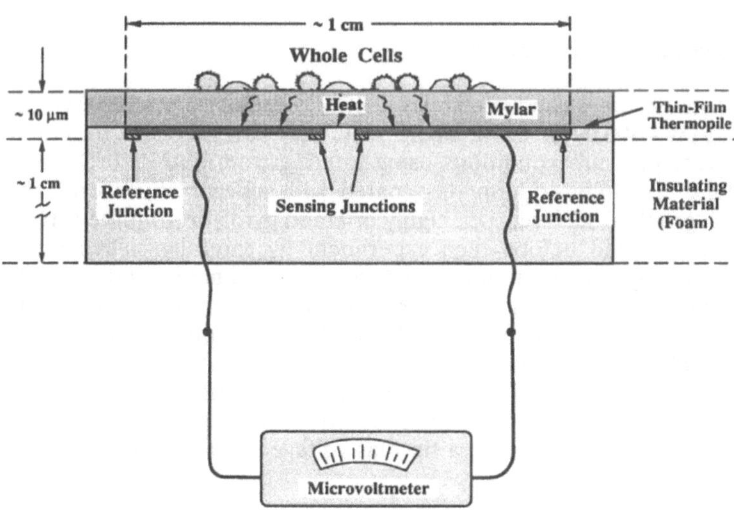

Figure 9. Schematic illustration of whole cell biosensor.

treated with sodium butyrate to promote cell maturation and were designated MC/9-2. The cellular adhesive proteins were applied as a solution (10 μgm/ml) of 20 μl drops to small areas on polystyrene culture plates. The drops were allowed to air dry for approximately 1 hour. All cell lines were spun down and resuspended in Hepes-buffered DMEM (H-DMEM) with 4% fetal calf serum at pH 7.3. A 20 μl drop of deionized water was placed on the dried protein deposit immediately before 450 μl of cell suspensions were added. This was done so that the dried salts in the protein deposit did not affect the osmolarity, since small drops of cell suspensions were used. The cell drops were humidified with plastic wells of D.I. water placed in the culture plates. The plates were then covered and placed in the incubator at 38°C for 100 minutes. Plates were then rinsed of non-adherent cells by gently squirting fine jets of H-DMEM across the plates. The plates were then tapped onto a paper towel to remove as much fluid as possible. The cell-bound surface was not disturbed due to the depth of the culture plate. The plates were then allowed to air dry for 2-3 minutes after which they were rinsed with 100% ethanol and dried as before. Toluidine blue dye solution was then placed in drops over the cell bound surfaces for 30 seconds and the surface was rinsed briefly with D.I. water. The plates were dried again and the adherent cells were visualized by light microscopy at 288x using a Universal Zeiss microscope.

Biosensor Cell Adhesion Studies

Additional ECM adhesion experiments were conducted whereby laminin or fibronectin was used to adhere MC/9 cells onto the Mylar® surface of a thin-film thermopile transducer prototype as depicted in Figure 9. Laminin (100 μgm/ml) and fibronectin (100 μgm/ml) solutions were used to precoat the substrate surface (~ one cm^2) with a minimal volume of solution and allowed to dry for one hour at room temperature. Suspensions of MC/9 cells (~10^6 cells) were then added to the ECM protein-coated surfaces and incubated for one hour at 37°C. Loose cells were gently drawn away with a pipette and the remaining adherent cells were visualized with light microscopy to assess cell immobilization.

Scanning Tunneling Microscopy Studies

STM studies were conducted on two ECM proteins (laminin, MW=900,000 and fibronectin, MW=500,000). STM experiments were carried out at room temperature and at ambient pressure conditions using an Angstrom TAK 3.0 STM instrument (Tempe, AZ). Platinum-iridium tips coated with apiezon wax were fabricated in house. Fresh graphite surfaces, i.e., highly oriented pyrolytic graphite (HOPG, Union Carbide) were created before each experiment by carefully cleaving the exposed graphite surface layer with an adhesive tape. Approximately 100 μl of Tris buffer solution containing 15 μgm/ml of ECM protein was placed into the microwell just prior to tunneling. The tunneling head was then carefully placed into the desired position and followed by a manual advancement of the tip through the liquid and within close proximity to the surface. At this point the instrument was placed into an automatic approach mode until it reached a constant tunneling current of approximately 1 nanoampere with a tip bias voltage of -0.1 volts.

RESULTS

Direct Cell/ECM Protein Binding Studies

Optical micrographs of the MC/9 and RBL-2H3 cell lines were taken to display the affects of ECM proteins on cell adhesion. More specifically, optical micrographs of the RBL-2H3 cell line are shown in Figures 10a-10c. As demonstrated by the polystyrene control (Figure 10a), these cells have a natural adhesion and spreading ability with many surfaces. The prevalent spreading morphology for control cells is a flattened-like appearance and having 1-4 filopodia (cell extension processes) per cell, termed 'protrusive' spreading.

RBL-2H3 cells in the presence of fibronectin (Figure 10b) and laminin (Figure 10c) displayed increased adhesion to the substrate, increased cell clumping and a drastically altered spreading morphology. In these cases filopodia were less common, virtually non-existent in the case of fibronectin and occurring rarely in the presence of laminin. Although the cells appear to have maintained a spherical shape from a two- dimensional perspective, they actually flattened out to a 'pancake-like' spreading conformation. The cell surfaces appeared venated with web-like patterns as a result of their flattened morphology and highly arranged cytoskeletons. Additionally, cell clumping was much more prevalent in the presence of ECM proteins probably due to cell-cell adhesion by resolubilized ECM protein which resulted in less defined cell boundaries. Similar results were obtained using the WEHI-3 cell line, also a natural adhering cell line, for both control and fibronectin experiments. Laminin, however, induced a decrease in cell adhesion.

MC/9 cells, a suspension cell line, were adhered to culture plates coated with laminin or fibronectin (Figures 11a-11c). Fibronectin-bound cells primarily displayed both the 'protrusive' and 'pancake-like' spreading as primary spreading morphologies in approximately equal populations. This distribution was shifted using butyrate treated MC/9 cells (MC/9-2) in which most cells were of the 'pancake-like' morphology. Fibronectin-bound cells resisted mild shear forces. However, laminin-bound cells detached as a cell-protein composite film when exposed to mild shear forces. This occurred in both the MC/9 and MC/9-2 cell lines.

The results of the in vitro binding studies of both the naturally-adherent (RBL-2H3 and WEHI-3) and suspension cell lines (MC/9 and MC/9-2) are summarized in Table 2. As can be seen from polystyrene control experiments shown in Table 2, both naturally-adherent and suspension cells behaved as expected, i.e., the naturally-adherent cells did bind and spread to the polystyrene surface. Specifically, the RBL cell line showed more attachment and spreading than the WEHI-3 cell line while the MC/9 cells remained in suspension.

Whole Cell Biosensor Adhesion

Preliminary results of the affect of the ECM protein immobilization method on whole cell attachment to biosensor substrates are shown in Figures 12a and 12b. Laminin and fibronectin were used to interface normally non-adherent MC/9 suspension cells onto the thermopile immunobiosensor. In these figures, highly dense cell assemblies of MC/9 cells can be seen immobilized on the protective Mylar® surface of the thermopile component of the whole cell hybrid biosensor. This is a promising result since, as mentioned previously, high cell population densities are

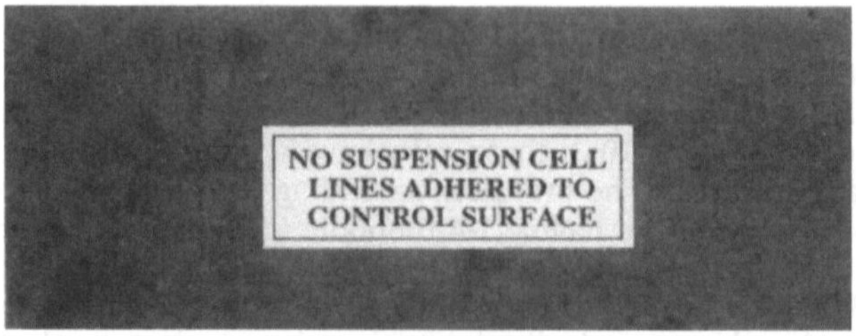

Figure 10a.

Figure 10b.

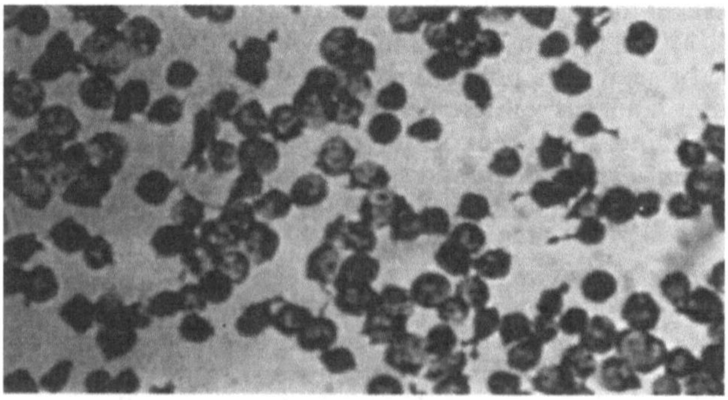

Figure 10c.

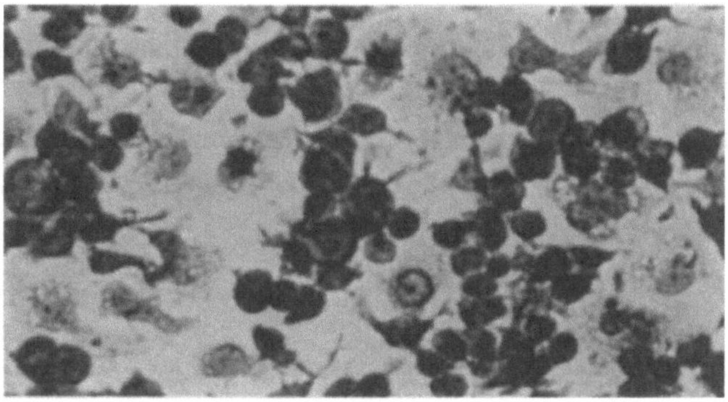

Figure 11a.

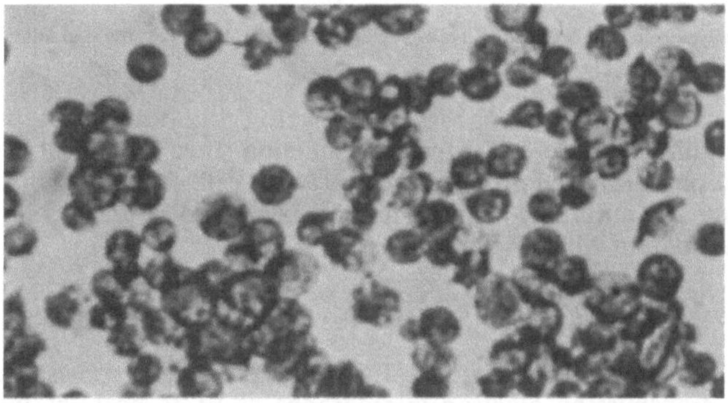

Figure 11b.

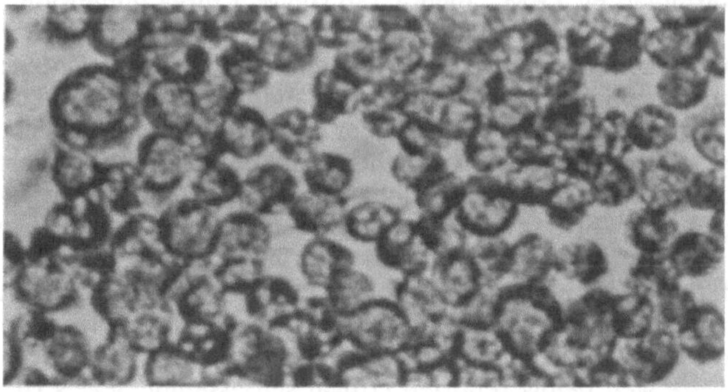

Figure 11c.

Table 2
Affect of ECM Protetins on Cell Attachment and Spreading

	POLYSTYRENE CONTROL		LAMININ		FIBRONECTIN	
	ADHERENCE	SPREADING	ADHERENCE	SPREADING	ADHERENCE	SPREADING
RBL	+++	++	+++	+++	+++	+++
MC/9*	--	--	+/-*	--	+++	++
MC/9-2*	--	--	+/-*	--	+++	+++
WEHI	++	++	+	+	+++	+++

Legend	Adherence	Spreading
- -	None	No spreading cells spherical in shape
- -	None	No spreading cells spherical in shape
-	1-29%	Poor spreading in some cells (1filopodia/cell) most cells spherical
+	30%-49%	Poor spreading in most cells (1 filopodia/cell) cells still somewhat spherical
++	50%-84%	Most cells spread, avg. 3 filopodia/cell; some cells highly flattened/contracted
+++	85%-100%	Most cells spread and highly flattened and contracted

* initially adhered but detached as a cell-protein film after prolonged exposure to solution

required to obtain adequate cell signal transduction since biological cells produce a metabolic heat output on the order of picowatts per cell. Note that the dark, equally spaced lines are the antimony-bismuth microthermocouples located on the opposite face of the transparent Mylar® film.

STM Studies of ECM Nanostructure

The STM images of individual ECM proteins were consistent in size and structure with transmission electron micrographs of ECM proteins but offer much greater molecular detail of these adhesive proteins. Figure 13a shows a three dimensional STM image of laminin which depicts a remarkably cruciform-like structure which is very similar to that reported in TEM studies as depicted in Figure 4b. In addition, the overall molecular dimensions are consistent with the dimensions of laminin as determined by TEM. Figure 13b shows a three dimensional STM image of fibronectin which portrays a characteristic 'cotter pin' like structure and molecular dimensions comparable to reported TEM studies. Of particular note in the laminin STM image are the highest topology regions indicated by lightest color intensity. These regions correspond to one of the major cell binding areas as depicted in Figure 4a. Similarly, the tallest regions (lightest areas) of the fibronectin molecule correspond to its reported binding domains (Figure 6a). Both individual and molecular aggregates of laminin and fibronectin were seen in all tunneling experiments. However, individual laminin and fibronectin molecules were seen less frequently and, thus, were less available to image. This was expected since there were no attempts made to control their self-association which is reported to occur at these room temperature conditions although the concentrations used in these experiments (10 μgm/ml) were well below reported critical self-assembly levels (0.1 mg/ml).[74,75] However, when individual molecules of laminin and fibronectin were located, high resolution STM images of their molecular structures were obtained in their hydrated states.

DISCUSSION

The adhesive utility of ECM proteins for applications, such as the engineering of cellular assemblies, depend on both their binding affinity to cells, as well as, to the desired substrate. In vivo, the specific high affinity sites in the ECM provide an optimum substrate for adhesive protein attachment. However, this condition certainly does not inherently exist in synthetic materials and careful consideration must be given to techniques which can achieve adequate immobilization to the polymer substrate. Our initial approach was to simply choose a material that exhibits a surface energy that would promote the nonspecific adsorption of the ECM proteins. In general, bioadhesion is favored when the critical surface tension is approximately 40 dynes/cm^2 or higher.[77] The substrate chosen in our study, Mylar®, has a reported[78] critical surface tension of 43 dynes/cm^2 which should enhance the bioadhesion for both proteins and cells. However, reliance on nonspecific adsorption processes to immobilize ECM proteins to synthetic materials may not be sufficient in the design of hybrid microstructures. For example, in this study, although MC/9 cells can be adhered to synthetic surfaces via laminin they tend to detach over time and more quickly under shear. This detachment usually occurred as films of cells and protein. Nonetheless, this also suggests that cell/ECM protein binding remained intact. On the other hand, good cell adhesion was obtained when using fibronectin which may be due to the relative orientation of the protein on the surface.[79] Overall, fibronectin provided superior MC/9 suspension cell line attachment (as seen in Figure 12b - higher cell densities immobilized) as compared to laminin for the same Mylar® substrate/sensor system shown in Figure 12a. In any event, a more efficient approach would be to use some form of surface modification technique, such as surface covalent immobilization or glow discharge, to enhance ECM protein immobilization, although these techniques may alter the protein conformation which can modulate the cell adhesion response in a unique way.[80] More recent approaches are incorporating the ECM binding sequences directly into synthetic substrate matrices.[60]

One noted difference between the results reported in this study and TEM results is the discrepancy in the absolute height and width dimensions i.e., the STM results indicate that these structures are larger in diameter but lower in thickness relative to the reported TEM values. This result is consistent with other STM results of lower than expected molecular thicknesses when imaging uncoated biological structures.[76] The molecular diameters of the individual chains appear to be significantly larger in the STM nanographs, suggesting that biological molecules may undergo deformation from local tip forces. This is an important issue to resolve for STM to realize its potential in characterizing biological structures. Also, a closer inspection of Figure 12a suggests that the long arm (A chain) may have folded back onto itself. If this scenario did occur, then one interpretation is that the molecular dimension of the A chain would be increased in this region. This notion is supported by the observation that it appears that two peaks appear adjacent to each other at the intersection point of these two chains as seen in Figure 13a.

The different responses observed between fibronectin and laminin are unclear but may be due to variations in their structure-function heuristics that are not well understood at this time. Thus, invaluable insight can be gained with STM (as well as AFM) as demonstrated in this study on fully hydrated protein structures presumably in their native conformational state. STM nanographs of laminin and fibronectin are consistent with TEM dimensions and additionally display important 3-D structural aspects which may prove to be functionally responsible for protein adsorption and cell binding properties.

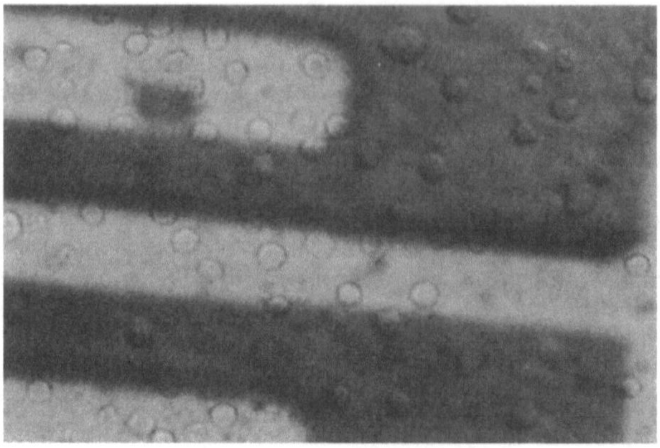

Figure 12a.

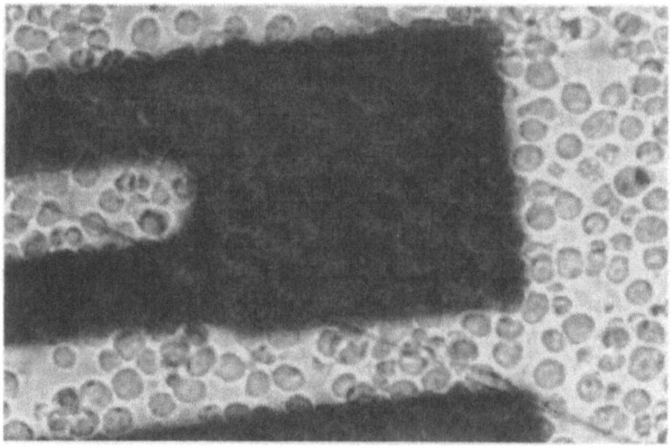

Figure 12b.

Figure 13a.

CONCLUSIONS

This study demonstrated that ECM proteins can serve as highly specific linkers useful for interfacing biological structures to synthetic substrates. A practical example of exploiting ECM proteins for this purpose was demonstrated with the use of the ECM proteins, fibronectin and laminin, for the fabrication of whole cell assemblies directly onto a thermopile transducer surface. A preliminary result of this study indicates that fibronectin is capable of anchoring cells firmly to synthetic substrate surfaces. The same may not be true with laminin based upon its inability to maintain good adherence to Mylar® surfaces at least when relying on physical adsorption forces alone. This circumstance emphasizes the need to acquire a much more fundamental level of understanding biointerfacial phenomena, such as protein adsorption, molecular recognition, and cell binding processes. In this light, SPM is a potentially valuable tool for gaining insight into complex biological structure-function heuristics essential for the rational design of molecular devices.

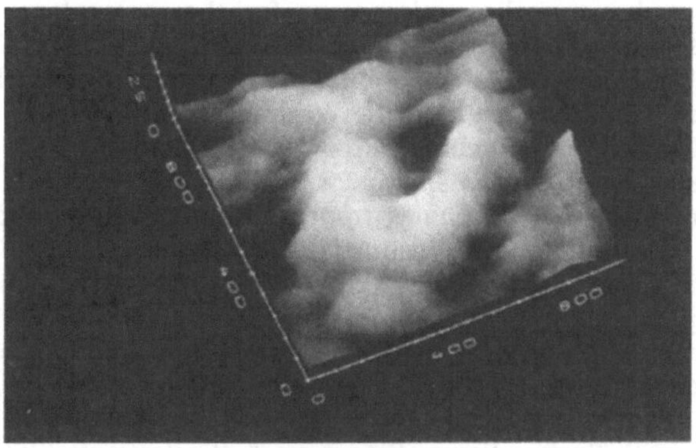

Figure 13b.

ACKNOWLEDGEMENTS

This work is supported in part by BRSG 2 S04 RR07112, Division of Research Resources, National Institutes of Health. The College of Engineering at ASU provided the support for the purchase of the STM instrument used in this study. Also, many fruitful discussions and helpful suggestions on STM were provided by my colleague Antonio Garcia. In addition, the support of the STM-IAP lab at ASU and the assistance of John Graham is appreciated. The thermopile devices, kindly provided by our colleagues Drs. Guilbeau and Towe, along with their helpful suggestions are gratefully acknowledged. Finally, special thanks to Sue Selkirk for her excellent graphical support.

REFERENCES

1. I. Karube and M. Suzuki, Microbial Biosensors, in: "Biosensors," A. E. G. Cass, ed., Oxford University Press, New York (1990).
2. J. W. Parce, J. O. Owicki, K. M. Kersco, G. B. Sigal, H. G. Wada, V. C. Muir, L. J. Bousse, K. L. Ross, B. I. Sikic, H. M. McDonnel, Detection of cell-affecting agents with a silicon biosensor, *Science* 246:243 (1989).
3. D. M. Rawson, A. J. Willmer, Whole-cell biosensors for environmental monitoring, *Biosensors* 4:299 (1989).
4. I. Karube, Hybrid biosensors for clinical analysis and fermentation control, in: "Methods in Enzymology: Immobilized Enzymes and Cells," Volume 137, Part D, K. Mosbach, ed., Academic Press Inc., San Diego (1988).
5. V. B. Pizziconi and D. L. Page, Whole cell biosensors in artificial organs, *Trans. ASAIO* (in preparation).
6. G. A. Rechnitz, Biosensors, *Chemical Engineering News* 66:24 (1988).
7. D. G. Parce, Light-addressable potentiometric sensor for biochemical systems, *Science* 240:1182 (1988).
8. E. J. Guilbeau, B. C. Towe, and M. J. Muelbauer, Potentilly implantable thermoelectric sensor for measuremenmt of glucose, *Trans. ASAIO* 10(3):329 (1987).
9. V. B. Pizziconi and D. L. Page, The elucidation of macromolecular assemblies using scanning tunneling microscopy for the molecular bioengineering of cellular self-assemblies in molecular device design, in: "ACS Symposium Series No. 493, Macromolecular Assemblies in Polymeric Systems," P. Stroeve and A. C. Balazs, eds., Chemical Society, Washington, DC (1992).
10. J. S. Sidwell and G. A. Rechnitz, Progress and challenges for biosensors using plant tissue materials, *Biosensors* 2:221 (1986).
11. A. Navaratne and G. A. Rechnitz, Improved plant tissue-based biosensor using in-vitro cultured tobacco callus tissue, *Analytica Chimica Acta* 257:59 (1992).
12. D. G. Margineau and H. Vais, Bioelectrochemical sensors with living cells, in: "Bioinstrumentation: Research, Developments and Applications," D. Wise ed., Butterworths, Boston (1990).
13. G. Xiuli, L. Yuqi, and Y. Guanghua, Study and application of a new adenosine electrode with thymus tissue, *Biosensors & Bioelectronics* 7:21 (1992).
14. D. Wijesuriya and G. A. Rechnitz, *Analytica Chimica Acta* 256:39 (1992).
15. R. M. Buch, T. Q. Barker, and G. A. Rechnitz, Intact chemoreceptor biosensors based on hawaiian aquatic species, *Analytica Chimica Acta* 243:157 (1991).
16. E. A. H. Hall, "Biosensors," London (1990).
17. J. Woodward, ed., "Immobilized Cells and Enzymes: A Practical Approach," IRL Press, Oxford (1985).
18. K. Mosbach, ed., "Methods in Enzymology: Immobilized Enzymes and Cells," Part B, Volume 135, Academic Press Inc., Orlando (1987).
19. K. Mosbach, ed., "Methods in Enzymology: Immobilized Enzymes and Cells," Part C, Volume 136, Academic Press Inc., Orlando (1987).
20. K. Mosbach, ed., "Methods in Enzymology: Immobilized Enzymes and Cells," Part D, Volume 137, Academic Press Inc., Orlando (1988).
21. R. F. Taylor, ed., "Protein Immobilization: Fundamentals and Applications," Marcel Dekker Inc., New York (1991).
22. J. Tampion and M. D. Tampion, "Immobilized Cells: Principles and Applications," Cambridge University Press, Cambridge (1987).

23. G. G. Guilbault, "Analytical Uses of Immobilized Enzymes," Marcel Dekker Inc., New York (1984).
24. B. Mattiasson, ed., "Immobilized Cells and Organelles," Volume I, CRC Press, Inc., Boca Raton (1983).
25. B. Mattiasson, ed., "Immobilized Cells and Organelles," Volume II, CRC Press, Inc., Boca Raton (1983).
26. I. Chibata, ed., "Immobilized Enzymes," Wiley, New York (1978).
27. J. W. Parce, G. B. Sigal, K. M. Kercso, J. C. Owicki, The silicon microphysiometer: Detection of biological effects of chemical and biochemical agents by alterations of celluar metabolic rate, in: "Biosensor Technology Fundamentals and Applications," R. P. Buck, W. E. Hatfield, M. Umana, and E. F. Bowden, eds., Marcel Dekker, Inc., New York (1990).
28. B. S. Jacobson and D. Branton, Plasma membrane: rapid isolation and exposure of the cytoplasmic surface by use of positively charged beads, *Science* 195:302 (1976).
29. B. C. Arkles, A. S. Miller, and W. S. Brinigar, Whole cell and cell organelle immobilization on siliceous surfaces, in: "Syliated Surfaces" D. E. Leyden and W. T. Collins, eds., Gorden and Breach, Inc. (1980).
30. B. A. Pethica, Microbial and cell adhesion, in: "Microbial Adhesion to Surfaces," R. C. W. Berkeley, J. M. Lynch, J. Melling, P. R. Rutter and B. Rutter, eds., Soc. Chem. Ind., London (1980).
31. E. Evans, Mechanics of cell deformation and cell-surface adhesion, in: "Physical Basis for Cell-Cell Adhesion," P. Bongrand, ed., CRC Press, Boca Raton (1988).
32. J. D. Andrade, L. M. Smith, and D. E. Gregonis, The contact angle and interface energetics, in: "Surface and Interfacial Aspects of Biomedical Polymers," J. Andrade, ed., Plenum Press, New York (1985).
33. J. M. Schakenraad, H. J. Busscher, C. R. H. Wildevuur, and J. Arends, The influence of substratum surface free energy on growth and spreading of human fibroblasts in the presence or absence of serum proteins, *J. Biomed. Mater. Res.* 20:773 (1986).
34. D. R. Absolom, L. A. Hawthorn, and G. Chang, Endothelialization of polymer surfaces, *J. Biomed. Mater. Res.* 22:271 (1988).
35. S. A. Klotz, Role of hydrophobic interactions in microbial adhesion to plastics used in medical devices, in: "Microbial Cell Surface Hydrophobicity," R. J. Doyle and M. Rosenberg, eds., American Society for Microbiology, Washington, DC (1990).
36. Y. Ito, Cell adhesion factor immobilized materials, in: "Synthesis of Biocomposite Materials," Y. Imanishi, ed., CRC Press, Boca Raton, (1992).
37. T. A. Horbett and B. Schway, Correlations between mouse 3T3 cell spreading and serum fibronectin adsorption on glass and hydroxyethylmethacrylate-ethylmethacrylate copolymers, *J. Biomed. Mater. Res.* 22:763 (1988).
38. E. Ruoslahti and M. D. Pierschbacher, New perspectives in cell adhesion: RGD and integrins, *Science* 238:491 (1987).
39. F. Grinnell, Cellular adhesivienes and extracellular substrates, *International Review of Cytology* 53:65 (1978).
40. K. M. Yamada, Cell surface interactions with extracellular materials, *Ann. Rev. Biochem.* 52:761 (1983).
41. E. Ruoslahti and M. D. Pierschbacher, ARG-GLY-ASP a versatile cell recognition signal, *Cell* 44:517 (1986).

42. R. Bar-Shavit, V. Sabbah, M. G. Lampugnani, P. C. Marchisio, J. W. Fenton II, I. Vlodavsky, and E. Dejena, ARG-GLY-ASPP sequence within thrombin promotes endothelial-cell adhesion, *J. Cell Biol.* 112(2):335 (1991).
43. E. C. Tsilibary, L. A. Reger, A. M. Vogel, G. G. Koliaakos, S. S. Anderson, A. S. Charonis, J. N. Alegre, and L. T. Furcht, Identification of a multifunctional, cell-binding peptide sequence from the a1(NC1) of type IV Collagen, *J. Cell Biol.* 111:1583 (1990).
44. D. W. Barnes and J. Silnutzer, Isolation of human serum spreading factor, *J. Biol. Chem.* 258(20):12548 (1983).
45. R. M. Senior, G. L. Griffin, R. M. Mecham, D. S. Wrenn, K. U. Prasad, and D. W. Urry, Val-Gly-Val-Ala-Pro-Gly, a repeating peptide in elastin, is chemotactic for fibroblasts and monocytes, *J. Cell Biol.* 99:870 (1984).
46. M. Hook, L. M. Switalski, T. Wadstrom, and M. Lindberg, Interactions of the pathogenic microorganisms with fibronectin, in: "Fibronectin," D. Mosher, ed., Biology of Extracellular Matrix: A Series, Academic Press, New York (1989).
47. H. L. Thompson, P. D. Burbelo, B. Segui-Real, Y. Yamada, and D. D. Metcalfe, Laminin promotes mast cell attatchment, *Immunology* 143(7):2323 (1989).
48. H. L. Thompson, P. D. Burbelo, Y. Yamada, H. K. Kleinman, and D. D. Metcalfe, Identification of an amino acid sequence in the laminin A chaim mediating mast cell attatchment and spreading, *Immunology* 72:144 (1991).
49. J. Dastych, J. J. Costa, H. L. Thompson, and D. D. Metcalfe, Mast cell adhesion to fibronectin, *Immunology* 73:478 (1991).
50. H. K. Kleinman, F. B. Cannon, G. W. Laurie, J. R. Hassell, M. Aumailley, V. P. Terranova, and M. DuBois-Dalcq, Biological activities of laminin, *J. Cell. Biochem.* 27:317 (1985).
51. J. Enenestein and L. T. Furcht, Epithelial and neural localization and heparin binding of the cell-substratum adhesion molecule, epinectin, *J. Invest. Dermatol.* 91(1):34 (1988).
52. K. S. Stenn, Epibolin: a protein of human plasma that supports epithelial cell movement, *Proc. Natl. Acad. Sci. USA* 78(11):6907 (1981).
53. W. G. Carter, M. C. Ryan, and P. J. Gahr, A new cell-adhesion ligand for integrin- Alpha-3-Beta-1 in epiithelial basement membranes, *Cell* 65:599 (1991).
54. C. R. Ruegg, R. Chiquet-Ehrismann, and S. S. Alkan, Tenascin, an extracellular matrix protein, exerts immunodulatory activities, *Proc. Natl. Acad. Sci. USA* 86:7437 (1989).
55. K. Tashiro, G. C. Sephel, B. Weeks, M. Sasaki, G. R. Martin, H. K. Kleinman, Y. Yamada, and G. R. Martin, A synthetic peptide containing the IKVAV sequence from the A chain of laminin mediates cell attatchment, migration, and neurite outgrowth, *J. Biol. Chem.* 264:16174 (1989).
56. A. Oldberg, A. Franzen, and D. Heinegard, Cloning and sequence analysis of rat bone sialoprotein (osteopontin) cDNA reveals an Arg-Gly-Asp cell binding sequence, *Pro. Natl. Acad. Sci. USA* 83:8819 (1986).
57. A. Oldberg, A. Franzen, and D. Heinegard, The primary structure of a cell-binding bone sialoprotein, *J. Biol. Chem.* 263(36):19430 (1988).
58. S. D. Wright, C3bi receptor (complement receptor type 3) recognizes a region of complement protein C3 containing the sequence of ARG-GLY-ASP, *Natl. Acad. Sci. USA* 84:1965 (1987).
59. E. Bell, Tissue engineering, an overview: reconstituting tissue and organ equivalents for grafting and for use as model systems, *J. Cellular Biochem. Supplement* 16F:116 (1992).

60. A. Nicol, D. C. Gowda, and D. W. Urry, Cell adhesion and growth on synthetic elastomeric matrices containing ARG-GLY-ASP-SER, *J. Biomed. Mater. Res.* 26:293 (1992).
61. S. P. Massia and J. A. Hubbell, Covalent surface immobilization of Arg-Gly-Asp-and Tyr-Ile-Gly-Ser-Arg-containing peptides to obtain well defined cell-adhesive substrates, *Analytical Biochemistry* 187:292 (1990).
62. L. G. Cima, J. P. Vacanti, C. Vacanti, D. Ingber, D, Mooney, and R. Langer, Tissue engineering by cell transplantation using degradable polymer substrates, *J. Biomech. Eng.* 113:143 (1991).
63. J. A. McDonald and R. P. Mecham, eds., "Receptors for Extracellular Matrix," Academic Press, Inc., San Diego (1991).
64. J. Engel, E. Odermatt, A. Engel, J. Madri, H. Furthmayer, H. Rohde, and R. Timpl, Shapes, domain organizations and flexibility of laminin and fibronectin, two multifunctional proteins of the extracellular matrix, *J. Mol. Biol.* 150:97 (1981).
65. D. R. Abramhamson, M. H. Irwin, P. L. St. John, M. A. Accavitti, L. W. Heck, and J. R. Couchman, Selective immunoreactivities of kidney basement membranes to monoclonal antibodies against laminin: localization of the end of the long arms to discrete microdomains, *J. Cell Biol.* 109:3477 (1989).
66. R. L. Trelsted, Matrix macromolecules: spatial relationships in two dimensions, *Ann. N.Y. Acad. Sci.* 580:391 (1990).
67. R. O. Hynes and K. M. Yamada, Fibronectins-multifunctional modular glycoproteins issue #2, *J. Cell. Biol.* 95:369 (1982).
68. R. B. Kemp, Heat dissipation and metabolism in isolated animal cells and whole tissues/organs, in: "Thermal and Energetic Studics of Cellular Biological Systems," A. M. James, ed., IOP Publishing Ltd., Bristol, Great Britain (1987).
69. G. Binnig and H. Rohrer, Scanning tunneling microscopy, *Surface Sci.* 126:236 (1983).
70. J. A. Golovchenko, The tunneling microscope: a new look at the atomic world, *Science* 232:48 (1986).
71. M. E. Welland and M. E. Taylor, in: "Modern Microscopies: Techniques and Applications," P. J. Duke and A. G. Michette, eds., Plenum Press, New York, N.Y. (1990).
72. R. J. Behm, N. Garcia, and H. Rohrer, "Scanning Tunneling Microscopy and Related Methods," Academic Publishers, Dordrecht, The Netherlands (1990).
73. D. Sarid, "Scanning Force Microscopy: With Applications to Electric, Magnetic, and Atomic Forces," Oxford University Press, New York, N.Y. (1991).
74. P. D. Yurencho, E. C. Tsilibary, A. S. Charonis, and H. Furthmayr, Laminin polymerzation in vitro: evidence for a two-step assembly with domain specificity, *J. Biol. Chem.* 260(12):7636 (1985).
75. P. D. Yurencho, Y. Cheng, and J. C. Schittny, Heparin modulation of laminin polymerzation, *J. Biol. Chem.* 265(7):3981 (1990).
76. Z. Wang, T. Hartmann, W. Baumiester, and R. Guckenberger, *Proc. N.Y. Acad. Sci., USA* 87:9343 (1990).
77. R. E. Baier, Role of surface energy in thrombogenesis, *Bull. N.Y. Acad. Med.* 48(2):257 (1972).
78. E. P. Plueddemann, "Silicone Coupling Agents," Plenum Press, New York (1991).
79. F. Grinnell, Fibronectin adsorption on material surfaces, in: "Blood in Contact with Natural and Artiifcial Surfaces," E. Leonard, V. T. Turitto, and L. Vroman, eds., Ann. N.Y. Acad. Sci. 516:280 (1987).

80. K. Lewandowska, N. Balachander, C. N. Sukenik, and L. A. Culp, Modulation of fibronectin adhesive proteins for fibroblasts and neural cells by chemically derivatized substrata, *J. Cell. Physiol.* 141:334 (1989).
81. S. R. Snyder and H. S. White, Scanning tunneling microscopy, atomic force microscopy, and related techniques, *Analytical Chemistry* 64(12):116R (1992).

IMPROVING THE ORIENTATION OF NERVE REGROWTH WITH A SIMPLE DEVICE

Luis de Medinaceli and Richard Jed Wyatt

National Institute of Mental Health, W.A.W. Bldg., St Elizabeths Hospital, Washington DC 20032

The synthetic microstructure described here is one of the simplest that can be imagined, and yet it partially solves one of the most complex problems currently encountered in reconstructive surgery. This microstructure is a simple rectangular sheet of biocompatible material, and it is used in the repair of divided nerves.

THE PROBLEM

A nerve can be compared to a telephone cable; it is comprised of thousands of individual fibers, stretching between the central nervous system and the periphery, and ceaselessly relaying information. A nerve, however, differs from a real telephone cable in that each line it contains is highly specialized and connects only two "stations" throughout life. When an injury severs the cable, the fibers heal and grow back, and connections between the central nervous system and periphery become reestablished. Unfortunately, the re-wiring is likely to be incorrect, providing only incoherent and often useless function. Therefore, extreme precision in the repair is needed. The problem is that current methods of nerve repair cannot be expected to attain any level of precision.

Inadequacy of Current Methods of Repair

Results of peripheral nerve surgery have always been quite variable, ranging from satisfactory to poor.[1,2] Increased use of the operating microscope stirred the hope that results could be improved. In fact, microsurgery has not changed fundamentally the prognosis of nerve lesions.[3,4] "Despite the application of meticulous technique to peripheral nerve repair, no adult with a major peripheral nerve transected has ever attained normal sensibility".[5] The reason for failure is found in the great multitude and the minuscule size of what is to be mended: the diameter of individual nerve fibers vary from 0.5 to 20 micrometers, and there are 3,000 to 100,000 of these fibers depending on the size of the nerve. The task of matching fibers seems impossible. At present, the surgeon aligns the nerve stumps as

Synthetic Microstructures in Biological Research, Edited by
J.M. Schnur and M. Peckerar, Plenum Press, New York, 1992

best he can; thousands of regenerating fibers wander at the suture line, interlacing occurs, and "Mother Nature" is blamed for the bad results.

Inadequacy of current methods has triggered research in many directions, and various solutions have been suggested, such as nerve intubulation, implants of growth-promoting substances, administration of growth-promoting drugs, and use of electric currents. The technique we suggested is quite different, and is based on a few theoretical observations.

Theoretical Basis for a Suggested Solution

What actually is the main reason for failure in nerve surgery? Although errors in the direction of the sprouts seem to be the major cause for persistent poor results after nerve repair, a number of other factors also have been suggested. Local scar or foreign bodies,[6,7] lesions of the protective sheaths,[8-11] disruption of local blood flow,[12,13] abnormalities of the fibers after healing and regrowth,[14-16] insufficiency of growth-promoting factors[17,18] have all been incriminated. We wanted to know if these factors had to be taken into consideration or could be considered negligible.

A probability model of nerve regeneration. In order to answer this question, we devised a probability model of nerve regeneration.[19] We tabulated the various types of fibers that are contained in a nerve, and also the various paths that are open to any regrowing fiber. We obtained 117 possibilities for the different types of regrowth; the probability for a healing fiber to enter a specific pathway was given by a formula of the general type:

$$p1p2(1\text{-}q)s1$$

where p = overall proportion of different kinds of fibers, q = probability of fibers that do not heal properly and are lost for function, and s = proportion of different kinds of sensory fibers. By varying all the parameters over reasonable ranges of values, their influence could be examined. A triplet of theoretical results was thus obtained for each of these combinations and a "reinnervation score" was attributed to each triplet (for a complete description of the model, see reference 19). The results predicted through this model were then compared to behavioral results obtained experimentally after various procedures of nerve repair.[20] A good correlation was observed between theoretical and experimental results; one of the conclusions of these studies was that the randomness of nerve regrowth and the resulting errors in connections sufficed to explain all clinical observations after nerve injury and repair. We considered it useful, therefore, to try and increase the precision of repair.

How to obtain a theoretically perfect match of fibers. When a nerve is transected, the two stumps are by definition a perfect mirror image of one another (Figure 1a). Given that nerve tissue is soft, each stump can be deformed in a random way with loss of the similitude as soon as the nerve is handled (Figure 1b). In fact, the perfect initial symmetry applies not only to the appearance of the nerve. The similitude applies also to the physical characteristics and therefore to the mechanical resistance or elasticity of each part of the nerve (Figure 1c). By pressing with a given force on a given point of one stump, a displacement of the structure is produced (Figure 1d). If an equal force were applied at the symmetrical point of the other stump, an equivalent displacement should be produced (Figure 1e). The symmetry of these displacements can be predicted because the fibers are enmeshed in a very thin, soft and ubiquitous net of jelly-like matrix that holds them together. Since the matrix secures the position of each fiber relative to the neighboring ones, it is not necessary to act individually on each fiber. It suffices to determine what global force applied to

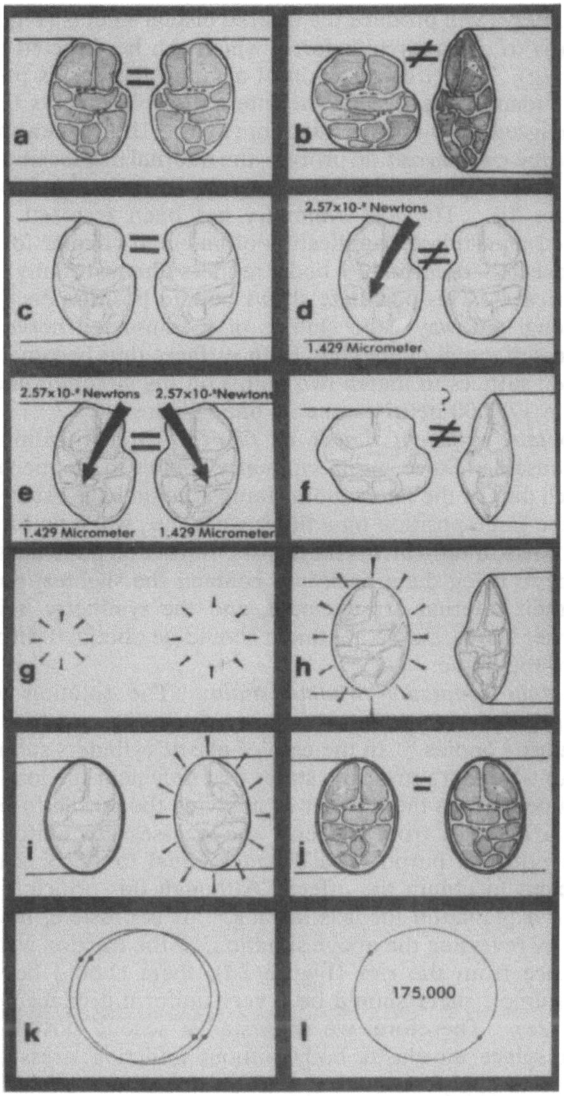

Figure 1. How to obtain theoretically perfect match of fibers
a) when a nerve is cut, both stumps surfaces are symmetrical; **b)** given that nerve tissue is soft, the symmetry is usually lost at once; **c)** the perfect initial symmetry applies also to the mechanical resistance or elasticity of each part of the stumps; **d)** application of a given force on a given point of one stump produces a corresponding displacement of the structure; **e)** application of an equal force at the symmetrical point of the other stump must produce an equivalent displacement; **f)** a simple way of acting on all neurites at the same time is.. **g)** ..to use a circular force, directed either inwards or outwards; **h)** circularization of a stump deforms it in a specific manner, linked to its mechanical characteristics; **i)** since the other stump has exactly the same mechanical characteristics, its circularization will deform it in precisely the same manner; **j)** the lost symmetry can thus be restored; **k)** adjusting two points on the circumference.. **l)** ..suffices to align all other points in the structure.

the exterior of the nerve will produce the desired displacements (Figure 1f). The most practical solution is to use a circular force, which can be directed either inwards or outwards (Figure 1g). The circularization of a stump flattens its prominent portions and levels off its indentations. While the internal displacements that ensue are too complex to be analyzed, they are not random (Figure 1h). When the other stump is submitted to the same circularization process, the internal displacements of its contents are totally predictable, in that it will be exactly the same as the displacements in the first stump (Figure 1i). The lost symmetry has been restored (Figure 1j). We therefore suggested a postulate: application of an equal circular force to each of the stumps of a transected deformable body renders the extremity of these stumps perfectly symmetrical.[21] This postulate shows how to position each fiber in front of its correct individual pathway: the stumps of a transected nerve are temporarily encased in cylinders of equal diameters. Both surfaces thus become mirror images of each other. It then suffices to match two points on the circumference (Figure 1k) in order to match the 175,000 neurites one by one (Figure 1l).

How to obtain sufficient match of fibers. Circularization of the stumps is not even indispensable. Since stump symmetry applies to the mechanical resistance or elasticity of each part of the nerve, both stumps should have exactly the same shape when they lie in the same position on a flat surface, i.e., when no force except gravity acts on them. A practical difficulty is that elastic retraction pulls the stumps apart and that it is necessary to bring them back into contact; the slightest pull applied to the stumps disrupts their internal arrangement, and the symmetry is immediately lost (Figure 2a). In other words, stump abutment should be obtained without applying any stress to the tip of the nerve.

How to obtain coaptation without strain. The solution to this step can be found in the principle of de Saint-Venant, which deals with the distribution of tensile forces in elastic bodies.[22] In the case of elastic cylinders submitted to an axial tension, the de Saint-Venant's principle states that only near the loaded ends does the stress distribution depend on the manner of applying the tensile force (Figure 2b); in cross sections at a distance from the end, the stresses are uniformly distributed[23] (Figure 2c). For practical purposes, a distance equal to 1.5 or 2 diameters of the structure is sufficient to obtain the effect. Although this principle holds whenever tested, a complete explanation for it is lacking.[24] It is possible, however, to draw a postulate from it, by reversing the above situation: if the traction were reported to an appropriate distance from the rim (Figure 2d), there should be no stress at the extremity of the stump: there should be a very uniform displacement of the whole structure (Figure 2e). Therefore, we suggested a second postulate: in order to manipulate and displace an elastic body without inflicting stress at the tip of the structure, it suffices to apply mechanical traction at a distance from this tip equal to at least one-and-a-half times the diameter of the structure.[21] This postulate suggests a simple method of reunion. A small sheet of any biocompatible material is sutured to the external sheath of nerve stumps at the appropriate distance from their tips. Thus, all mechanical stress is diverted from the zone of repair, and the stumps are brought smoothly in contact (Figure 2f).

Two Technical Prerequisites for Success

Proper trimming. Truly precise abutment of nerve stumps can be attempted only if the cut-surfaces are flat. In clinical practice, this is never the case. Nerves do not cut well no matter how sharp the wounding agent. Histologic studies of freshly severed nerves show that the surface of the stumps is strikingly uneven.[25,26] A number of guillotine-type techniques have been devised to obtain a clean-cut surface before suturing.[25,27,28]

Inadequacy of current trimming techniques. Strict immobilization of nerve stumps in a holder during transection yields cut-surfaces that are passably flat. This technique of trimming, however, falls short of its ultimate goal because it does not take into consideration a major mechanical factor;[29] the strength and elasticity of the nerve envelopes is in sharp contrast with the softness and fragility of their contents. Therefore, the sheaths will not divide before the cells they contain are destroyed by the applied pressure. This damage is not strictly local: the crush-, twist- and tear-damage extends longitudinally, i.e. proximally and distally to the site of trimming. The presence of bruised axons at the tip of nerve stumps was demonstrated as early as 1852.[30] More than a century later, however, local damage of stumps still is not taken into consideration during nerve repair. The presence of a damaged zone at the tip of the stumps nullifies the precision of repair and increases the probability of misdirected sprouting.

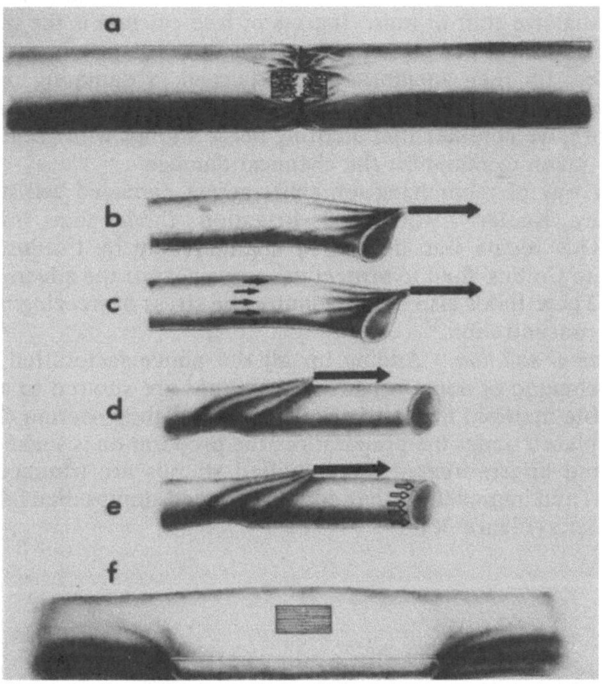

Figure 2. How to obtain coaptation without strain. **a)** Suturing nerve stumps submits the tips to non-uniform stress that disrupts the soft structure and disorganizes the longitudinal alignment of the fibers. **b)** in the case of elastic cylinders submitted to an axial tension, only near the loaded ends does the stress distribution depend on the manner of applying the tensile force; **c)** in cross sections at a distance from the end, the stresses are uniformly distributed (de Saint-Venant's principle); **d)** if the traction were transferred to an appropriate distance from the rim, there should be no stress at the end of the stump: **e)** there should be a very uniform displacement of the whole structure; **f)** it is therefore possible to abut soft nerve stumps without disorganizing the longitudinal alignment of the fibers (From reference 21, with permission).

Difficulty in eliminating physical damage of the stump. Local damage of stumps seems difficult to eliminate. Any attempt to shave off the unhealthy portion with a blade will simply spread the damage further along the structure, and this iatrogenic lesion is not precluded by immobilizing the stumps in a holder. Using a laser beam eliminates pressure damage but is not an improvement because the heat produces denaturation of proteins. The charred zone left at the tip of each stump by a laser beam can be as wide as 200 micrometers, leads to formation of scar tissue,[31] and does nothing to improve the precision of sprout regrowth.

Of all currently available techniques, hardening the nerve is the only one that provides a clean, smooth and non-traumatic cut. Such hardening can be obtained through temporary freezing.[32,33] Cold damage can be avoided by monitoring depth and duration of freezing, and by using appropriate irrigation fluids.[34]

Chemical protection. At the site of an injury, fibers are not only physically bruised, they also are chemically burned. The ionic composition of the fiber substance, the cytoplasm, differs greatly from that of the surrounding medium, the extracellular fluid. Therefore, any breach in the fiber membrane must result in an immediate and massive shift of ions. Ingress of free calcium is the most deleterious of these ion shifts but it is not the only one. Because of their elongated shape, nerve fibers quickly seal up their membrane and succeed in damming up the chemical disaster. The extremity of the divided fiber, however, is "burned" along a significant length. Therefore, we consider that abutting nerve stumps with great care is useless unless steps are taken to minimize the chemical damage.

A simple way of minimizing ion shifts across damaged cell membranes was demonstrated by Keeler[35] who used irrigation fluids that mimic axoplasm composition. This recipe was developed commercially by Collins.[36] We added chlorpromazine to Collins' fluid to protect the nerve against the adverse effects of free calcium ions.[34] These fluids also help minimize the stress of freezing because of their high potassium concentration.[37]

The practical solution. Adding up all the above factors led us to propose the following technique of nerve repair: The stumps are sutured to a support made of a bioabsorbable material in such a way that they slightly overlap (Figure 3a); the tip of a cryod is placed under the preparation; the preparation is soaked in a modified Collins fluid[34] and briefly frozen; the solidified stumps are trimmed with a sharp blade (Figure 3b) and immediately thawed; the trimmed stumps then fall automatically in correct abutment (Figure 3c).

RESULTS

Experimental Results

Histologic examination of such reunions always shows a striking difference compared with those obtained by ordinary microsutures. With this new method, the fibers stay in excellent alignment; there is no disruption of the longitudinal direction and no buckling. Also, the gap between the stumps is dramatically reduced.[26] Most importantly, the functional outcome is consistently favorable. Satisfactory functional recovery is observed in all experimental animals when the procedure is used.[38] These satisfactory functional results, which we consider the most valuable criterion of technical efficacy, have been confirmed in independent experimental studies.[39-41]

Clinical results. We are currently testing the method in clinical practice. Preliminary results are encouraging.

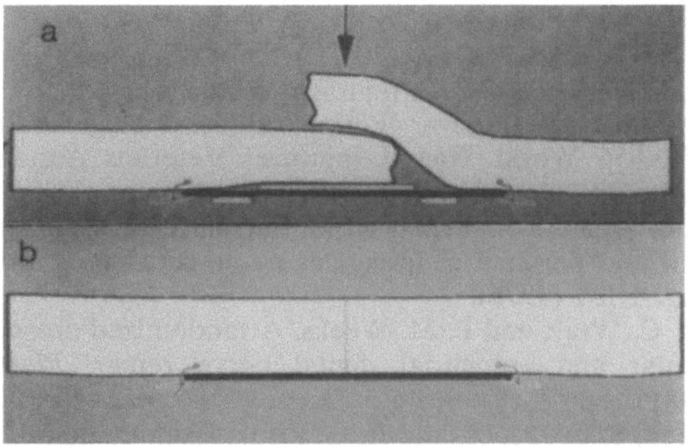

Figure 3. The practical solution. **a)** The stumps are sutured to a support device with a slight overlap; **b)** both stumps are transected at the same time; **c)** the trimmed stumps fall automatically into satisfactory abutment.

Limits of the method. Injuries cannot be expected to produce a smooth perpendicular cut in every case. This difficulty, however, can be circumvented because it is possible to shave off a considerable length of the nerve without significantly losing the similitude. As a rule of thumb, a length equal to one diameter of the nerve can be removed from each stump without significant modification of the internal pattern and disposition. In most cases, such a distance is ample to obtain smooth perpendicular stump surfaces.

Evidently, the method cannot be used in the case of true loss of substance. The adverse effects of tension at the suture line have long been described,[42,43] and even moderate defects require grafting.[44,45] In order to know how frequent this situation is in civilian practice, we recently reviewed a series of 2,181 fresh nerve injuries of the upper limb treated by conventional methods. In 59% of all cases, two or more nerves were injured in a same patient. We found that 1,482 injuries (68%) were located in the digits and that 387 times, the injured limb segment was lost or beyond repair (20% of all cases). In the 1,794 remaining nerve injuries, primary treatment was accomplished by end-to-end suture 1,568 times (87%), by graft 33 times (<2%), and was impossible in 193 cases (11%). This indicates that extensive loss of substance is not the most frequent situation in clinical practice.

Another characteristic of this method is that it is applicable exclusively to recent injuries. Healing processes bring minute modifications to the internal pattern of the nerve and therefore render futile any secondary attempt to match neurites one by one. This is one of the many reasons why we advocate immediate repair of nerve injuries.

The Future?

We think that this type of surgical approach may have various applications, and that repair of other organs might benefit from the principles of "Cell Surgery." Also, we believe that this work indicates the possibility of developing new techniques in biology and medicine by taking into consideration physical, chemical and mathematical factors.

REFERENCES

1. S. Sunderland, "Nerve and Nerve Injuries," Williams & Wilkins, Baltimore (1968).
2. B. Woodhall and G. W. Beebe, "Peripheral Nerve Regeneration: A Follow-up Study of 3,656 World War II Injuries," Veterans Administration medical monograph, U.S. Govt. Printing Office, Washington, DC (1956).
3. H. E. Cabaud, W. G. Rodkey, H. R. McCarrol, S. B. Mutz, and J. J. Niebauer, Epineurial and perineurial fascicular nerve repair; a critical comparison, *J. Hand Surg.* 1:131 (1976).
4. L. Young, R. C. Wray, and P. M. Weeks, A randomized prospective comparison of fascicular and epineurial digital nerve repair, *Plast. Reconstr. Surg.* 68:89 (1981).
5. A. L. van Beek, S. C. Jacobs, and E. G. Zook, Examination of peripheral nerves with the scanning electron microscope, *Plast. Reconstr. Surg.* 63:509 (1979).
6. J. M. Lane, F. W. Bora, and D. Pleasure, Neuroma scar formation following peripheral nerve transection, *J. Bone Joint Surg. (Am)* 60:197 (1978).
7. S. Ramon and Y. Cajal, "Degeneration and Regeneration of the Nervous System, (R. May transl. & ed.), reprint 1968," Hafner, New York (1928).
8. P. H. Spencer, H. J. Weinberg, C. S. Raine, and J. W. Prineas, The perineurial window - a new model for the study of demyelination and remyelination, *Brain Res.* 96:329 (1975).
9. R. S. Mellick and J. B. Cavanagh, Changes in blood vessel permeability during degeneration and regeneration in peripheral nerves, *Brain* 91:141 (1968).
10. Y. Olsson, Studies on vascular permeability in peripheral nerves. I. Distribution of circulating fluorescent serum albumin in normal, crushed and sectioned rat sciatic nerve, *Acta Neuropathol.* 7:1 (1968).
11. M. Schwartz, B. A. Sela, and N. Eshhar, Antibodies to gangliosides and myelin autoantigens are produced in mice following sciatic nerve injury, *J. Neurochem.* 38:1192 (1982).
12. J. P. Kendall, I. A. Stokes, and J. P. O'Hara, Tension and creep phenomena in peripheral nerve, *Acta Orthop. Scand.* 50:721 (1979).
13. Y. Miyamoto, Experimental studies on repair for peripheral nerves; Effects of tension at suture site on intraneural circulation, *Hiroshima J. Med. Sci.* 28:13 (1979).
14. B. G. Cragg and P. K. Thomas, The conduction velocity of regenerated peripheral nerve fibers, *J. Physiol.* 171:103 (1964).
15. W. A. Rushton, A theory of the effects of fibre size in medullated nerve, *J. Physiol.* 115:101 (1951).
16. J. M. Schroder, Altered ratio between axon diameter and myelin sheath thickness in regenerated nerve fibers, *Brain Res.* 45:49 (1972).
17. T. Ebendal, L. Olson, and A. Seiger, The level of nerve growth factor (NGF) as a function of innervation, *Exp. Cell Res.* 148:311 (1983).
18. N. Miki, Y. Hayashi, and H. Higashida, Characterization of chick gizzard extract that promotes neurite outgrowth in cultured ciliary neurons, *J. Neurochem.* 37:627 (1981).
19. L. de Medinaceli and R. R. Rawlings, Is it possible to predict the outcome of peripheral nerve injuries? A probability model based on prospects for regenerating neurites, *BioSystems* 20:243 (1987).
20. L. de Medinaceli, Functional consequences of experimental nerve lesions: effects of reinnervation blend, *Exp. Neurol.* 100:166 (1988).

21. L. de Medinaceli, How to correctly match 175,000 neurites: two postulates for a quick solution, *BioSystems* 20:307 (1987).
22. A. de Saint-Venant, Memoire sur la torsion des prismes, avec des considerations sur leur flexion ainsi que sur l'equilibre interieur des solides elastiques en general, et des formules pratiques pour le calcul de leur resistance a divers efforts s'exercant simultanement, *Mem. Sav. Acad. Sci. Paris* 14:233 (1856).
23. S. Timoshenko and J. N. Goodier, Saint-Venant's Principle, in: "Theory of Elasticity," 2nd Edn., p. 33, McGraw-Hill (1951).
24. R. D. Schile and R. L. Sierakowski, On the Saint-Venant problem for a nonhomogeneous elastic material, *Quat. Appl. Math.* 23:19 (1965).
25. S. Edshage, Peripheral nerve suture, *Act. Chir. Scand. Suppl.* 331:2 (1964).
26. L. de Medinaceli and W. J. Freed, Peripheral nerve reconnection: immediate histologic consequences of distributed mechanical support, *Exp. Neurol.* 81:459 (1983).
27. G. L. Clark, A method of preparation of nerve ends for suturing, *Plast. Reconstr. Surg.* 34:233 (1964).
28. V. E. Meyer and J. Smahel, The surgical cut-surface of peripheral nerves, *Int. J. Microsurg.* 2:187 (1980).
29. L. de Medinaceli, R. J. Wyatt, and W. J. Freed, Peripheral nerve reconnection: mechanical, thermal and ionic conditions that promote the return of function, *Exp. Neurol.* 81:469 (1983).
30. A. Waller, Examen des alterations qui ont lieu dans les filets d'origine du nerf pneumo-gastrique et des nerfs rachidiens, par suite de la section de ces nerfs au-dessus de leurs ganglions, *C. R. Hebd. Acad. Sci. Paris* 34:842 (1852).
31. M. A. Badalamente, L. C. Hurst, and D. Blum, Carbon dioxide laser transection of peripheral nerve, *Orthop. Res. Soc. Abstr.* 30:21 (1984).
32. L. de Medinaceli and M. de Medinaceli, "Reflexions sur les Lesions Traumatiques des Nerfs Peripheriques," Foulon, Imprimeur, Paris (1970).
33. S. Edshage, Evaluation of freezing as a method to improve cut surfaces in peripheral nerves preparatory to suturing, *Plast. Reconstr. Surg.* 37:196 (1966).
34. L. de Medinaceli and A. C. Church, Peripheral nerve reconnection: inhibition of early degenerative processes through the use of a novel fluid medium, *Exp. Neurol.* 84:396 (1984).
35. R. Keeler, J. Swinney, R. M. Taylor, and P. R. Uldall, The problem of renal preservation, *Br. J. Urol.* 38:653 (1966).
36. G. M. Collins, M. Bravo-Shugarman, and P. I. Terasaki, Kidney preservation for transportation. Initial perfusion and 30 hours'ice storage, *Lancet* II:1219 (1969).
37. L. van den Berg, The effect of addition of sodium and potassium chloride to the reciprocal system: KH2PO4-Na2HPO4-H2O on pH and composition during freezing, *Arch. Biochem. Biophys.* 84:305 (1959).
38. L. de Medinaceli, W. J. Freed, and R. J. Wyatt, Peripheral nerve reconnection: improvement of long term functional effects under simulated clinical conditions in the rat, *Exp. Neurol.* 81:488 (1983).
39. R. P. Wikholm, J. E. Swett, Y. Torigoe, and H. I. Blanks, Repair of severed peripheral nerve: a superior anatomic and functional recovery with a new "reconnection" technique, *Otolar. Head Neck Surg.* 99:353 (1988).
40. R. T. Zellem, D. W. Miller, J. A. Kenning, E. M. Hoenig, and W. A. Bucheit, Experimental peripheral nerve repair: environmental control directed at the cellular level, *Microsurgery* 10:290 (1989).

41. A. V. Seaber, J. M. Boswick, W. H. Joyner, and J. R. Urbaniak, Sutureless reconnection of peripheral nerves, *Orthop. Res. Soc. Abstr.* 30:(1984).
42. B. Stookey, Surgical considerations of peripheral nerve injuries, Surg.Gynec.Obst. 27:362 (1918).
43. B. Stookey, The technic of nerve suture, *J. Am. Med. Ass.* 74:1380 (1920).
44. C. A. Elsberg, Technic of nerve suture and nerve grafting, *J. Am. Med. Ass.* 73:1422 (1919).
45. G. C. Huber, Repair of peripheral nerve injuries, *Surg. Gynec. Obst.* 30:464 (1920).

CONTRIBUTORS

Breckenridge, L. Clark, P. Connolly, P. Curtis, A.S.G. Dow, J.A.T. Wilson, R. Lind, R. Wilkinson, C.D.W.	Department of Electronics and Electrical Engineering and Department of Cell Biology, The University, Glasgow, Scotland, United Kingdom
Christie, P. Ennis, D.J. DiLosa, V.R.	Electrical Engineering Department, University of Delaware, Newark DE 19716-3130, USA
de Medinaceli, L. Wyatt, R.J.	NIMH Neuroscience Center, Neuropsychiatry Branch, W.A.W. Building, St. Elizabeths Hospital, Washington DC 20032, USA
Furuno, T. Sato, A. Sasabe, H.	Frontier Research Program, RIKEN, Hirosawa 2-1, Wako, Saitama 351-01, Japan
Garcia, A.A. Druyor, R.	Department of Chemical, Bio, and Materials Engineering, Arizona State University, Tempe AZ 85287-6006
Oden, P. Knipping, U.	Department of Physics, Arizona State University, Tempe AZ 85287-1504, USA
Ostroff, G.	Alpha-Beta Technology, Two Biotech Park, 373 Plantation Street, Worcester MA 01605

Hallmark, V.M. Chiang, S.	Almaden Research Center, Physical Science, 650 Harry Road, San Jose CA 995120-6099, USA
Brown, J.K.	UCSF Physics Research Laboratory, University of California at San Francisco, 389 Oyster Point Boulevard,South San Francisco CA 94080, USA
Wöll, Ch.	Institut für Angewandte Physikalische Chemie, Universität Heidelberg, Im Neuenheimer Feld 253, D-6900 Heidelberg, Germany
Hui, S.W. Yu, H.	Biophysics Department, Roswell Park Cancer Institute, Elm and Carlton Streets, Buffalo NY 14263, USA
Xu, Z. Bittman, R.	Department of Chemistry and Biochemistry, Queens College of the City University of New York, Flushing NY 11367, USA
Krishnasastry, M. Walker, B. Bayley, H.	Worcester Foundation for Experimental Biology, 222 Maple Avenue, Shrewsbury MA 01545, USA
Zorn, L.	Graduate School of Biomedical Sciences, University of Massachusetts Medical Center, 55 Lake Avenue North, Worcester MA 01655, USA
Kasianowicz, J.	NIDDK-LBM, National Institutes of Health, Bethesda MD 20892, USA
Ligler, F.S.	Center for Bio/Molecular Science and Engineering, Naval Research Laboratory, Washington DC 20375, USA
Rabbany, S.Y.	Bioengineering Program, Department of Engineering, Hofstra University, Hempstead NY 11550, USA
Lösche, M.	Universität Mainz, Institut für Physikalische Chemie, Welder-Weg 11, D-6500 Mainz, Germany
Markowitz, M.A. Schnur, J.M. Singh, A.	Center for Bio/Molecular Science and Engineering, Naval Research Laboratory, Washington DC 20375, USA

Peckerar, M. Rebbert, M.	Nanoelectronics Processing Facility, Naval Research Laboratory, Washington DC 20375, USA
Pizziconi, V.B. Page, D.L.	Department of Chemical, Bio, and Materials Engineering, Arizona State University, Tempe AZ 85287-6006, USA
Schmitt, F.-J. Knoll, W.	Max-Planck-Institut für Polymerforschung, Ackermannweg 10, D-6500 Mainz, Germany
Blankenburg, R. Häußling, L. Ringsdorf, H.	Johannes-Gutenberg-Universität Mainz, Institut für Organische Chemie, J.-J. Becherweg 18-20, D-6500 Mainz, F.R.G.
Weisenhorn, A.L. Hansma, P.K.	Department of Physics, University of California, Santa Barbara CA 93106, USA
Leckband, D.E. Israelachvili, J.N.	Department of Nuclear & Chemical Engineering, University of California, Santa Barbara CA 93106, USA
Seeger, S. Bierbaum, K. Dahint, R. Mantar, M. Grunze, M.	Institut für Angewandte Physikalische Chemie, Universität Heidelberg, Im Neuenheimer Feld 253, 6900, Heidelberg, Germany
Feng, C.L.	Faculty of Chemistry, University of Liaoning, Huanghe Lu 850, 116022 Dalian, China
Stayton, P.S. Olinger, J.M. Wollman, S.T. Thurman, E.M. Bohn, P.W. Sligar, S.G.	Beckman Institute, University of Illinois at Urbana-Champaign, 405 N. Mathews, Urbana IL 61801, USA
Todorov, A.T. Petrov, A.G. Brandt, M.O. Picard, G. Denicourt, N. Fendler, J.H.	Département de Chimie, Université de Montréal, C.P. 6128, succursale A, Montréal, Québec, Canada H3C 3J7

Turin, L.	Laboratory of Cellular and Molecular Pharmacology, NIEHS, Research Triangle Park NC 22709, USA / Unité Associée 671 du CNRS, Station Marine, F-06230 Villefranche/Mer, France
Williams, S. Jacobsen, C. Kirz, J. Zhang, X.	Brookhaven National Labs, NSLS Building, 725, Beamline X1A, Upton NY 11973, USA

INDEX

MIX
Papier aus verantwortungsvollen Quellen
Paper from responsible sources
FSC® C105338

If you have any concerns about our products,
you can contact us on
ProductSafety@springernature.com

In case Publisher is established outside the EU,
the EU authorized representative is:
Springer Nature Customer Service Center GmbH
Europaplatz 3, 69115 Heidelberg, Germany

Printed by Libri Plureos GmbH
in Hamburg, Germany